Lecture Notes in Artificial Intelligence 9426

Subseries of Lecture Notes in Computer Science

Antonis Bikakis · Xianghan Zheng (Eds.)

Multi-disciplinary Trends in Artificial Intelligence

9th International Workshop, MIWAI 2015
Fuzhou, China, November 13–15, 2015
Proceedings

 Springer

Editors
Antonis Bikakis
University College of London
London
UK

Xianghan Zheng
Fuzhou University
Fuzhou
China

ISSN 0302-9743   ISSN 1611-3349   (electronic)
Lecture Notes in Artificial Intelligence
ISBN 978-3-319-26180-5   ISBN 978-3-319-26181-2   (eBook)
DOI 10.1007/978-3-319-26181-2

Library of Congress Control Number: 2015953001

LNCS Sublibrary: SL7 – Artificial Intelligence

Printed on acid-free paper

Springer International Publishing AG Switzerland is part of Springer Science+Business Media
(www.springer.com)

Preface

MIWAI has evolved from an annual series of international workshops, which was initiated in 2007 by Mahasarakham University in Thailand as the Mahasarakham International Workshop on Artificial Intelligence. In 2011, it was renamed as the Multi-disciplinary International Workshop on Artificial Intelligence, and was for the first time held outside Thailand, in Hyderabad, India. This year, the 9th workshop in this series, MIWAI 2015, took place in Fuzhou, China during November 13–15, and was hosted by one of the biggest Chinese universities, Fuzhou University.

MIWAI covers a wide range of research areas such as cognitive science, computational philosophy, computational intelligence, computer vision, evolutionary computing, game theory, knowledge representation and reasoning, machine learning, multi-agent systems, natural language processing, pattern recognition, planning and scheduling, robotics, uncertainty in artificial intelligence and others. It also deals with applications of artificial intelligence research in domains such as bio-informatics, e-commerce, Internet of Things, knowledge management, privacy and security, recommender systems, social networks, software engineering, surveillance, telecommunications, the Web and others. Its aims are to provide a meeting place where artificial intelligence (AI) researchers and practitioners can present and discuss new research ideas and results from all fields of AI, and to promote synergies between researchers working in different countries and research fields.

MIWAI 2015 was one of the most successful workshops in this series with respect to the number and quality of submissions. It received 67 submissions from 21 different countries in Asia, Europe, America, and Oceania, and on several different topics ranging from more theoretical ones, such as cognitive science and logic-based knowledge representation, to more applied topics, such as smart health services and traffic sensing in vehicular networks. Each submission was carefully reviewed by at least two members of an international Program Committee (PC) through a double-blind review process, as well as by the chairs of the PC. From the 67 submissions, 31 were accepted as full papers (acceptance rate: 46 %) and were allocated 12 pages in the proceedings. There were also 15 submissions that received positive reviews; these were accepted as short papers and were allocated eight pages in the proceedings. From the accepted papers, one full paper and three short ones were withdrawn. This volume contains the remaining 30 full and 12 short papers, presented in eight topical sections: Knowledge Representation, Reasoning, and Management; Multi-agent Systems; Data Mining and Machine Learning; Computer Vision; Robotics; AI in Bioinformatics; AI in Security and Networks; and Other AI Applications.

The technical program of MIWAI 2015 included the presentations of all accepted papers, but also two keynote talks from two prominent members of the AI community: Prof. Luc De Raedt from Katholieke Universiteit Leuven, Belgium, and Prof. Fangzhen Lin, from the Hong Kong University of Science and Technology. We would like to thank the keynote speakers for accepting our invitation and for their very interesting talks.

We also thank the authors for submitting their papers, responding to the reviewers' comments, and abiding by our production schedule, and especially those travelling from far away to come to Fuzhou and participate in the workshop. We thank the excellent Program Committee for their hard work in reviewing the submitted papers. Their criticism and very useful comments and suggestions were instrumental in achieving a high-quality technical program and a first-class publication.

A special thanks goes to the local organizers in Fuzhou for the excellent hospitality and for making all the necessary arrangements for the workshop. Last but not least, we would like to thank our sponsors and supporters. Their contribution and support were crucial to the success of MIWAI 2015.

September 2015

Antonis Bikakis
Xianghan Zheng

Organization

Steering Committee

Arun Agarwal	University of Hyderabad, India
Rajkumar Buyya	University of Melbourne, Australia
Patrick Doherty	University of Linkoping, Sweden
Jrme Lang	University of Paris-Dauphine, France
James F. Peters	University of Manitoba, Canada
Srinivasan Ramani	IIIT Bangalore, India
C. Raghavendra Rao	University of Hyderabad, India
Leon Van Der Torre	University of Luxembourg, Luxembourg

Conveners

Richard Booth	Mahasarakham University, Thailand
Chattrakul Sombattheera	Mahasarakham University, Thailand

General Chairs

Shuying Cheng	Fuzhou University, China
Weixing Wang	KTH, Sweden
Grigoris Antoniou	University of Huddersfield, UK

Program Chairs

Antonis Bikakis	University College London, UK
Xianghan Zheng	Fuzhou University, China

Publicity Chairs

Jielin Xu	Fuzhou University, China
Wenzhong Guo	Fuzhou University, China
Rapeeporn Chamchong	Mahasarakham University, Thailand

Workshop Administrator

Zhicong Chen	Fuzhou University, China

Web Administrator

Panich Sudkhot	Mahasarakham University, Thailand

Program Committee

Arun Agarwal	University of Hyderabad, India
Costin Badica	University of Craiova, Romania
Sotiris Batsakis	University of Huddersfield, UK
Antonis Bikakis	UCL, UK
Hima Bindu	University of Hyderabad, India
Veera Boonjing	King Mongkut's Institute of Technology Ladkrabang, Thailand
Richard Booth	Mahasarakham University, Thailand
Darko Brodic	University of Belgrade, Serbia
Patrice Caire	University of Luxembourg, Luxembourg
Rapeeporn Chamchong	Mahasarakham University, Thailand
Dehua Chen	Donghua University, China
Zhicong Chen	Fuzhou University, China
Pham Cong Thien	Ho Chi Minh City - Nong Lam University, Vietnam
Broderick Crawford	Pontificia Universidad Catolica de Valparaiso, Chile
Tiago De Lima	University of Artois and CNRS, France
Patrick Doherty	Linköping University, Sweden
Vladimir Estivill-Castro	Griffith University, Australia
Giorgos Flouris	FORTH-ICS, Greece
Chun Che Fung	Murdoch University, Australia
Ulrich Furbach	University of Koblenz, Germany
Guido Governatori	NICTA, Australia
Jingzhi Guo	University of Macau, SAR China
Wenzhong Guo	University of Fuzhou, China
Christos Hadjinikolis	King's College London, UK
Sachio Hirokawa	Kyushu University, Japan
Bingquan Huang	University College Dublin, Ireland
Xinyi Huang	Fujian Normal University, China
Zhisheng Huang	Vrije University of Amsterdam, The Netherlands
Jason Jung	Yeungnam University, South Korea
Manasawee Kaenampornpan	Mahasarakham University, Thailand
Mohan Kankanhalli	National University of Singapore, Singapore
Jérôme Lang	Université Paris-Dauphine, France
Kittichai Lavangnananda	King Mongkut's University of Technology Thonburi (KMUTT), Thailand
Jimmy Lee	The Chinese University of Hong Kong, SAR China
Changlu Lin	Fujian Normal University, China
Pawan Lingras	Saint Mary's University, Canada
Martin Lukac	Nazarbayev University, Kazakhstan
Dickson Lukose	MIMOS Berhad, Knowledge Technology, Malaysia
Michael Maher	UNSW Canberra, Australia
Bm Mehtre	IDRBT, India
Jérôme Mengin	IRIT - Université de Toulouse, France

Rajeev Wankar   University of Hyderabad, India
Paul Weng   LIP6, France
Kevin Kok Wai Wong   Murdoch University, Australia
Jingtao Yao   University of Regina, Canada
Wai Yeap   Auckland University of Technology, New Zealand
Lim Tek Yong   Multimedia University, Malaysia
Cheah Yu-N   Universiti Sains Malaysia, Malaysia
Xianghan Zheng   Fuzhou University, China
Dominik Ślezak   University of Warsaw and Infobright Inc., Poland

Additional Reviewers

Alex Becheru   Francesco Kriegel   Yu Wang
Filippos Gouidis   Thien Pham   Yuexin Zhang
Nattiya Khaitiyakun   Hiran V. Nath

Keynote Abstracts

Probabilistic Programming
and Its Applications
(Keynote Abstract)

Luc De Raedt

Department of Computer Science, KU Leuven, Belgium
luc.deraedt@cs.kuleuven.be

Abstract. Probabilistic logic programs [4] combine the power of a programming
language with a possible world semantics; they are typically based on Sato's
distribution semantics [9, 8], and it is possible to learn their parameters and to
some extent also their structure. They have been studied for over twenty years
now. In this talk, I shall introduce the state of the art in probabilistic logic
programs and report on some recent progress in applying this paradigm to
challenging applications. The first application domain will be that of robotics,
where we have developed extensions of the basic distribution semantics to cope
with dynamics as well continuous distributions [5]. The resulting representations
are now being used to learn multi-relational object affordances, which specify
the conditions under which actions can be applied on particular objects [6, 7].
The second application is in a biological domain, where a decision theoretic
extension of the distribution semantics [10] is the underlying inference engine
of the PheNetic system [2], which extracts from an interactome, the sub-network
that best explains genes prioritized through a molecular profiling experiment.
Finally, I shall report on our results in applying ProbFOIL [3] to the problem of
machine reading in CMU's Never Ending Language Learning system [1].
ProbFOIL is an extension of the traditional rule-learning system FOIL for use
with the distribution semantics.

References

[1] Carlson, A., Betteridge, J., Kisiel, B., Settles, B., Hruschka, E.R. Jr., Mitchell, T.M.: Toward
an architecture for never-ending language learning. In: AAAI, vol. 5, p. 3 (2010)

[2] De Maeyer, D., Renkens, J., Cloots, L., De Raedt, L., Marchal, K.: Phenetic: network-based
interpretation of unstructured gene lists in E. Coli. Mol. Biosyst. 9(7), 1594–1603 (2013)

[3] De Raedt, L., Dries, A., Thon, I., Van den Broeck, G., Verbeke, M.: Inducing probabilistic
relational rules from probabilistic examples. In: Proceedings of 24th International Joint
Conference on Artificial Intelligence (IJCAI) (2015)

[4] De Raedt, L., Kimmig, A.: Probabilistic (logic) programming concepts. Mach. Learn. 1–43
(2015)

[5] Gutmann, B., Thon, I., Kimmig, A., Bruynooghe, M., De Raedt, L.: The magic of logical
inference in probabilistic programming. Theory Pract. Logic Program. (TPLP) 11(4–5), 663–
680 (2011)

[6] Moldovan, B., Moreno, P., van Otterlo, M., Santos-Victor, J., De Raedt, L.: Learning relational affordance models for robots in multi-object manipulation tasks. In: 2012 IEEE International Conference on Robotics and Automation (ICRA), pp. 4373–4378. IEEE (2012)

[7] Nitti, D., De Laet, T., De Raedt, L.: A particle filter for hybrid relational domains. In: 2013 IEEE/RSJ International Conference on Intelligent Robots and Systems (IROS), pp. 2764–2771. IEEE (2013)

[8] Poole, D.: Probabilistic Horn abduction and Bayesian networks. Artif. Intell. **64**, 81–129 (1993)

[9] Sato, T.: A statistical learning method for logic programs with distribution semantics. In: Proceedings of the 12th International Conference on Logic Programming (ICLP 1995) (1995)

[10] Van den Broeck, G., Thon, I., van Otterlo, M., De Raedt, L.: DTProbLog: a decision-theoretic probabilistic prolog. In: Proceedings of the 24th AAAI Conference on Artificial Intelligence (AAAI 2010) (2010)

A Formalization of Programs
in First-Order Logic with a Discrete Linear Order
(Keynote Abstract)

Fangzhen Lin

Department of Computer Science
The Hong Kong University of Science and Technology
Clear Water Bay, Kowloon, Hong Kong

The Voice: If you build it, he will come.
— *Field of Dreams (1989)*

Computer programs are among the most complex man-made systems. Given their widespread uses, many of them in critical applications, their reliability is of utmost importance. There have been many formalisms and methods proposed for reasoning about computer programs, and many techniques and methodologies for designing and debugging programs. In this talk, I will present my recent work on translating computer programs to first-order logic with quantification over natural numbers. I describe how this can be done for a core non-concurrent procedural language with loops. The key feature of this approach is that the translated first-order theory captures all the behaviors of the input program (under a standard execution model) so that whatever one wants to know about the program, one can find out using this first-order theory. My goal is to eventually extend this to include all computer programs, large or small.

This talk is based on the following two papers:

1. Fangzhen Lin. A Formalization of Programs in First-Order Logic with a Discrete Linear Order. In *Proceedings of KR-2014*. An extended version of this paper is available at http://www.cs.ust.hk/faculty/flin/papers/program-fol.pdf
2. Fangzhen Lin and Bo Yang. Reasoning about Mutable Data Structures in First-Order Logic with Arithmetic: Lists and Binary Trees Technical Report, 2015. http://www.cs.ust.hk/faculty/flin/papers/dsw2015.pdf.

Contents

Computer Vision

Robotics

AI in Bioinformatics

AI in Security and Networks

Other AI Applications

Knowledge Representation, Reasoning and Management

Representing Time for the Semantic Web

Sotiris Batsakis, Ilias Tachmazidis[✉],
and Grigoris Antoniou

Department of Informatics, University of Huddersfield,
Queensgate, Huddersfield, West Yorkshire HD1 3DH, UK
{s.batsakis,ilias.tachmazidis,g.antoniou}@hud.ac.uk

Abstract. Representation of temporal information for the Semantic
Web often involves qualitative defined information (i.e., information
described using natural language terms such as "before"), since precise
dates are not always available in addition to quantitative defined temporal information. This work proposes several representations for time
points and intervals in ontologies by means of OWL properties and reasoning rules in SWRL embedded into the ontology. Although qualitative representations for interval and point relations exist, in addition to
quantitative ones, this is the first work proposing representations combining qualitative and quantitative information for the Semantic Web.
In addition to this, several existing and proposed approaches are compared using different reasoners and experimental results are presented
in detail. Experimental results illustrate that reasoning performance dif-
fers greatly between different representations and reasoners. To the best
of our knowledge this is the first such experimental evaluation of both
qualitative and quantitative Semantic Web temporal representations.

Keywords: Temporal representation and reasoning · Semantic Web ·
Rules

1 Introduction

Understanding the meaning of Web information requires formal definitions of
concepts and their properties, using the Semantic Web Ontology definition language OWL[1]. This language provides the means for defining concepts, their
properties and their relations, allowing for reasoning over the definitions and the
assertions of specific individuals using reasoners such as Pellet [12] and HermiT
[13]. Furthermore, reasoning rules can be embedded into the ontology using the
SWRL rule language.

Temporal information is an important aspect of represented objects in many
application areas involving change. Temporal information in turn can be defined
using quantitative (e.g., using dates) and qualitative terms (i.e., using natural
language expressions such as "During"). Quantitative approaches are used for

---

[1] http://www.w3.org/TR/owl-features/.

© Springer International Publishing Switzerland 2015
A. Bikakis and X. Zheng (Eds.): MIWAI 2015, LNAI 9426, pp. 3–15, 2015.
DOI: 10.1007/978-3-319-26181-2_1

example in [10] and in OWL-Time [7]. Qualitative temporal terms have specific semantics which can be embedded into the ontology using reasoning rules. In previous work [2,11] such a representation is proposed for qualitative defined temporal information in OWL, but combining qualitative and quantitative information was not supported.

Current work deals exactly with the case of combined qualitative and quantitative information which is more expressive than the representation proposed in [2]. In addition, reasoning performance using different point and temporal representations and reasoners is evaluated. Each point in time can be represented quantitatively using a date or qualitatively using relations with other points. These relations are *before*, *after* or *equals*. Intervals can be defined using their end-points, which in turn can be defined using dates or point relations. Alternatively intervals can be defined using qualitative interval relations. Specifically, between each pair of intervals qualitative relations are asserted (e.g., "Before" or "During"). These relations represent the relative placement of intervals along the axis of time [3].

Reasoning can be applied for interfering point relations using either dates or qualitative relations or both. In case of dates, SWRL[2] rules are used combined with support for date datatypes by the reasoner. In case of qualitative point relations both OWL axioms and SWRL rules can be used. Both approaches and their combination with dates are evaluated. Intervals with specific end points can be representing by attaching two dates (start and end) directly to an interval as datatype properties or by attaching the dates to points which in turn are associated with intervals. When end-points are not defined using dates, qualitative point relations such as *after* can be used. Alternatively, interval relations can be inferred using directly reasoning over Allen relations [3] between intervals, instead of reasoning over points and then extracting the interval relations. Again all approaches are evaluated using different reasoners.

Although many different representations are proposed and evaluated, all of them are based on existing standards and recommendations using OWL and SWRL. Embedding reasoning rules into the ontology makes sharing of data easier since all SWRL compliant reasoners (such as Pellet and HermiT) can be used for temporal reasoning. To the best of our knowledge, this is the first work proposing combined qualitative and quantitative representations for the Semantic Web, and provides an evaluation of these approaches.

Current work is organized as follows: related work in the field of temporal knowledge representation is discussed in Sect. 2. The proposed representations are presented in Sect. 3 and the corresponding reasoning mechanisms in Sect. 4. The combined interval-point reasoning mechanism is presented in Sect. 5. Evaluation is presented in Sect. 6 and conclusions and issues for future work in Sect. 7.

2   Background and Related Work

Definition of ontologies for the Semantic Web is achieved using the Web Ontology Language OWL. Specifically, the current W3C standard is the OWL 2[3] language,

---

[2] http://www.w3.org/Submission/SWRL/.
[3] http://www.w3.org/TR/owl2-overview/.

offering increased expressiveness while retaining decidability of basic reasoning tasks. Reasoning tasks are applied both on concept and property definitions into the ontology (TBox), and on assertions of individual objects and their relations (ABox). Reasoners include among others Pellet[4] and HermiT[5]. Reasoning rules can be embedded into the ontology using SWRL. To guarantee decidability, the rules are restricted to *DL-safe rules* that apply only on named individuals in the ontology ABox.

Qualitative temporal reasoning (i.e., inferring implied relations and detecting inconsistencies in a set of asserted relations) typically corresponds to Constraint Satisfaction problems which are NP, but tractable sets (i.e., solvable by polynomial algorithms) are known to exist [4]. These tractable sets (i.e., sets of qualitative relations between intervals) form the basis of the current work. Relations between dynamic (i.e. evolving in time and having time dependent properties) entities in ontologies are typically represented using Allen temporal relations of Fig. 1.

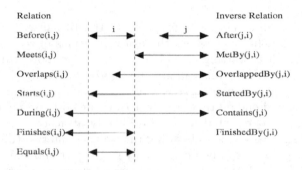

Fig. 1. Allen temporal relations

Embedding temporal reasoning into the ontology, by means of SWRL rules applied on temporal intervals, forms the basis of the SOWL model proposed in [1] and the CHRONOS system [9]. CHRONOS and the underlying SOWL model were both not addressing the issue of combined qualitative and quantitative representation using different approaches and reasoners. Thus, the selection of the most efficient representation with respect to the type of available data remained an open issue [14]. Current work addresses this issue and to the best of our knowledge is the first such work for temporal representations for the Semantic Web.

3   Temporal Representation

This work deals with qualitative relations between points in addition to interval Allen relations. Qualitative relations of two points are represented using an

---

[4] http://clarkparsia.com/pellet/.
[5] http://hermit-reasoner.com/.

object property specifying their relative position on the axis of time. Specifically between two points three relations can hold, these relations are "<", ">", "=" also referred to as *before, after* and *equals* respectively. If date/time is available then a corresponding datatype property can be used. Qualitative an quantitative representations can be combined (Fig. 2).

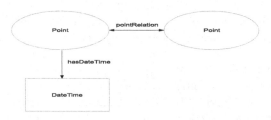

Fig. 2. Point representations

An interval temporal relation can be one of the 13 pairwise disjoint Allen relations [3] of Fig. 1. In cases where the exact durations of temporal intervals are unknown (i.e., their starting or ending points are not specified), their temporal relations to other intervals can still be asserted qualitatively by means of temporal relations (e.g., "interval i_1 is before interval i_2" even in cases where the exact starting and ending time of either i_1, i_2, or both are unknown).

Intervals can be represented using two directly attached datatype properties, corresponding to starting and ending time of each interval (Fig. 3(a)). This straightforward approach can be applied only when start and end time of intervals are known. Interval relations can be inferred using comparisons of starting/ending dates using SWRL rules. Another more flexible and more complex approach is presented in Fig. 3(b). In this case intervals are related with starting and ending points, and not directly with dates. These points can be associated with dates, as in Fig. 2, and/or with other points using point relations (such as *after*). Point relations can be inferred using comparisons of dates and/or reasoning rules over asserted point relations. When point relations are inferred, then Allen relations between intervals can be inferred using SWRL rules implementing the definitions of Fig. 1. Finally, reasoning over qualitative defined Allen relations can be applied directly without using dates or points as in Fig. 3(c).

Besides temporal property definitions, additional OWL axioms are required for the proposed representation; basic relations are pairwise disjoint i.e., "<", ">" and "=" point relations are pairwise disjoint and all Allen relations of Fig. 1 are pairwise disjoint as well. In addition, "<" is inverse of ">", while "=" is symmetric and transitive. Also *Before* is *inverse* of *After*, *Meets* is *inverse* of *MetBy*, *During* is *inverse* of *Contains*, *Finishes* is *inverse* of *FinishedBy*, *Starts* is *inverse* of *startedBy* and *Overlaps* is *inverse* of *OverlappedBy*. *Equals* is symmetric and transitive.

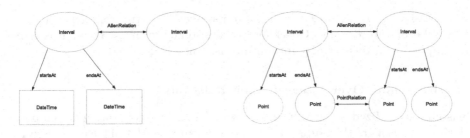

(a) Direct Interval Representation   (b) Point-Interval Representation

(c) Allen-Based Interval Representation

Fig. 3. Example of (a) Direct (b) Point Based (c) Allen Based Interval Representations

4   Temporal Reasoning

Inferring implied relations and detecting inconsistencies are handled by a reasoning mechanism. In the case of qualitative relations, assertions of relations holding between temporal entities (i.e., intervals) restrict the possible assertions holding between other temporal entities in the knowledge base. Then, reasoning on qualitative temporal relations can be transformed into a *constraint satisfaction problem*, which is known to be an *NP-hard* problem in the general case [4]. Inferring implied relations is achieved by specifying the result of *compositions* of existing relations. Specifically, when a relation (or a set of possible relations) R_1 holds between intervals i_1 and i_2 and a relation (or a set of relations) R_2 holds between intervals i_2 and i_3 then, the *composition* of relations R_1, R_2 (denoted as $R_1 \circ R_2$) is the set (which may contain only one relation) R_3 of relations holding between i_1 and i_3.

Qualitative relations under the intended semantics may not apply simultaneously between a pair of individuals. For example, given the time intervals i_1 and i_2, i_1 can not be simultaneously *before* and *after* i_2. Typically, in temporal representations (e.g., using Allen relations), all basic relations (i.e., simple relations and not disjunctions of relations) are pairwise disjoint. When disjunctions of basic relations hold true simultaneously, then their set intersection holds true as well. For example, if i_1 is *before or equals* i_2 and simultaneously i_1 is *after or equals* i_2 then i_1 *equals* i_2. In case the intersection of two relations is empty, these relations are disjoint. Checking for consistency means, whenever asserted and implied relations are disjoint, an inconsistency is detected.

4.1 Reasoning Over Interval Allen Relations

Reasoning is realized by introducing a set of SWRL rules operating on temporal relations. Reasoners that support DL-safe rules such as HermiT can be used for inference and consistency checking over Allen relations. The temporal reasoning rules for Allen relations are based on the composition of pairs of the basic Allen relations of Fig. 1 as defined in [3]. Specifically, if relation R_1 holds between intervals i_1 and i_2, and relation R_2 holds between intervals i_2 and i_3, then the composition table defined in [3] denotes the possible relation(s) holding between intervals i_1 and i_3. Not all compositions yield a unique relation as a result. For example, the composition of relations *During* and *Meets* yields the relation *Before* as a result, while the composition of relations *Overlaps* and *During* yields three possible relations namely *Starts, Overlaps and During*.

A series of compositions of relations may yield relations which are inconsistent with existing ones (e.g., if i_1 *before* i_2 is inferred using compositions, a contradiction arises if i_1 *after* i_2 has been also asserted into the ontology). Reasoning over temporal relations is known to be an NP-hard problem and identifying tractable cases of this problem has been in the center of many research efforts over the last few years [4]. The notion of *k-consistency* is very important in this research. Given a set of n intervals with relations asserted between them imposing certain restrictions, *k-consistency* means that every subset of the n intervals containing at most k intervals does not contain an inconsistency. Notice that, checking for all subsets of n entities for consistency is exponential on the n.

There are cases where, although *k-consistency* does not imply *n-consistency* in general, there are specific sets of relations R_t (which are subsets of the set of all possible disjunctions of basic relations R), with the following property: if asserted relations are restricted to this set, then *k-consistency* implies *n-consistency* and R_t is a *tractable set* of relations or a *tractable subset of R* [4]. Tractable sets of Allen interval algebra have been identified in [8] and tractable subsets for Point Algebra have been identified in [6]. Additional tractable sets for Allen relations have been identified in [1]. Consistency checking is achieved by ensuring path consistency by applying formula:

$$\forall x, y, k \; R_s(x, y) \leftarrow R_i(x, y) \cap (R_j(x, k) \circ R_k(k, y))$$

representing intersection of compositions of relations with existing relations (symbol $\cap$ denotes intersection, symbol $\circ$ denotes composition and R_i, R_j, R_k, R_s denote temporal relations). The formula is applied until a fixed point is reached (i.e., the application of the rules above does not yield new inferences) or until the empty set is reached, implying that the ontology is inconsistent. Implementing path-consistency formula (using SWRL in this work) requires rules for both compositions and intersections of pairs of relations.

Compositions of relations R_1 and R_2 yielding a unique relation R_3 as a result are expressed in SWRL using rules of the form:

$$R_1(x, y) \wedge R_2(y, z) \rightarrow R_3(x, z)$$

The following is an example of such a composition rule:

$$Before(x, y) \wedge Contains(y, z) \rightarrow Before(x, z)$$

Rules yielding a set of possible relations cannot be represented directly in SWRL since, disjunctions of atomic formulas are not permitted as a rule head. Instead, disjunctions of relations are represented using new relations whose compositions must also be defined and asserted into the knowledge base. For example, the composition of relations $Overlaps$ and $During$ yields the disjunction of three possible relations ($During$, $Overlaps$ and $Starts$) as a result:

$$Overlaps(x, y) \wedge During(y, z) \rightarrow During(x, z) \vee Starts(x, z) \vee Overlaps(x, z)$$

If the relation DOS represents the disjunction of relations $During$, $Overlaps$ and $Starts$, then the composition of $Overlaps$ and $During$ can be represented using SWRL as follows:

$$Overlaps(x, y) \wedge During(y, z) \rightarrow DOS(x, z)$$

The set of possible disjunctions over all basic Allen's relations contains 2^{13} relations, and complete reasoning over all temporal Allen relations has exponential time complexity. However, tractable subsets of this set that are closed under composition (i.e., compositions of relation pairs from this subset yield also a relation in this subset) are also known to exist [6,8]. In this work, we use the subset identified in [1].

An additional set of rules defining the result of intersection of relations holding between two intervals is also required. These rules are of the form:

$$R_1(x, y) \wedge R_2(x, y) \rightarrow R_3(x, y),$$

where R_3 can be the empty relation. For example, the intersection of relation DOS (represents the disjunction of $During$, $Overlaps$ and $Starts$) with relation $During$, yields relation $During$ as a result:

$$DOS(x, y) \wedge During(x, y) \rightarrow During(x, y).$$

The intersection of relations $During$ and $Starts$ yields the empty relation, and an inconsistency is detected:

$$Starts(x, y) \wedge During(x, y) \rightarrow \perp.$$

Thus, path consistency is implemented by defining compositions and intersections of relations using SWRL rules and OWL axioms for inverse relations as presented in Sect. 3.

4.2 Reasoning Over Point Relations

Possible relations between points are *before*, *after* and *equals*, denoted as "$<$", "$>$", "$=$" respectively. Table 1 illustrates the set of reasoning rules defined on the composition of existing relation pairs. The three temporal relations are declared as pairwise disjoint, since they cannot simultaneously hold between two points. Not all compositions yield a unique relation as a result. For example, the composition of relations *before* and *after* yields all possible relations as a result. Because such compositions do not yield new information these rules are discarded. Rules corresponding to compositions of relations R_1 *and* R_2 yielding a unique relation R_3 as a result are retained (7 out of the 9 entries of Table 1 are retained), and are directly expressed in SWRL. The following is an example of such a temporal inference rule:

$$before(x, y) \wedge equals(y, z) \rightarrow before(x, z)$$

Table 1. Composition Table for point-based temporal relations.

Relations	$<$	$=$	$>$
$<$	$<$	$<$	$<, =, >$
$=$	$<$	$=$	$>$
$>$	$<, =, >$	$>$	$>$

Therefore, 7 out of 9 entries in Table 1 can be expressed using SWRL rules, while the two remaining entries do not convey new information. A series of compositions of relations may imply relations which are inconsistent with existing ones. In addition to rules implementing compositions of temporal relations, a set of rules defining the result of intersecting relations holding between two instances must also be defined in order to check consistency.

For example the intersection of relations *before* and *after* yields the empty relation, and an inconsistency is detected (i.e., they cannot hold simultaneously between two points). As shown in Table 1, compositions of relations may yield one of the following four relations: *before, after, equals* and the disjunction of these three relations. Intersecting the disjunction of all three relations with any of these leaves existing relations unchanged. Intersecting any one of the three basic (non disjunctive) relations with itself also leaves existing relations unaffected. Only intersections of pairs of different basic relations affect the ontology by yielding the empty relation as a result, thus detecting an inconsistency. By declaring the three basic relations *before, after, equals* as pairwise disjoint, all intersections that can affect the ontology are defined. Thus, checking consistency of point relations is implemented by defining compositions of relations using SWRL rules and by declaring the three basic relations as disjoint (no intersection rules are needed). In case of quantitative relations (i.e., using dates) then qualitative relations are extracted by comparing dates using SWRL rules.

Compositions of relations can be also expressed using OWL *Role Inclusion Axioms* [5] instead of SWRL rules. For example, the composition of *before* and *equals* can be expressed using the following axiom:

$$before \circ equals \sqsubseteq before$$

Properties involved in Role Inclusion Axioms (RIA) cannot be combined with disjointness axioms in OWL, but in case the consistency of point relations is guaranteed, then this approach can be used instead of SWRL rules. In addition to this, Role Inclusion Axioms can be used in conjunction with date/times in a combined qualitative/quantitative approach. In total, based on the reasoning mechanism, 5 different representations for points have been implemented:

- Quantitative Point Representation (**P1**): Relations are extracted by comparing date/time values using SWRL.
- Qualitative Only using SWRL (**P2**): Only qualitative point relations are asserted and reasoning using Path Consistency implemented in SWRL is applied.
- Qualitative Only using Role Inclusion Axioms (**P3**): Only qualitative point relations are asserted and reasoning using Path Consistency is implemented using OWL 2 Role Inclusion Axioms.
- Combined representation using SWRL (**P4**): Both dates and qualitative relations are asserted and reasoning mechanism combines rules from representations P1 and P2.
- Combined representation using OWL Role Inclusion Axioms (**P5**): Both dates and qualitative relations are asserted and reasoning mechanism combines SWRL rules from representations P1 and OWL axioms from P3.

5   Combining Interval and Point Representation and Reasoning

In addition to the Allen based Interval representation (see Fig. 3(c)) and the corresponding reasoning mechanism of Sect. 4.1, interval relations can be extracted using endpoint relations and/or comparisons of dates of endpoints. Using the direct representation of Fig. 3(a), Allen relations are extracted using end-point date comparisons. There are 13 SWRL rules, one for each basic Allen relation. For example the Allen *intervalMeets* (or *Meets*) relation is inferred using the following rule:

$$ProperInterval(a) \wedge ProperInterval(x) \wedge endValue(x, z1) \wedge startValue(a, b1) \wedge$$
$$lessThanOrEqual(b1, z1) \wedge lessThanOrEqual(z1, b1) \rightarrow intervalMeets(x, a)$$

In case of the representation of Fig. 3(b) both date comparisons and point algebra reasoning from Sect. 4 are applied for inferring qualitative relations between end-points. Allen relations between intervals are then inferred using relations of end-points. There are 13 SWRL rules, one for each basic Allen relation. For example, the rule for extracting the interval *Meets* relation is:

$$ProperInterval(a) \land ProperInterval(x) \land equals(z, b) \land hasBeginning(a, b) \land$$
$$hasEnd(x, z) \rightarrow intervalMeets(x, a)$$

Based on the above rules and the representations of Sect. 3 (see Fig. 3), five different interval representations have been implemented[6].

- Allen-based Interval Representation (**I1**): Qualitative Allen relations only are asserted directly between intervals (points are not used, see Fig. 3(c)) combined with the reasoning mechanism of Sect. 4.1.
- Quantitative Only-direct intervals (**I2**): Only dates/times are asserted attached directly to intervals (see Fig. 3(a)) and Allen relations are extracted by date/time comparisons.
- Quantitative Only using Points (**I3**): Only dates/times are asserted attached to Points representing end-points of intervals (see Fig. 3(b)) and Allen relations are extracted by date/time comparisons.
- Qualitative Only Point Based Interval representation (**I4**): Only qualitative relations between points (see Fig. 3(b)) are asserted and reasoning mechanism is based on Point reasoning rules from Sect. 4.2 and Allen extraction rules from Sect. 5).
- Combined qualitative/quantitative Interval representation (**I5**): Both dates and qualitative relations between points are asserted (see Fig. 3(b)) and date/time comparisons are combined with SWRL rules of Sect. 4.2 and Allen extraction rules from Sect. 5).

Table 2. Comparison of point and interval representations

Representation	P1	P2	P3	P4	P5	I1	I2	I3	I4	I5
Qualitative	No	Yes	Yes	Yes	Yes	Yes	No	No	Yes	Yes
Quantitative	Yes	No	No	Yes	Yes	No	Yes	Yes	No	Yes
Reasoning support: HermiT (H), Pellet (P)	P	H, P	H, P	P	P	H, P	P	P	H, P	P
Consistency checking	N/A	Yes	No	Yes	No	Yes	N/A	N/A	Yes	Yes

6 Evaluation

The required expressiveness of the proposed representations is within the limits of OWL 2 expressiveness combined with SWRL and date/time datatypes.

---

[6] We have made all point and interval representations available on the Web at: https://github.com/sbatsakis/TemporalRepresentations.

Thus, reasoners such as Pellet and HermiT can be used for reasoning. Reasoning mechanism is tractable since it consists of Date/time comparisons and/or path consistency using SWRL [8]. A summary of all proposed representations is presented in Table 2.

Notice that quantitative only approaches don't need to perform consistency checking since date/time assertions represent a valid instantiation of such values, while qualitative assertions my impose restrictions that cannot be satisfied. Also to the best of our knowledge HermiT and Pellet are the only reasoners currently supporting SWRL, while only Pellet currently supports date/time comparisons needed for SWRL rules used by quantitative approaches.

6.1 Experimental Evaluation

Measuring the efficiency of the proposed representations requires temporal intervals and points as defined in Sect. 3 containing instances. Thus, datasets of various sizes containing points and intervals, both qualitative (using relations) and quantitative (using dates) generated randomly were used for the experimental evaluation. Reasoning response times of the temporal reasoning rules are measured as the average over 10 runs. HermiT 1.3.8 and Pellet 2.3.0 running as a Java library were the reasoners used in the experiments. All experiments where run on a PC, with Intel Core CPU at 2.4 GHz, 6 GB RAM, and Windows 7.

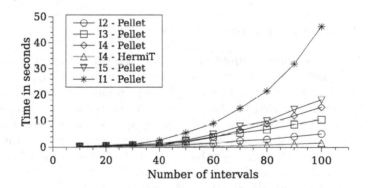

Fig. 4. Average reasoning time as a function of the number of intervals

Measurements illustrate that there are major differences in performance between various approaches, and reasoners. Interval representations can be used for reasoning over 100 intervals, while qualitative representation combined with HermiT reasoner (representation I1 with HermiT, not presented in Fig. 4) can reason over 500 intervals in 133.1 s when using Allen relations directly (representation I1). For 100 intervals corresponding time is 2.1 s respectively, clearly outperforming representations of Fig. 4.

Point representations can be used for reasoning over 500 points efficiently (with the exception of qualitative representations using SWRL -P2 and P4- and

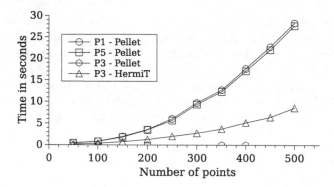

Fig. 5. Average reasoning time as a function of the number of points

Pellet, that can be practically used for at most 100 points and they are not presented in Fig. 5, while reasoning time for representation P2 using HermiT over 500 points is 286 s, thus slower than all measurements presented in Fig. 5). An interesting case is the representation based on Role Inclusion Axioms (P3) that can be used for reasoning over 100 K points in less than 3 s when using Pellet (but not when using HermiT, see Fig. 5) being orders of magnitude faster than all other approaches. This illustrates that there is clearly room for optimization on SWRL implementations of current reasoners.

7   Conclusions and Future Work

In this work, several representations for handling points and interval relations in OWL ontologies are presented. The proposed representations are fully compliant with existing Semantic Web standards and specifications, which increases their applicability. Being compatible with W3C specifications the proposed representations can be used in conjunction with existing editors, reasoners and querying tools such as Protégé and HermiT without requiring specialized software. Therefore, information can be easily distributed, shared and modified. Directions of future work include the development of real-world applications based on the proposed mechanism and optimizations of reasoning engines. Furthermore, parallelizing our rule based reasoning mechanisms is another promising direction of future research.

References

1. Batsakis, S., Petrakis, E.G.M.: SOWL: a framework for handling spatio-temporal information in OWL 2.0. In: Bassiliades, N., Governatori, G., Paschke, A. (eds.) RuleML 2011 - Europe. LNCS, vol. 6826, pp. 242–249. Springer, Heidelberg (2011)
2. Batsakis, S., Stravoskoufos, K., Petrakis, E.G.M.: Temporal reasoning for supporting temporal queries in OWL 2.0. In: 15th International Conference on Knowledge-Based and Intelligent Information and Engineering Systems (KES 2011), pp. 558–567 (2011)

3. Allen, J.F.: Maintaining knowledge about temporal intervals. Commun. ACM **26**(11), 832–843 (1983)
4. Renz, J., Nebel, B.: Qualitative spatial reasoning using constraint calculi. In: Handbook of Spatial Logics, pp. 161–215, Springer, Netherlands (2007)
5. Horrocks, I., Kutz, O., Sattler, U.: The even more irresistible SROIQ. In: Proceedings of the KR 2006, Lake District, UK (2006)
6. van Beek, P., Cohen, R.: Exact and approximate reasoning about temporal relations. Comput. Intell. **6**(3), 132–147 (1990)
7. Hobbs, J.R., Pan, F.: Time ontology in OWL. W3C Working Draft, September 2006
8. Nebel, B., Burckert, H.J.: Reasoning about temporal relations: a maximal tractable subclass of Allen's interval algebra. J. ACM (JACM) **42**(1), 43–66 (1995)
9. Preventis, A., Makri, X., Petrakis, E., Batsakis, S.: CHRONOS Ed: a tool for handling temporal ontologies in protégé. Int. J. Artif. Intell. Tools **23**(4) (2014)
10. Noy, N., Rector, A., Hayes, P., Welty, C.: Defining N-ary relations on the semantic web. W3C Working Group Note, 12 April 2006
11. Batsakis, S., Antoniou, G., Tachmazidis, I.: Integrated representation of temporal intervals and durations for the semantic web. In: Bassiliades, N., Ivanovic, M., Kon-Popovska, M., Manolopoulos, Y., Palpanas, T., Trajcevski, G., Vakali, A. (eds.) New Trends in Database and Information Systems II. AISC, vol. 312, pp. 147–158. Springer, Heidelberg (2015)
12. Sirin, E., Parsia, B., Grau, B.C., Kalyanpur, A., Katz, Y.: Pellet: a practical owl-dl reasoner. Web Semant.: Sci. Serv. Agent. World Wide Web **5**(2), 51–53 (2007)
13. Shearer, R., Motik, B., Horrocks, I.: HermiT: a highly-efficient OWL reasoner. In: OWLED, vol. 432 (2008)
14. Ermolayev, V., Batsakis, S., Keberle, N., Tatarintseva, O., Antoniou, G.: Ontologies of time: review and trends. Int. J. Comput. Sci. Appl. **11**(3), 57–115 (2014)

Construction of P-Minimal Models Using Paraconsistent Relational Model

Badrinath Jayakumar$^{(\boxtimes)}$ and Rajshekhar Sunderraman

Department of Computer Science, Georgia State University, Atlanta, GA, USA
{bjayakumar2,raj}@cs.gsu.edu
http://www.cs.gsu.edu/

Abstract. Positive extended disjunctive deductive databases are those that contain explicit negation both in the head and body of the clauses. For such databases, paraconsistent minimal models (p-minimal models) have been proposed based on multi-valued logic (four-valued logic). Moreover, the paraconsistent relational model is also based on four-valued logic. In this paper, we propose an algorithm, which converts clauses to equations and solves it, to find p-minimal models using the paraconsistent relational model. In order to accomplish that, we use disjunctive paraconsistent relation model.

Keywords: Inconsistency · Paraconsistent relational model · Fixed-point semantics · Four-valued logic

1 Introduction

The paraconsistent relational model moves a step forward and completes the relational model by representing both positive and negative information for any given relation. The model was first proposed by Bagai and Sunderraman [5]. The authors have given two applications for the paraconsistent relational model: weak well-founded semantics [5] and well-founded semantics [7] for general deductive databases. Bagai and Sunderraman find the models by constructing a system of algebraic equations for the clauses in the database. There are two advantages for this approach: it operates on a set of tuples instead of "tuple-at-a-time" basis and the algebraic expression in the algebraic equation can be optimized based on various laws of equality. The optimizations are similar to the ordinary relations case where selections and projections are pushed deeper into expressions whenever possible [5].

Paraconsistent logic [4,10,13] does not trivialize the result in the presence of inconsistent information. Four-valued logic [8], which is a type of paraconsistent logic, was introduced in logic programming by Blair and Subrahmanian [9]. Three prominent works have been done in positive extended disjunctive deductive databases with respect to inconsistencies: The first, answer set semantics, by Gelfond and Lifschitz [12], trivialize the results in the presence of inconsistencies. The second, p-minimal models, by Sakama and Inoue [17], which is based

© Springer International Publishing Switzerland 2015
A. Bikakis and X. Zheng (Eds.): MIWAI 2015, LNAI 9426, pp. 16–28, 2015.
DOI: 10.1007/978-3-319-26181-2_2

on four-valued logic [8], tolerates inconsistencies. In addition to that, for both logic programs and disjunctive logic programs many works have been proposed [1–3,11], where all of the approaches are based on four-valued logic. The third, the quasi-classic models, by Zhang et al. [19], has stronger inference power than p-minimal models because the quasi-classic models support disjunctive syllogism and disjunction introduction. Moreover, the quasi-classic models are based on quasi-classic logic [14].

In this paper, we use the paraconsistent relational model and propose an algorithm to find p-minimal models for positive extended disjunctive deductive databases. The central idea in arriving at p-minimal models for a given positive extended disjunctive deductive database is to associate paraconsistent relations with the predicate symbols. We then construct a system of algebraic equations for the clauses in positive extended disjunctive deductive databases. The equations are then used to incrementally construct p-minimal models with the help of disjunctive paraconsistent relations.

The rest of this paper is organized as follows: in Sect. 2, we discuss preliminaries to understand the paper; in Sect. 3, we explain the disjunctive paraconsistent relational model; in Sect. 4, we propose an algorithm to find p-minimal models; in Sect. 5, we state the conclusion and future work for the paper.

2   Preliminaries

Before we explain the details of the actual contribution of this paper, we must briefly review positive extended disjunctive deductive databases [16,17] and the paraconsistent relation model [5,6] which help to understand the paper.

Given a first order language $\mathcal{L}$, a disjunctive deductive database P [16] consists of logical inference rules of the form: r (rule) $= l_0 \vee \cdots \vee l_n \leftarrow l_{n+1}, \ldots, l_m$. A rule is called a positive disjunctive rule if the rule has both head (disjunction of literals) and body (conjunction of literals). Concretely, the rule r is called positive extended disjunctive rule if $l_0, \ldots, l_n, l_{n+1}, \ldots, l_m$ are literals which are either positive or negative ($\neg$) atoms. For the given syntax of positive extended disjunctive deductive databases, we reproduce the fixed-point semantics of P [17].

Fixed-Point Semantics. Let P be a positive extended disjunctive deductive database and $\mathcal{I}$ be a set of interpretations, then $T_P(\mathcal{I}) = \bigcup_{I \in \mathcal{I}} T_P(I)$

$$
T_P(I) = \begin{cases} \emptyset, \text{ if } l_{n+1}, \ldots, l_m \subseteq I \text{ for some} \\ \quad \text{ground constraint } \leftarrow l_{n+1} \ldots l_m \text{from } P; \\ \{J \mid \text{ for each ground clause} \\ \quad r_i \colon l_0 \vee \cdots \vee l_n \leftarrow l_{n+1}, \ldots, l_m \text{ such that} \\ \quad \{l_{n+1}, \ldots, l_m\} \subseteq I, J = I \cup \bigcup_{r_i} \{l_j\} (0 \leq j \leq n)\}, \text{otherwise.} \end{cases}
$$

In the definition of $T_P(I)$, $\{l_j\}(1 \leq j \leq n)$ is a collection of sets where every set in the collection contains a disjunct. For any positive extended disjunctive deductive database P, $\mathcal{T}_P$ is finite and $\mathcal{T}_P \uparrow n = \mathcal{T}_P \uparrow \omega$ where n is a successor ordinal and ω is a limit ordinal. For any positive extended disjunctive deductive database P, p-minimal models $= min(\mu(\mathcal{T}_P \uparrow \omega)^1)$ where $min(\mathcal{I}) = \{I \in \mathcal{I} \mid \nexists J \in \mathcal{I}$ such that $J \subset I\}$.

Unlike normal relations where we only retain information believed to be true of a particular predicate, we also retain what is believed to be false of a particular predicate in the paraconsistent relational model. Let a relation scheme Σ be a finite set of attribute names, where for any attribute name $A \in \Sigma$, $dom(A)$ is a non-empty domain of values for A. A tuple on Σ is any map $t: \Sigma \to \bigcup_{A \in \Sigma} dom(A)$, such that $t(A) \in dom(A)$ for each $A \in \Sigma$. Let $\tau(\Sigma)$ denote the set of all tuples on Σ. An ordinary relation on scheme Σ is thus any subset of $\tau(\Sigma)$. The paraconsistent relation on a scheme Σ is a pair $< R^+, R^- >$ where R^+ and R^- are ordinary relations on Σ. Thus R^+ represents the set of tuples believed to be true of R, and R^- represents the set of tuples believed to be false.

Algebraic Operators. Two types of algebraic operators are defined here: (i) Set Theoretic Operators, and (ii) Relational Theoretic Operators.

Set Theoretic Operators. Let R and S be two paraconsistent relations on scheme Σ.

Union. The union of R and S, denoted $R \dot{\cup} S$, is a paraconsistent relation on scheme Σ, given that $(R \dot{\cup} S)^+ = R^+ \cup S^+$, $(R \dot{\cup} S)^- = R^- \cap S^-$.

Complement. The complement of R, denoted $\dot{-}R$, is a paraconsistent relation on scheme Σ, given that $\dot{-}R^+ = R^-$, $\dot{-}R^- = R^+$.

Intersection. The intersection of R and S, denoted $R \dot{\cap} S$, is a paraconsistent relation on scheme Σ, given that $(R \dot{\cap} S)^+ = R^+ \cap S^+$, $(R \dot{\cap} S)^- = R^- \cup S^-$.

Difference. The difference of R and S, denoted $R \dot{-} S$, is a paraconsistent relation on scheme Σ, given that $(R \dot{-} S)^+ = R^+ \cap S^-$, $(R \dot{-} S)^- = R^- \cup S^+$.

Relation Theoretic Operators. Let Σ and Δ be relation schemes such that $\Sigma \subseteq \Delta$ and let R and S be paraconsistent relations on schemes Σ and Δ.

Join. The join of R and S, denoted $R \dot{\bowtie} S$, is a paraconsistent relation on scheme $\Sigma \cup \Delta$, given that $(R \dot{\bowtie} S)^+ = R^+ \bowtie S^+$, $(R \dot{\bowtie} S)^- = (R^-)^{\Sigma \cup \Delta} \cup (S^-)^{\Sigma \cup \Delta}$.

Projection. The projection of R onto Δ, denoted $\dot{\pi}_\Delta(R)$, is a paraconsistent relation on Δ, given that $\dot{\pi}_\Delta(R)^+ = \pi_\Delta(R^+)^{\Sigma \cup \Delta}$, $\dot{\pi}_\Delta(R)^- = \{t \in \tau(\Delta) \mid t^{\Sigma \cup \Delta} \subseteq (R^-)^{\Sigma \cup \Delta}\}$ where π_Δ is the usual projection over Δ of ordinary relations.

Selection. Let F be any logic formula involving attribute names in Σ, constant symbols, and any of these symbols $\{==, \neg, \wedge, \vee\}$. Then, the selection of R by F, denoted $\dot{\sigma}_F(R)$, is a paraconsistent relation on scheme Σ, given that $\dot{\sigma}_F(R)^+ = \sigma_F(R^+)$, $\dot{\sigma}_F(R)^- = R^- \cup \sigma_{\neg F}(\tau(\Sigma))$, where σ_F is a usual selection of tuples satisfying F from ordinary relations.

The following example is taken from Bagai and Sunderraman's paraconsistent relational data model [5].

---

[1] $\mu(\mathcal{T}_P \uparrow \omega) = \{I \mid I \in \mathcal{T}_P \uparrow \omega \text{ and } I \in \mathcal{T}_P(I)\}$.

Example 1. *Strictly speaking, relation schemes are sets of attribute names. However, in this example we treat them as ordered sequence of attribute names, so tuples can be viewed as the usual lists of values. Let $\{a, b, c\}$ be a common domain for all attribute names, and let R and S be the following paraconsistent relations on schemes $\langle X, Y \rangle$ and $\langle Y, Z \rangle$ respectively:*

$$R^+ = \{(b,b),(b,c)\}, R^- = \{(a,a),(a,b),(a,c)\}$$
$$S^+ = \{(a,c),(c,a)\}, S^- = \{(c,b)\}.$$

Then, $R \bowtie S$ is the paraconsistent relation on scheme $\langle X, Y, Z \rangle$:

$(R \bowtie S)^+ = \{(b,c,a)\}$

$(R \bowtie S)^- = \{(a,a,a),(a,a,b),(a,a,c),(a,b,a),(a,b,b),(a,b,c),(a,c,a),$
$\qquad (a,c,b),(a,c,c),(b,c,b),(c,c,b)\}$

Now, $\dot{\pi}_{\langle X,Z \rangle}(R \bowtie S)$ becomes the paraconsistent relation on scheme $\langle X, Z \rangle$:

$$\dot{\pi}_{\langle X,Z \rangle}(R \bowtie S)^+ = \{(b,a)\}$$
$$\dot{\pi}_{\langle X,Z \rangle}(R \bowtie S)^- = \{(a,a),(a,b),(a,c)\}$$

Finally, $\dot{\sigma}_{\neg X=Z}(\dot{\pi}_{\langle X,Z \rangle}(R \bowtie S))$ becomes the paraconsistent relation on scheme $\langle X, Z \rangle$:

$$\dot{\sigma}_{\neg X=Z}(\dot{\pi}_{\langle X,Z \rangle}(R \bowtie S))^+ = \{(b,a)\}$$
$$\dot{\sigma}_{\neg X=Z}(\dot{\pi}_{\langle X,Z \rangle}(R \bowtie S))^- = \{(a,a),(a,b),(a,c)(b,b),(c,c)\}$$

□

In the rest of the paper, relations mean paraconsistent relations. In order to find p-minimal models easily in our algorithm, we create a copy for a given relation. For any given relation R, the copy of R is R'. Both R and R' are different relations with the same attributes and the same tuples. R is called an exact relation and R' is called a copy relation. In addition to this, the replica of the relation R is R, where replica R has the same name, the same tuples, and the same attributes. We assume that a relation and its replica can not appear in the same set, but can appear in different sets. If two relations (a relation and its replica) appear in the same set, then we merge the tuples of them and write it as one relation.

In the next section, we explain the disjunctive relation model, which is an adaptation of disjunctive relational model introduced by Jayakumar and Sunderraman [15].

3   Disjunctive Relation

Let a disjunctive relation scheme 2^Σ be a finite set of attribute sets, where for any attribute set $A \in 2^\Sigma$, dom(a) is a non-empty domain of values for each

$a \in A$. Let $\tau(2^{\Sigma})$ denote the set of all tuples on 2^{Σ}. A disjunctive relation on scheme 2^{Σ} is thus any subset of $\tau(2^{\Sigma})$. A disjunctive relation, DR, over the scheme 2^{Σ} consists of two components $\langle DR^{+}, DR^{-} \rangle$, where $DR^{+} \subseteq \tau(2^{\Sigma})$ and $DR^{-} \subseteq \tau(2^{\Sigma})$. DR^{+} is the component that consists of a set of tuples. Each tuple in this component represents a disjunction of facts. In the case where the tuple is a singleton, we have a definite fact. DR^{-} is the component that consists of a set of tuples. Each tuple in this component represents a conjunction of facts. In the case where the tuple is a singleton, we have a definite fact. Let T be a tuple in DR, then for all $t \in T$, $Att(t)$ is an attribute set that represents the element in the tuple of the disjunctive relation DR, and let $Att(R)$ be an attribute set of relation R over the scheme Σ.

In the remainder of the section, we define a rename operator, mapping, and necessary definitions, which play a key role in constructing the p-minimal models.

Rename Operator. Rename operator renames the attributes for any relations.

Attribute Rename (Θ). Let R be a relation over scheme Σ and $\Sigma = \{A_1 \ldots A_m, R.A_1 \ldots R.A_m\}$. Then,

$$\Theta_{A_1 \ldots A_m \to R.A_1 \ldots R.A_m}(R) \text{ and } \Theta_{R.A_1 \ldots R.A_m \to A_1 \ldots A_m}(R).$$

This operator (Θ) is used to maintain uniqueness of attributes between any two relations.

Tuple Mapping to Disjunctive Relation. The algebraic equivalent for disjunction ($\vee$) is union. So, we represent the disjunctive information in P as paraconsistent unions ($\dot{\cup}$) of relations. However, it is not very flexible to construct p-minimal models with paraconsistent unions ($\dot{\cup}$) of relations. So, we map the information in relations to a disjunctive relation DR. Let $R_1 \ldots R_n$ be relations over schemes $\Sigma_1 \ldots \Sigma_n$ where every $\Sigma_i \subseteq \Sigma$ and $1 \leq i \leq n$. Then a set of attribute sets for any DR that refers $R_1 \dot{\cup} \ldots \dot{\cup} R_n$ is $\{\Sigma_1 \ldots \Sigma_n\}$. Next, we map the tuples of relations containing paraconsistent unions to a disjunctive relation. For each $t \in T$, T is a tuple for any disjunctive relation (DR). Then $t: \Sigma \to \cup_{A \in Att(R_i)} dom(A)$ such that $t(A) \in dom(A)$ for every i in $R_1 \dot{\cup} \ldots \dot{\cup} R_n$ where $Att(t) = Att(R_i)$. Informally, a disjunctive relation can be considered as a collection of relations. It is intuitive to map each disjunctive relation back to base relations because every $t \in T$ of the disjunctive relation represents the corresponding tuple in the relation.

The following example [15] is very specific, but helps to understand the algorithm clearly.

Example 2. *Let R_1, R_2 and C are relations over schemes $\{X\}$, $\{Y, Z\}$ and $\{X, Y, Z\}$ and domain for every attribute is $\{a, b, c\}$. Then, we have the following equation:*

$$(\dot{\pi}_{\{X,Y,Z\}}(R_1(X) \dot{\cup} \dot{-} R_2(Y, Z)))[X, Y, Z] = (\dot{\pi}_{\{X,Y,Z\}}(C(X, Y, Z)))^{+}[X, Y, Z]$$

where $C^{+} = \{(a, b, c)\}$, $R_1^{-} = \{(b)\}$ and $R_2^{+} = \{(a, c), (b, c)\}$

Solution. Before the tuples of C are distributed to R_1 and R_2, it is impera-
tive to note that R_1 and R_2 contain definite tuples, which are not disjunctive
(conjunctive). The first step is to map the definite tuples of R_1 and R_2 to a disjunctive relation. The definite tuples have no disjunction (conjunction) in any
disjunctive relation. So, we rename the attributes (Θ) of R_1 and R_2. Then we
map the definite tuples to DR.

In the rest of the paper, we differentiate positive and negative parts of a
relation (disjunctive relation) with a double line in every relation (disjunctive
relation) diagram. Also, we call relations in left hand side of the equation as base
relations.

$$DR = \begin{array}{|c|c|} \hline \{R_1.X\} & \{R_2.Y, R_2.Z\} \\ \hline (b) & \\ \hline\hline & (a,c) \\ \hline & (b,c) \\ \hline \end{array}$$

The next step is to distribute the tuples from
C to each individual relation in any union after applying Θ to R_1 and R_2. It
is necessary to apply Θ before the distribution of tuples from C because we
changed the attributes of R_1 and R_2 before we map the definite tuples.

$$R_1 = \begin{array}{|c|} \hline \{X\} \\ \hline (a) \\ \hline\hline (b) \\ \hline \end{array} \text{ and } \dot-R_2 = \begin{array}{|c|} \hline \{Y, Z\} \\ \hline (b,c) \\ \hline\hline (a,c) \\ \hline (b,c) \\ \hline \end{array}$$

The next step is to again rename (Θ) the attributes.

$$R_1 = \begin{array}{|c|} \hline \{R_1.X\} \\ \hline (a) \\ \hline\hline (b) \\ \hline \end{array} \text{ and } \dot-R_2 = \begin{array}{|c|} \hline \{R_2.Y, R_2.Z\} \\ \hline (b,c) \\ \hline\hline (a,c) \\ \hline (b,c) \\ \hline \end{array}$$

Then we map the newly added tuples of $R_1 \dot\cup \dot-R_2$ to DR.

$$DR = \begin{array}{|c|c|} \hline \{R_1.X\} & \{R_2.Y, R_2.Z\} \\ \hline (a) \quad \vee & (b,c) \\ \hline (b) & \\ \hline\hline & (a,c) \\ \hline & (b,c) \\ \hline \end{array}$$

$\square$

This state of DR is base DR. To reiterate, DR^+ contains tuples which in
turn contain disjunction. From the base DR, multiple DR can be obtained by
applying disjunction in tuples. Each newly created DR from the base DR should
not lose any tuple set; otherwise, it leads to incorrect models. The following
definition addresses the issue.

Proper Disjunctive Relation (PDR). Let DR be a base disjunctive relation.
A proper disjunctive relation is a set, which contains all disjunctive relations that
can be formed from DR by applying disjunction in tuples. Concretely, for every
disjunctive relation (DR_i), which is obtained from DR by applying disjunction,
$\tau(DR^+) = \tau(DR_i^+)$ where $1 \leq i \leq (2^n - 1)^{\tau(DR^+)}$ such DR_i is a PDR^i.

Example 3. *Continuing from Example 2.*

Solution. The next step is to create a set of proper disjunctive relation from DR.

$PDR = \{PDR^1, PDR^2, PDR^3 \}$ where $PDR^1 =$

$\{R_1.X\}$	$\{R_2.Y, R_2.Z\}$
(a)	
(b)	
	(a, c)
	(b, c)

$PDR^2 =$

$\{R_1.X\}$	$\{R_2.Y, R_2.Z\}$
	(b, c)
(b)	
	(a, c)
	(b, c)

$PDR^3 =$

$\{R_1.X\}$	$\{R_2.Y, R_2.Z\}$
(a) $\vee$	(b, c)
(b)	
	(a, c)
	(b, c)

The size of PDR is 3. Correspondingly, there should be three replicas of base relations. We sometimes superscript the set with a number in order to show the difference between any two sets that looks the same.

$$\{\{(\dot{\pi}_{\{X,Y,Z\}}(R_1(R_1.X)\dot{\cup}\dot{-}R_2(R_2.Y, R_2.Z)))[X, Y, Z]\}^1,$$
$$\{(\dot{\pi}_{\{X,Y,Z\}}(R_1(R_1.X)\dot{\cup}\dot{-}R_2(R_2.Y, R_2.Z)))[X, Y, Z]\}^2,$$
$$\{(\dot{\pi}_{\{X,Y,Z\}}(R_1(R_1.X)\dot{\cup}\dot{-}R_2(R_2.Y, R_2.Z)))[X, Y, Z]\}^3\}.$$

For every p in PDR, reverse map tuples to a set of base relations.

Finally, rename (Θ) each attribute name of every relation back to its old name in every replica. Hence, R_1 attribute is $< X >$ and R_2 attribute is $< Y, Z >$. □

To individualize the relation, we have the following definition.

Relationalize. Let $R_1\dot{\cup}R_2\dot{\cup}\ldots\dot{\cup}R_n$ and $R_1, R_2 \ldots R_n$ be relations on scheme Σ.

$$Relationalize(\dot{\pi}_{\{\Sigma\}}(R_1\dot{\cup}R_2\dot{\cup}\ldots\dot{\cup}R_n)[\Sigma]) := \{R_1, R_2 \ldots R_n\}$$

The relationalize operator removes the unions from relations and the projection for the expression. By doing so, the operator produces a set of relations. If there is a select operation associated with the expression, then apply the operation before *Relationalize* is applied. *Relationalize* is in accordance to $\{l_i\}$ (defined in Preliminaries section) [17].

Example 4. *Continuing from Example 3. In this example, we relationalize only one replica* $(\{(\dot{\pi}_{\{X,Y,Z\}}(R_1(X)\dot{\cup}\dot{-}R_2(Y, Z)))[X, Y, Z]\}^1)$.

Solution. $Relationalize(\dot{\pi}_{\{X,Y,Z\}}(R_1 \dot{\cup} R_2)[X,Y,Z]) = \{R_1, R_2\}$ where

$$R_1 = \begin{array}{|c|} \hline \{X\} \\ \hline (a) \\ \hline (b) \\ \hline \end{array} \text{ and } \dot{-}R_2 = \begin{array}{|c|} \hline \{Y,Z\} \\ \hline (a,c) \\ \hline (b,c) \\ \hline \end{array} \text{ or } R_2 = \begin{array}{|c|} \hline \{Y,Z\} \\ \hline (a,c) \\ \hline (b,c) \\ \hline \end{array}$$

$\square$

During p-minimal models construction, we encounter a set of redundant relation sets. In order to remove it, we define the following.

Minimize. Let $\{R1_1 \ldots R1_m\}$ and $\{R2_1 \ldots R2_n\}$ be two sets of relations where $m \leq n$.

$$Minimize(\{\{R1_1 \ldots R1_n\}, \{R2_1 \ldots R2_m\}\}) := \{\{R1_1 \ldots R1_m\} \mid R1_i = R2_j$$
$$\wedge \; Att(R1_i) - Att(R2_j) \wedge \tau(R1_i) = \tau(R2_j) \text{ such that } \forall i, 1 \leq i \leq m \wedge \exists j, 1 \leq i \leq n\}.$$

By using the definitions and operators in this section, we propose an algorithm in the following section.

4   P-Minimal Models for Positive Extended Disjunctive Deductive Databases

By using the algebra of the relational model, we present a bottom up method for constructing p-minimal models for the positive extended disjunctive deductive database. The algorithm that we present in this section is an extension of the algorithm proposed by Bagai and Sunderraman [5]. The reader is requested to refer to p-minimal models [17] and paraconsistent logics [18] in order to supplement additional knowledge. P-minimal models construction involves two steps. The first step is to convert P into a set of relation definitions for the predicate symbols occuring in P. These definitions are of the form

$$U_r = D_{U_r}$$

where U_r is the paraconsistent union of disjunctive head predicate symbols of P, and D_{U_r} is an algebraic expression involving predicate symbols of P. Here r refers to the equation number, $1 \leq r \leq N$, where N refers to a total number of equations. The second step is to iteratively evaluate the expressions in these definitions to incrementally construct the relations associated with the predicate symbols. The first step is called SERIALIZE and the second step is called Model Construction.

Algorithm. SERIALIZE
Input. A positive extended disjunctive deductive database clause $l_0 \vee \cdots \vee l_n \leftarrow l_{n+1} \ldots l_m$. For any i, $0 \leq i \leq m$, l_i is either of the form $p_i(A_{i1} \ldots A_{ik_i})$ or $\neg p_i(A_{i1} \ldots A_{ik_i})$, and let V_i be the set of all variables occurring in l_i.

Output. An algebraic expression involving paraconsistent relations.

Method. The expression is constructed by the following steps:

1. For each argument A_{ij} of literal l_i, construct argument B_{ij} and condition C_{ij} as follows:
   (a) If A_{ij} is a constant a, then B_{ij} is any brand new variable and C_{ij} is $B_{ij} = a$.
   (b) If A_{ij} is a variable, such that for each k, $1 \leq k < j$, $A_{ik} \neq A_{ij}$, then B_{ij} is A_{ij} and C_{ij} is true.
   (c) If A_{ij} is a variable, such that for some k, $1 \leq k < j$, $A_{ik} = A_{ij}$, then B_{ij} is a brand new variable and C_{ij} is $A_{ij} = B_{ij}$.
2. Let $\hat{l}_i$ be the atom $p_i(B_{i1} \ldots B_{ik_i})$, and F_i be the conjunction $C_{i1} \wedge \cdots \wedge C_{ik_i}$. If l_i is a positive literal, then Q_i is the expression $\dot{\pi}_{V_i} \dot{\sigma}_{F_i}(\hat{l}_i)$. Otherwise, let Q_i be the expression $\dot{-}\dot{\pi}_{V_i}(\dot{\sigma}_{F_i}(\hat{l}_i))$.
   As a syntatic optimisation, if all conjuncts of F_i are true (i.e. all arguments of l_i are distinct variables), then both $\dot{\sigma}_{F_i}$ and $\dot{\pi}_{V_i}$ are reduced to identity operations, and are hence dropped from the expression $\dot{\sigma}_{F_i}$.
3. Let U be the union ($\dot{\cup}$) of the Q_i's thus obtained, $0 \leq i \leq n$. The output expression is $(\dot{\sigma}_{F_1}(\dot{\pi}_{DV}(U)))$ $[B_{01} \ldots B_{n_{kn}}]$ where DV is the set of distinct variables occurring in all l_i.
4. Let E be the natural join ($\bowtie$) of the Q_i's thus obtained, $n + 1 \leq i \leq m$. The output expression is $(\dot{\sigma}_{F_1}(\dot{\pi}_{DV}(E)))$ $[B_{01} \ldots B_{n_{kn}}]$. As in step 2, if all conjuncts are true, then $\dot{\sigma}_{F1}$ is dropped from the output expression.

From the algebraic expression of the algorithm, we then construct a system of equations.

For any positive extended disjunctive deductive database P, EQN (P) is a set of all equations of the form $U_r = D_{U_r}$, where U_r is a union of the head predicate symbols of P, and D_{U_r} is the paraconsistent union ($\dot{\cup}$) of all expressions obtained by the algorithm *SERIALIZE* for clauses in P with the same U_r in their head. If all literals in the head are the same for any two rules, then U_r is the same for the two rules.

The final step is then to construct the model by incrementally constructing the relation values in P. For any positive extended disjunctive deductive database, P_E is the non disjunctive-facts (clauses in P without bodies), and P_B is the disjunctive rules (clauses in P with bodies). P_E^* refers to a set of all ground instances of clauses in P_E. Then, $P_I = P_E^* \cup P_B$.

The following algorithm finds p-minimal models for P.

ALGORITHM. Model Construction

Input. A positive extended disjunctive deductive database (P)

Output. P-minimal models for P.

Method: The values are computed by the following steps:

1. (Initialization)
   (a) Compute EQN(P_I) using the algorithm *SERIALIZE* for each clause in P_I.

(b) SModel $= \emptyset$, For each predicate symbol p in P_E, set

$$p^+ = \{(a1 \ldots ak) \mid p(a1 \ldots, ak) \in P_E^*\}, \text{ and } p^- = \emptyset \text{ or}$$
$$p^- = \{(a1 \ldots ak) \mid \neg p(a1, \ldots ak) \in P_E^*\} \text{ and } p^+ = \emptyset$$

SModel $= p$
End for.

2. (Rule Application)

(a) For every SModel (SModel $\neq \emptyset$), create copies of the relations in SModel and replace the SModel with the copies. DModel $= \emptyset$.

(b) For every equation r of the form $U_r = D_{U_r}$, create DR_r and insert the tuples from the copies in SModel into the corresponding exact relation in the equation r. Apply Θ to every relation in U_r and map the definite tuples for the relations in U_r to DR_r. Again, apply Θ to every relation in U_r. Compute the expression D_{U_r} and set the relations in U_r with $D_{U_r}^+$.

(c) Apply Θ to every relation in U_r, map the newly added tuples of U_r to DR_r and create a set of proper disjunctive relations (PDR_r) from the DR_r.

(d) Delete all tuples for the relations in U_r and create multiple replicas of U_r, which is denoted by the set C_r, where $|C_r| = |PDR_r|$.

(e) Re-map each p in PDR_r to C where $C \in C_r$.
For every $C \in C_r$,
$C = Relationalize(C)$
For every $R \in C$
$R = \Theta(R)$
End For.
End For.
DModel $=$ DModel $\bigcup C_r$ /* Merging relations of every equation */

(f) Once all equations are evaluated for the current SModel, perform the following: (i) for every $M \in$ DModel and for every exact relation for SModel that is not in M, create the exact relation in M; and (ii) for every $M \in$ DModel and for every exact relation for SModel that is in M, insert the tuples from the copy relation in SModel into the exact relation of M. Then add DModel to TempModel.

(g) Once every SModel is applied, start from step 2 (a) with SModel $= Minimize$(TempModel) and stop when there is no change in SModel.

3. P-models: rewrite the set of relations in SModel as a set of literals. P-minimal models $= min$(P-models) (min() is defined in Preliminaries).

It is very intuitive from the algorithm that if the computation of D_{U_r} is empty for any SModel, then discard the SModel. We found that the algorithm should be extended a little to accommodate for disjunctive facts, duplicate variables in disjunctive literals, and constants in disjunctive literals.

The following example shows that how the algorithm works.

Example 5. *Let P be a positive extended disjunctive deductive database. It has the following facts and rules:*

$$r(a,c), p(a), p(c), \neg f(a,b), s(c)$$
$$g(X) \vee \neg p(X) \leftarrow r(X,Y), s(Y)$$
$$g(X) \vee \neg p(X) \leftarrow \neg f(X,Y)$$

Solution. After step 1 (a) in initialization, EQN(P_I) returns:

$$(U_1)(\dot{\pi}_{\{X\}}(g(X)\dot{\cup}\bar{\,}p(X))[X] = (\dot{\pi}_{\{X\}}(r(X,Y)\dot{\bowtie}s(Y)))^+[X]\dot{\cup}(\dot{\pi}_{\{X\}}(\bar{\,}f(X,Y)))^+[X]$$

After step 1 (b) in initialization, SModel $= \{r,p,s,f\}$ where r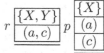

$$s \begin{array}{|c|} \hline \{Y\} \\ \hline (c) \\ \hline \end{array} \quad f \begin{array}{|c|} \hline \{X,Y\} \\ \hline (a,b) \\ \hline \end{array}$$

After step 2 (a), SModel $= \{r',p',s',f'\}$ (COPIES) where

$$r' \begin{array}{|c|} \hline \{X,Y\} \\ \hline (a,c) \\ \hline \end{array} \quad p' \begin{array}{|c|} \hline \{X\} \\ \hline (a) \\ \hline (c) \\ \hline \end{array} \quad s' \begin{array}{|c|} \hline \{Y\} \\ \hline (c) \\ \hline \end{array} \quad f' \begin{array}{|c|} \hline \{X,Y\} \\ \hline (a,b) \\ \hline \end{array}$$

In step 2 (b), there is only one SModel and one equation. It is necessary to insert the tuples from the copies in SModel to the corresponding relations in the equation. DModel $= \emptyset$. Then map the definite tuples to DR_1 for the current SModel. Compute the expression and assign it to U_1.

$$DR_1 \begin{array}{|c|c|} \hline \{g.X\} & \{p.X\} \\ \hline & (a) \\ \hline & (c) \\ \hline \end{array}$$

By step 2 (c), map the newly added (disjunctive) tuples to DR_1.

$$DR_1 \begin{array}{|c|c|} \hline \{g.X\} & \{p.X\} \\ \hline (a) & \vee \quad (a) \\ \hline & (a) \\ \hline & (c) \\ \hline \end{array}$$

$PDR_1 = \{PDR_1^1, PDR_1^2, PDR_1^3\}$

$$PDR_1^1 \begin{array}{|c|c|} \hline \{g.X\} & \{p.X\} \\ \hline (a) & \vee \quad (a) \\ \hline & (a) \\ \hline & (c) \\ \hline \end{array} \quad PDR_1^2 \begin{array}{|c|c|} \hline \{g.X\} & \{p.X\} \\ \hline & (a) \\ \hline & (a) \\ \hline & (c) \\ \hline \end{array} \quad PDR_1^3 \begin{array}{|c|c|} \hline \{g.X\} & \{p.X\} \\ \hline (a) & \\ \hline & (a) \\ \hline & (c) \\ \hline \end{array}$$

We skip a step (2 (d)) here. Map every p in PDR_1 back to a set of base relation. We write after relationalizing the set of relations and applying Θ (step 2 (e)).

$C_1= \{\{g,p\}^1, \{p\}^2, \{g,p\}^3\}$

$$\{g,p\}^1 \quad g \begin{array}{|c|} \hline \{X\} \\ \hline (a) \\ \hline \end{array} \quad p \begin{array}{|c|} \hline \{X\} \\ \hline (a) \\ \hline (c) \\ \hline (a) \\ \hline \end{array} \quad \{p\}^2 \quad p \begin{array}{|c|} \hline \{X\} \\ \hline (a) \\ \hline (c) \\ \hline (a) \\ \hline \end{array} \quad \{g,p\}^3 \quad g \begin{array}{|c|} \hline \{X\} \\ \hline (a) \\ \hline \end{array} \quad p \begin{array}{|c|} \hline \{X\} \\ \hline (a) \\ \hline (c) \\ \hline \end{array}$$

DModel = DModel $\bigcup C_1$

By step 2 (f), DModel = $\{\{g,p,r,s,f\}^1, \{p,r,s,f\}^2, \{g,p,r,s,f\}^3\}$

$$\{g,p,r,s,f\}^1 \quad g\;\begin{array}{|c|}\hline \{X\} \\\hline (a) \\\hline\end{array}\; p\;\begin{array}{|c|}\hline \{X\} \\\hline (a) \\\hline (c) \\\hline (a) \\\hline\end{array}\; r\;\begin{array}{|c|}\hline \{X,Y\} \\\hline (a,c) \\\hline\end{array}\; s\;\begin{array}{|c|}\hline \{Y\} \\\hline (c) \\\hline\end{array}\; f\;\begin{array}{|c|}\hline \{X,Y\} \\\hline (a,b) \\\hline\end{array}$$

$$\{p,r,s,f\}^2 \quad p\;\begin{array}{|c|}\hline \{X\} \\\hline (a) \\\hline (c) \\\hline (a) \\\hline\end{array}\; r\;\begin{array}{|c|}\hline \{X,Y\} \\\hline (a,c) \\\hline\end{array}\; s\;\begin{array}{|c|}\hline \{Y\} \\\hline (c) \\\hline\end{array}\; f\;\begin{array}{|c|}\hline \{X,Y\} \\\hline (a,b) \\\hline\end{array}$$

$$\{g,p,r,s,f\}^3 \quad g\;\begin{array}{|c|}\hline \{X\} \\\hline (a) \\\hline\end{array}\; p\;\begin{array}{|c|}\hline \{X\} \\\hline (a) \\\hline (c) \\\hline\end{array}\; r\;\begin{array}{|c|}\hline \{X,Y\} \\\hline (a,c) \\\hline\end{array}\; s\;\begin{array}{|c|}\hline \{Y\} \\\hline (c) \\\hline\end{array}\; f\;\begin{array}{|c|}\hline \{X,Y\} \\\hline (a,b) \\\hline\end{array}$$

Add DModel to TempModel.

By step 2 (g), SModel = $Minimize$(TempModel). The algorithm stops when there is no change in SModel. We skip further iterations and go to the final step (3). In the final step, we first rewrite the relation in the form of literals,

$$\text{P-models} = \{\{g(a), p(a), p(c), \neg p(a), r(a,c), s(c), \neg f(a,b)\}, \{p(a), p(c), \neg p(a),$$
$$r(a,c), s(c), \neg f(a,b)\}, \{g(a), p(a), p(c), r(a,c), s(c), \neg f(a,b)\}\}.$$

Then, p-minimal models = $\{\{p(a), p(c), \neg p(a), r(a,c), s(c), \neg f(a,b)\}, \{g(a), p(a), p(c), r(a,c), s(c), \neg f(a,b)\}\}$. This result is the same for **fixed-point semantics** defined in Preliminaries section.

5 Conclusion

In this paper, we proposed an algorithm to find p-minimal models for any positive extended disjunctive deductive database. We also used a disjunctive relational model to represent the relations containing paraconsistent unions [15].

Though we find the model for any given positive extended disjunctive deductive database, the algorithm does not find models for the databases with recursions and constraints, which could be a good future work for this algorithm. It would be very interesting to analyze the algorithm by allowing default negation in program P. We observe that we have not proven the correctness and complexities of the algorithm. We have also left that for future work. In query-intensive applications, this precomputation of the model enables efficient processing of subsequent queries. The creation of many proper disjunctive databases are expensive, given the p-minimal models computation, and are probably not worth the extra computation.

References

1. Alcântara, J., Damásio, C.V., Pereira, L.M.: A declarative characterisation of disjunctive paraconsistent answer sets. In: ECAI, vol. 16, p. 951. Citeseer (2004)
2. Alcântara, J., Damásio, C.V., Moniz, L.M.: Paraconsistent logic programs. In: Flesca, S., Greco, S., Leone, N., Ianni, G. (eds.) JELIA 2002. LNCS (LNAI), vol. 2424, pp. 345–356. Springer, Heidelberg (2002)
3. Arieli, O.: Paraconsistent declarative semantics for extended logic programs. Ann. Math. Artif. Intell. **36**(4), 381–417 (2002)
4. Arieli, O.: Distance-based paraconsistent logics. Int. J. Approximate Reasoning **48**(3), 766–783 (2008)
5. Bagai, R., Sunderraman, R.: A paraconsistent relational data model. Int. J. Comput. Math. **55**(1–2), 39–55 (1995)
6. Bagai, R., Sunderraman, R.: Bottom-up computation of the fitting model for general deductive databases. J. Intell. Inf. Syst. **6**(1), 59–75 (1996)
7. Bagai, R., Sunderraman, R.: Computing the well-founded model of deductive databases. Int. J. Uncertainty Fuzziness Knowl. Based Syst. **4**(2), 157–175 (1996)
8. Belnap Jr., N.D.: A useful four-valued logic. In: Michael Dunn, J., Epstein, G. (eds.) Modern Uses of Multiple-Valued Logic, pp. 5–37. Springer, Netherlands (1977)
9. Blair, II.A., Subrahmanian, V.: Paraconsistent logic programming. Theor. Comput. Sci. **68**(2), 135–154 (1989)
10. Carnielli, W., Coniglio, M.E., Marcos, J.: Logics of formal inconsistency. In: Gabbay, D.M., Guenthner, F. (eds.) Handbook of Philosophical Logic, pp. 1–93. Springer, Netherlands (2007)
11. Damásio, C.V., Pereira, L.M.: A survey of paraconsistent semantics for logic programs. In: Besnard, P., Hunter, A. (eds.) Reasoning with Actual and Potential Contradictions, pp. 241–320. Springer, Netherlands (1998)
12. Gelfond, M., Lifschitz, V.: Classical negation in logic programs and disjunctive databases. New Gener. Comput. **9**(3–4), 365–385 (1991)
13. Hunter, A.: Paraconsistent logics. In: Besnard, P., Hunter, A. (eds.) Reasoning with Actual and Potential Contradictions, pp. 11–36. Springer, Netherlands (1998)
14. Hunter, A.: Reasoning with contradictory information using quasi-classical logic. J. Logic Comput. **10**(5), 677–703 (2000)
15. Jayakumar, B., Sunderraman, R.: Paraconsistent relational model: a quasi-classic logic approach. In: IJCAI Workshop 13 Ontologies and Logic Programming for Query Answering, Buenos Aires, Argentina, pp. 82–90, July 2015
16. Minker, J., Seipel, D.: Disjunctive logic programming: a survey and assessment. In: Kakas, A.C., Sadri, F. (eds.) Computational Logic: Logic Programming and Beyond. LNCS (LNAI), vol. 2407, pp. 472–511. Springer, Heidelberg (2002)
17. Sakama, C., Inoue, K.: Paraconsistent stable semantics for extended disjunctive programs. J. Logic Comput. **5**(3), 265–285 (1995)
18. Subrahmanian, V.: Paraconsistent disjunctive deductive databases. In: Proceedings of the Twentieth International Symposium on Multiple-Valued Logic, pp. 339–346. IEEE (1990)
19. Zhang, Z., Lin, Z., Ren, S.: Quasi-classical model semantics for logic programs – a paraconsistent approach. In: Rauch, J., Raś, Z.W., Berka, P., Elomaa, T. (eds.) ISMIS 2009. LNCS, vol. 5722, pp. 181–190. Springer, Heidelberg (2009)

The Study Trend and Application Case of Research and Development Integrated Information Provision System for Small and Medium-sized Companies

Youngkon Lee[1(✉)] and Ukhyun Lee[2(✉)]

[1] e-Business Department, Korea Polytechnic University,
2121 Jeongwangdong, Siheung City, Korea
yklee777@kpu.ac.kr
[2] School of IT Convergence Engineering, Shinhan University,
233-1 Sangpae dong, Dongducheon City, Korea
uhlee@shinhan.ac.kr

Abstract. In order to enhance the success rate of research and development, various systems, which analyze and verify the value of technologies developed by private and public companies, have been developed. Recently, the methodologies analyzing the trend of future technology are being actively studied. Especially, the studies, which explore future technology in the huge patent documents, which can be said a treasure house of technology information, are increasing. However, the existing systems for it have limitation issues in performance and accuracy. In order to complement the existing methods, this study approaches patent documents by technology-unit instead of utilizing related information by patent-unit. This study draws core keywords through data mining and suggests a method detecting user defined technology trend discovery (UDTTD). As the result, this study developed a research and development integrated information provision system for small and medium-sized companies and an actual application case of the system on a medium-sized company is presented herewith.

Keywords: Future technology · Patent document · UDTTD · IPC · TF-IDF · SKOS · Ontology RDF · Technology trend analysis · Prediction by artificial neural network

1 Introduction

Research and development in a company suggests a direction, along which the company would go on and it is the most important activity for a company to survive in the competition. At present, the success rate of domestic research and development of private companies and government does not reach 1 % of project planning stage. In order to enhance the success rate of research and development, various systems, which analyze and verify the value of the technology developed by private and public companies, have been developed. Recently, the methodologies analyzing the trend of future technology are being actively studied in major advanced countries including

© Springer International Publishing Switzerland 2015
A. Bikakis and X. Zheng (Eds.): MIWAI 2015, LNAI 9426, pp. 29–38, 2015.
DOI: 10.1007/978-3-319-26181-2_3

United States, Japan and Germany. Especially, the studies, which explore future technology in huge patent documents, which can be said a treasure house of technology information, are increasing [1].

The M&S (modeling and simulation) development team in this study has been working on the development of 'research and development integrated information provision system for small and medium-sized companies' during the past 4 years. Here, the meaning of 'for small and medium-sized companies' is that a small and medium-sized company can get research and development related information from the system produced for this study at low cost. The word 'integrated' is used because the system provides with all information required at all steps in the lifecycle of research and development. Since there is limitation in paper space to introduce the whole system, this study paper will give explanation focusing on the technology analysis and trend provision system. In Sect. 2, Related Studies, information regarding Aureka, the patent analysis solution being commonly used globally, and PIAS, which was developed by Korean Intellectual Property Office and commonly used in Korea, will be given to explain the difference between Aureka/PIAS and the system produced for this study. In Sect. 3, the development methodology and system application case of the system produced for this study is stated. Section 4 will close this study paper by suggesting the expected results from the system produced for this study.

2 Related Studies

2.1 Trend of Related Systems

Patent analysis service is advancing with United States in the center. At present, the competition to develop new technology or service has virtually disappeared after Thompson Scientific of United States had merged major patent data based information companies. Especially, no new patent information service has appeared and the patent search market disappeared in United States since google.com had opened free patent search in 2005 (which is under suspension now).

Globally, there are methods to predict technology trend by utilizing various predicting methods such as specialist investigation or biblio-metrical analysis; however, the results given by current technology prediction methods are all on macro level; while the medium and large scale future technology prediction is under fierce competition by some big companies; therefore, it is difficult for small and medium-sized companies to predict technology with good prospect.

Aureka of Thompson Company in United States and True-Teller, vatagepoint.com technology analysis system of Nomura Research Institute (NRI) in Japan are well known in Korea. They are technical keyword based cluster analysis methods based on the distance information between the technical keywords; therefore, they can analyze technology owned by others but they cannot analyze the whole technology area. Aureka shows what patents exist in certain technology field by performing cluster analysis on the technical keyword by text mining [2]. However, as seen in the Aureka system analysis result screen in Fig. 1, the coordinate space does not have any technical meaning. The visual result is just the expression of technical keywords with short distance among them on the coordinate; while it is not possible to tell what technical meaning the blank area has. In addition, since it only visually shows the pre-occupied technology area, it does not provide a company with direct information, by which the company can make certain choice. The analysis result of True-Teller analysis software of NRI is same with Aureka in this regard.

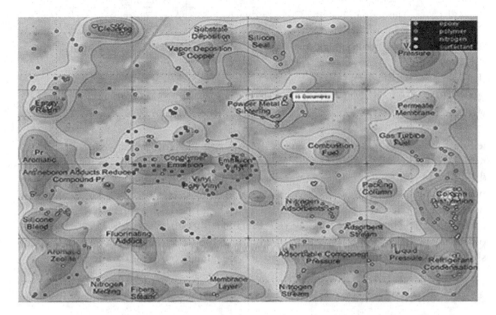

Fig. 1. The Aureka system analysis result screen

Meanwhile, the patent ranking service of Ocean Tomo Company in United States provides with a service in which they give the names of the companies with relation to an individual patent together with the ranking of individual patents which makes a patent population. They also give the ranking of the inventors. The patent ranking service of IPB in Japan provides with class evaluation service of a patent right; however, it does not provide with effective information on the company related to the technology (Table 1).

The PIAS (patent information analysis system) provided by KIPO of Korea is a patent data analysis solution. It is a patent information analysis system, which systematically collects various patent information and knowledge, provides with data and supports various analyses in the field of science and technology [3]. It was developed by KIPO and companies can use it free-of-charge; however, it is not commonly used because of following issues; though it has multiple merits. First, technology analysis from the viewpoint of a user is difficult. The system analyzes patent applicants based on subclass step, which is the topmost step of IPC class code, which is provided by the system as default. However, the scope of patent data technology lower than the subclass is too diverse that the subclass cannot be defined as a technology level. A small and medium-sized company, which is doing research and development on a specialized

Table 1. The status of patent technology analysis system products and services (company names and technology name)

Country name	Company name	Technology name
United States	Thompson Scientific Company www. delphion.com	Patent search and analysis technology
United States	Thompson Scientific Company www. micropatent.com	Aureka service
United States	www.vantagepoint.com	Patent keyword converged analysis service
United States	Ocean Tomo	Patent ranking service
Japan	IPB	Patent ranking service

technology, wants to know the trend of that specific technology; however, PIAS does not meet this needs. It is believed that this issue should be improved. Second, there is a difficulty in securing latest patent data in real time. PIAS was produced as a client-side program and it does patent data search and analysis by user's search word; therefore, it does not guarantee the quality of service when a user queries huge patent data. In order to set up a direction for a company or an institution to go forward by analyzing latest patent data trend from the viewpoint of a research and development planner, the system should provide with service by securing service quality on latest data in real-time.

3   Research Trend

3.1   Research and Development Integrated Information Provision System for Small and Medium-sized Companies

We used the patent documents of United States, which country is more technically advanced than Korea, as the raw data for technology trend analysis. We defined technology as a set of patents utilizing the international IPC (International Patent Classification) classification system and assumed following two for technology analysis [4]. First, the classification system for subclass or below which is mapped with patent data by industry should not be biased for the number of lower classification systems at specific technology level. Second, the classification name of IPC classification system should include a technical word which can be defined by technology.

The research and development information provision system we developed depends on ontology, simulation and natural word based keyword extraction system. Ontology is utilized in the keyword extraction included in a patent and relation analysis between technologies; while simulation is used in the prediction of technology trend utilizing a keyword. The keyword extraction algorithm consists of following major 3 steps.

- Keyword analysis utilizing TF-IDF (term frequency - inverse document frequency) algorithm
- Refining of equivalence and synonym utilizing a thesaurus
- Handling and refining of stop words

TF-IDF is a representative keyword extracting algorithm of text mining. It expresses the importance of an arbitrary core keyword in a document group made of N ea of document as a statistical value through weight based on vector space model [2]. When drawing keywords from diverse documents, grouping is required to classify the documents in accordance with their characters [5].

Since the keyword drawn from patent claim is the keyword having same character with the word drawn from the patent document which has same IPC class code, the synonym/equivalence refining process by word is essential. A thesaurus is absolutely necessary for the vocabulary refining process between the words. The raw data to be entered into vocabulary dictionary database is English version vocabulary dictionary in the format of SKOS (simple knowledge organization system) downloaded from Wikipedia. In addition, stop words removal process is essentially required during extracting a keyword; because general words or words required for grammar purpose are substantially included [6]. The major function of technology analysis system consists of "lower level technology trend analysis", "technology trend analysis between technologies" and "user group technology trend analysis". Figure 2 shows the process and modules for analyzing big patent data to obtain technology value and trend. The system is composed of patent DB, ontology translator, inference engine, and simulator. The system extracts the technology relationship data by calculating the technology distance based on patent ontology and neural network. We borrowed the keyword reference group from SKOS framework, by which we classified the keywords used frequently in patents and calculate grades by technology distance by weight factor [7, 8].

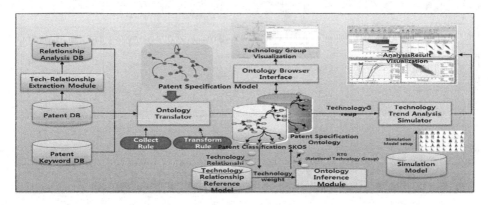

Fig. 2. System configuration diagram of technology analysis and trend prediction

Technology analysis system extracts RTG (related technology group) with high correlation to a technology and visually shows it. It shows the trend of major technology development by time and information of related persons at a glance.

In addition, we developed a neural network based system to predict future technology trend based on the result of analysis. The trend of future technology can be measured by the transition of technology value and the value of a technology was judged by technology value indices such as spread ability, convergence ability and prospect ability. The prospect ability tells the future value of certain technology while the convergence ability tells the utilization possibility of certain technology in other technologies. The spread ability tells the impact of a technology on other technologies. The technology value index was estimated by the patent increasing rate, the number of industrial areas with the patent which belongs to certain technology and the number of quotation of the patent. Figure 3 below shows the technology value prediction process of TVNS (technology value neural network system) developed by this study. We developed a data refining and arrangement program to estimate efficient technology value indices (spread ability, convergence ability and prospect ability) in large capacity database environment. In order to enhance the accuracy of technology value utilizing artificial neural network, we went through artificial neural network learning process utilizing the technology value indices of 3 years (t + 1, t + 2, t + 3). The weight of each index was estimated by a specialist on technology value and the technology value trend prediction was possible through it.

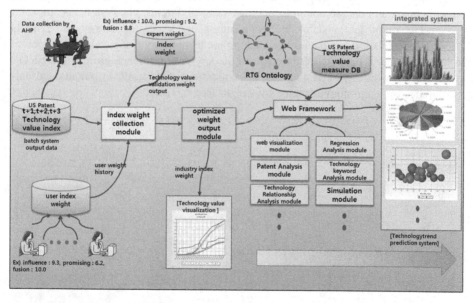

Fig. 3. Technology trend prediction by artificial neural network

3.2   Application Cases of Research and Development Integrated Information Provision System

We applied the research and development integrated information provision system we developed on the 'AutoGen' Company Limited, which is a medium-sized company. The 'AutoGen' Company located in Shihwa Industrial Complex is specialized in manufacturing automotive body parts and molds since it was registered as a cooperating company of Daewoo Motor Company in 1980. The company is quite active in research and development by establishing a technology research center to become a global small hidden champion. We applied the system produced for this study on the information analysis and feasibility securing of the technology, which the company is developing now.

First, the analysis object technology item is 'Other parts for motor vehicle' (IPC: B21D 53/88), which includes process, knowhow and component manufacturing process required during the manufacturing of various automotive parts. These are essential technologies required for the production of a motor vehicle which warrants stable and trustworthy quality including patent information on the core materials required for the manufacturing.

According to the technology trend analysis, the number of patent rapidly increased more than 2 times since year 2005 and there were sudden increases in the number of patent in 2009 and 2010 (please refer to Fig. 4). The reason is believed the boom in automotive industry, restructuring of domestic/overseas car makers and the opening of domestic car market to overseas makers.

I Number of patent by year

This is the chart for the total number of patent relevant to the technology code. The bar is the total number of patent relevant to the technology of the year and the broken line is the cumulative total.

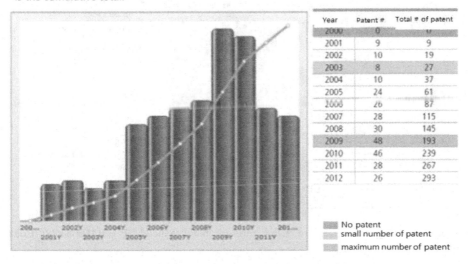

Year	Patent #	Total # of patent
2000	0	0
2001	9	9
2002	10	19
2003	8	27
2004	10	37
2005	24	61
2006	26	87
2007	28	115
2008	30	145
2009	48	193
2010	46	239
2011	28	267
2012	26	293

No patent
small number of patent
maximum number of patent

Fig. 4. Technology development trend of 'Other parts for motor vehicle' (IPC: B21D 53/88)

The prospect ability tells the future value of certain technology. We allocated 40 % weight to prospect ability because we think it is an important element of the technology. The convergence ability tells the utilization possibility of certain technology in other technology. We allocated 30 % weight of the technology value index to it. The spread ability tells the impact of a technology on other technologies. We allocated 30 % weight of the technology value index to it.

According to the weight graph analysis result, the future value of analysis object technology does not have big change as time passes by. It is judged that the technology will keep certain value regardless of the prosperity or recession in the automotive industry (Please refer to Fig. 5). The convergence ability of the technology shows continuous increasing pattern; though the increasing rate is not big. It means that the technology is also necessary in other mechanical industries other than automotive industry. The spread ability rapidly increases from 2012. It means that the technologies, which quote this technology are rapidly increasing.

| The technology value factor weight graph by year
This graph predicts the future value of a technology by each technology value factor and the 4 lines of technology value indices inclusive of weights by year.

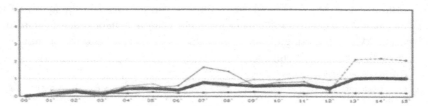

| Technology value factor value by year

Year	Prospect ability	convergence ability	spread ability	technology value
2000	0			0
2001	0.18511	0.33333	0	17.4043
2002	0.19796	0.4	0.2	25.9184
2003	0.16053	0.25	0	13.9212
2004	0.1776	0.6	0.6	43.104
2005	0.24553	0.70833	0.5	46.0711
2006	0.20564	0.30769	0.61538	35.9177
2007	0.21487	0.71429	1.67857	80.3806
2008	0.22999	0.56667	1.43333	69.1996
2009	0.26526	0.97917	0.64583	59.3604
2010	0.22462	0.97826	0.80435	62.4631
2011	0.20127	1.10714	0.85714	66.9792
2012	0.23569	0.96154	0.34615	48.6583
2013	0.2143	1.08188	2.14501	105.3787
2014	0.21469	1.08819	2.19729	107.152
2015	0.20967	1.13526	2.09505	105.2961

□ Colors and weights by factor

Prospect ability   40%
convergence ability 30%
spread ability   30%
technology value

□ Meaning of each indicator

▨ No value
▨ lowest value
▨ highest value
☐ prediction index

Fig. 5. Technology value trend of 'Other parts for motor vehicle' (IPC: B21D 53/88) (Color figure online)

According to the prediction result by the technology analysis system, the technology value of the analysis object technology will increase as a whole. Especially, the technology will become more important in the aspect of spread ability. It means that there is substantial number of other industries which require this technology. The technology value will also increase in the aspect of convergence ability. It is judged that it would be more advantageous to develop this technology as universal technology for general mechanical industry instead of a technology exclusive for automotive industry.

It is possible to see that the increasing trend of patents by selection patent applicants appears about 2 years later than the increasing trend of total number of patents. It means that the selection patent applicants have a trend of following the technology trend instead of leading the technology in the related industries (Refer to Fig. 6). Regarding the number of patents by applicant and by year, the patent application by selection patent applicant is concentrated in the 3 years from 2009 to 2011; however, it shows a state of lull again from 2012. The reason for this is believed that the effect of the patent application was less than expectation or the necessity of the patent in the market was not so big.

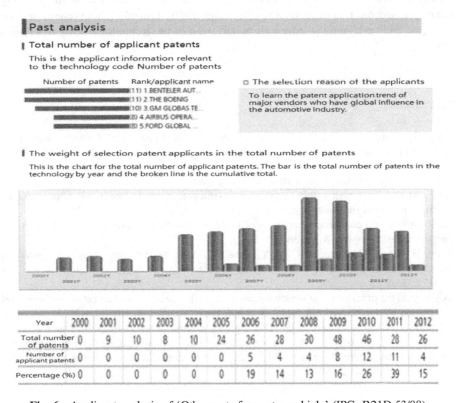

Year	2000	2001	2002	2003	2004	2005	2006	2007	2008	2009	2010	2011	2012
Total number of patents	0	9	10	8	10	24	26	28	30	48	46	28	26
Number of applicant patents	0	0	0	0	0	0	5	4	4	8	12	11	4
Percentage (%)	0	0	0	0	0	0	19	14	13	16	26	39	15

Fig. 6. Applicant analysis of 'Other parts for motor vehicle' (IPC: B21D 53/88)

It was possible to quantitatively verify the patent analysis information and technology trend in United States market of the 'Other parts for motor vehicle', which is being developed in 'AutoGen' Company, by applying the research and development integrated information provision system. The 'AutoGen' Company is now actively reflecting the information obtained by the system produced for this study in its research and development activities.

4   Conclusion

This study introduced a system, which analyzes technology related information and provides with technology related information, and gave the utilization possibility of the developed system by way of a substantiation case. This study proved that the system produced for this study is a 3D technology trend analysis system with high accuracy, which greatly improves the existing fragmental patent analysis system, provides with technology analysis information from multilateral aspects and combines ontology, keyword extraction and artificial neural network skill. It is expected that the web based system produced for this study will enable small and medium-sized companies greatly enhance the success rate of technology development by utilizing patent related service at low cost and verifying the information and trend of the technology being developed by them.

References

1. Jeon, Y., Kim, Y., Jeong, Y., Ryu, J., Maeng, S.: Language model for technology trend detection from patent document text and clue-based mechanical study method. Softw. Appl. **5** (2009). Book 36
2. http://thomsonreuters.com/scholarly-scientific-research/
3. Kim, M.: Major patent information analysis solution status at home and abroad. Korea Institute of Patent Information (2006)
4. Jo, J.: A study on convergence analysis of IPC and utilization plan of it in the evaluation and judgment. Ind. Property Right **38** (2012)
5. Jeon, H., Jeong, C., Song, S., Choi, Y., Choi, S., Jeong, H.: Building of technology-related word bundle for technology trend analysis. Korean Internet Inf. Assoc. Summer Conf. J. **1** (2012). Book 13
6. A study on the efficient noise removal during patent data analysis and patent technology level evaluation with improved reliability. Technol. Innov. Acad. J. **1**, 105–128 (2012). Book 15
7. Son, G., Lee, S.: Compound noun weight giving method in patent literature search (2005)
8. Buehrer, G., Chellapilla, K.: A scalable pattern mining approach to web graph compression with communities. In: Proceedings of the 2008 International Conference on Web Search and Data Mining. ACM (2008)

iCurate: A Research Data Management System

Shuo Liang$^{(\boxtimes)}$, Violeta Holmes, Grigoris Antoniou,
and Joshua Higgins

University of Huddersfield, Huddersfield, UK
{shuo.liang,v.holmes,g.antoniou,joshua.higgins}@hud.ac.uk

Abstract. Scientific research activities generate a large amount of data, which varies in format, volume, structure and ownership. Although there are revision control systems and databases developed for data archiving, the traditional data management methods are not suitable for High-Performance Computing (HPC) systems. The files in such systems do not have semantic annotations and cannot be archived and managed for public dissemination.

We have proposed and developed a Research Data Management (RDM) system, 'iCurate', which provides easy-to-use RDM facilities with semantic annotations. The system incorporates Metadata Retrieval, Departmental Archiving, Workflow Management System, Meta data Validation and Self Inferencing. The 'i' emphasises the user-oriented design. iCurate will support researchers by annotating their data in a clearer and machine readable way from its production to publication for the future reuse.

Keywords: Big data · Data curation · Linked Open Data · Research Data Management · Research support · Semantic web · Human-computer interaction

1   Introduction

At many research institutions, scientific research activities generate a large amount of data, which vary in format, volume, structure and ownership. The data could also be generated on various media. The major challenge in the research community is converting the research data into a generally machine readable format to enable further reuse of the content. Researchers often store their data without proper consideration of security, availability and curation of data. The data includes simulation data in HPC (High-Performance Computing) facilities, raw data from scientific equipment, video and image, text document, etc. Currently, the researchers store their data on various media, often large portable disks. Many research institutions have RDM policies in compliance with the research funding body requirements.

Several tools were suggested by DCC (Digital Curation Centre) [6] to help digital preservation. However, learning methods of preservation and usage of

© Springer International Publishing Switzerland 2015
A. Bikakis and X. Zheng (Eds.): MIWAI 2015, LNAI 9426, pp. 39–47, 2015.
DOI: 10.1007/978-3-319-26181-2_4

these tools is costly and consumes time. Combining these tools into institutions' existing ways of the research process is also expensive and time consuming.

In the current situation, HPC users' data can not be directly categorised and integrated into the curation process. The HPC file storage system is often maintained at a department level, and not integrated into an institutional storage and cloud storage. The disparate data storage systems are hindering the management process.

Therefore, it is demanded to develop a tool or service in order to improve the current situation of RDM (Research Data Management) and comply with the institutions' RDM policies. However, the research institutions' RDM solutions are not tailored particularly for HPC RDM or enhanced with artificial intelligence to reduce users' effort.

In this paper, we present an overview of current methods of RDM, and evaluate their suitability in the context of the HPC systems. We propose an RDM system, 'iCurate', which provides easy-to-use RDM facilities, combined with state-of-the-art semantic access. We will focus on preservation and curation services of research data, and explore the feasibility of linking research data to related publications maintained within a research institution or university repository. The system aims to integrate multiple storage media and university repository services to serve the digital curation purpose. We have designed a number of components for the existing HPC systems to facilitate easy access across these infrastructures.

2 Background

2.1 Defining Research Data

In a recent public consultation on open research data of the European Commission, the question 'what is research data' was addressed by researchers, funders, industry, etc. In the researchers' point of view, "research data are all data from an experiment, study or measurement, including the metadata and processing details" [4]. This reflects that researchers focus more on the data they collected and processed. One view from a funder, The Open Knowledge Foundation, argued that "research data is extremely heterogeneous and that it takes a variety of forms" [4].

2.2 Ontology and Linked Open Data

One way of increasing data reusability is to publish them with a schema mapped to LOV (Linked Open Vocabularies). LOV is the set of ontologies behind LOD (Linked Open Data). The PREMIS (Preservation Metadata: Implementation Strategies) Ontology is a digital preservation standard based on the OAIS (Open Archival Information System) reference model. It is published as part of LOV by Library of Congress. OAIS is designed to support digital libraries. It is an international standard of archiving [5]. OAIS functional model is composed of six functional entities and related interfaces: Ingest, Archival Storage, Access, Data Management, Preservation Planning and Administration.

Table 1. A comparison of current RDM solutions

	DataStage	DataBank	DataUp [10]	DMPOnline [3]	Taverna [8]
Metadata retrieval	✓	✓	✓		✓
Workflow evaluation				✓	✓
Workflow management					
Departmental archiving		✓	✓		
Repository	✓				
OAIS model					
Metadata validation					

2.3   Current and Past Projects for Data Repository

In Table 1, some of the key features of existing RDM tool are compared.

The DataStage and DataBank are departmental data depository tool and University Repository tool developed by DataFlow team. Both tools are open-source. DataStage is designed as a customisable network-attached storage, which could be mounted on end-user's PC as "a mapped drive" for research group level. It has been claimed web-accessible, metadata enrich-able, backup enabled, group shared, password protected, etc. [1]. DataBank is a data repository designed for institutions. It provides "a definitive, sustainable, reference-able location for research datasets" [2]. DataBank assigns DOIs (Digital Object Identifiers) to datasets in the data repository.

The DataBank projects did not consider the various data storage structure and flexibility for a fragmented storage system. In the application level, Data-Bank can convert deposited data into publications as a repository. Some other applications such as DataUP and Taverna focus on a specific discipline.

Regardless of compatibility, there are various applications designed for every stage of RDM process. Where a required feature is missing, institutions can combine the applications together, just as the Data Management Toolkits in Penn State University [9].

However, a whole RDM process will change the researchers' habits. This change may not be suitable to each individual. Furthermore, it affects the quality of the whole RDM plan. The SWORD protocol [7] that DataBank used has set a foundation of a few RDM programs, while other Web service formats such as WSDL and REST have already been used in Digital Preservation area. The co-existence of these formats have made fragmentations in RDM domain.

3   Motivation

The RDM is considered an extension of traditional library databases, thus are built around library repositories because this is the traditional way of publishing. RDM as a part of the research process is preferred ideally to support record-ing the missing meta-data associated with the data that has been generated or

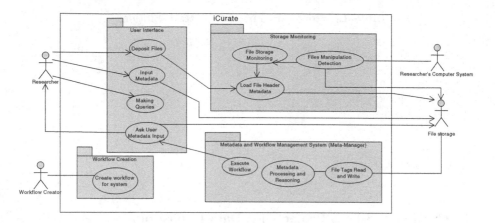

Fig. 1. The use case diagram of iCurate

archived that is related to a future publication. Nevertheless, a standard has not yet been proposed, which defines how detailed the meta-data should be, how to grade the meta-data, and how to improve the process on the scale of time.

The meta-data collection and evaluation tools seen in the previous section are intrusive to a user's own working routine, which makes them difficult to adopt. It is an extra work to manually categorise and group data in a separate system or repository to facilitate annotation, even more so to repeat this process.

Moreover, especially in the HPC field, the computational power is not as fast as the generation rate of the data. The data that has been preserved today needs to be pre-processed before being archived, in order to serve the further re-use. The pre-processed data from an HPC system will have more reusability with rich meta-data.

4 System Design - iCurate

4.1 Main Features

Based on our investigation outlined in Sect. 2, it is evident that the existing solutions do not completely satisfy the needs of RDM for users of HPC. The current solutions do not generate and summarise the information of research activities. Hence, we design and develop an RDM system (iCurate) that enables curation process from research data's production to its publication. The 'i' emphasises the user-oriented design philosophy. This system will overcome some limitations of the existing systems and have following features: metadata retrieval, departmental archiving, workflow system, ontology-driven self-inferencing.

1. *Metadata Retrieval Collecting.* Metadata is an essential component of an RDM system. Some meta-data which are generated by particular applications has already been stored in the header of a file. The file headers form the

basic meta-data of iCurate. Furthermore, researchers' other activities such as file manipulation, editing and work schedule are recorded as part of meta-data as well.

2. *Departmental Archiving.* iCurate will be appearing on researcher's computer or HPC facilities as a virtual folder or drive. Researcher's files appears on iCurate in structured view, e.g. tree structure or tag cloud. regardless how they were stored on the drive. iCurate compresses files in the order of curation.

3. *Workflow System.* In order to overcome the extra contributing efforts in RDM, workflow management is suggested to be added into an RDM system. The workflow system is used in the business process and many computer programs. By sharing workflows of the processing methods which are employed by researchers or anyone who works in similar discipline through the community, people will not be planning and designing their own RDM plan alone, and will contribute more effort on their research or workflow improvement.

4. *Metadata Validation and Self Inferencing.* Being compatible to OAIS and also PREMIS Ontology, will give the RDM system the ability to expose easily not only the public but also Linked Open Data. Inferencing is a step further to workflow management. The inferencing, which is based on the pre-configured workflow and ontologies, gives iCurate error checking, auto-suggestions and other new features. Inferencing requires all meta-data to be mapped onto RDM ontologies to enable the inferencing system. Once the RDM system have collected the data, the workflow system can be programmed to check if there is any contradiction among the data and give suggestions. Users should be able to log-in to a web front-end to perform daily error checking or between any period which are considered appropriate.

In the use case of iCurate, the user interface is depicted in Fig. 1. There are two human actors making inputs to the system. A workflow creator defines the workflow of RDM process, it could be adopted by researchers, who are actually recording their metadata. A workflow creator can be a project director, an external user who created and shared his/her RDM workflow, or the researchers themselves. Once the workflow is defined in a particular user's case, the system extracts meta-data from file storage actor, and generating questionnaires based on the pre-defined workflows, extracts further information about the research work. In the HPC environment it is common for storage to be centralised, such as an external NAS (Network-Attached Storage). This is convenient to monitor files for changes as they are created or modified, unlike existing systems which require the user to manually upload changes to the RDM system.

4.2   Meta-data and Workflow Management

As mentioned above, a workflow creator can be a researcher or another person. In order to cater for particular cases, the workflow can be continuously amended for the new requirement. In Fig. 2, an example of a workflow is given. The File Sets are determined by the folder structure and editing time. By grouping files into

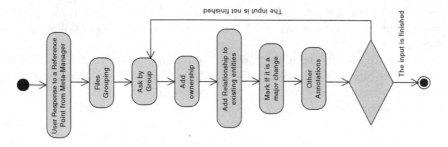

Fig. 2. A use case diagram of iCurate

Fig. 3. iCurate web change form

File Sets, the minimal intellectual entity can be identified. Requested meta-data will be asked after the File Sets are defined by the user.

A database, which collects and stores user inputed meta-data, is convenient but not yet smart. Based on the meta-data set obtained from researchers and reasoning engine, iCurate infers information inconsistency and gives research feedback to correct. Eventually, it will benefit not only the data management but also research activity itself.

4.3   Web Interface

The web interface of iCurate has two functions: to give a summary of the health-iness of researcher's meta-data in the dash board, and to collect meta-data. The dash board gives user an overview of their metadata inputs, project progresses, to-do lists.

Figure 3 shows a meta-data collection form of iCurate. The user is notified at the time which is predefined by their RDM workflow, e.g. every end of the day, after activity threshold has been reached. The questions are dynamically arranged depending on user's previous inputs. The activity period is highlighted at the top. The files are changed during this period is shown and can be checked to group up as a set of a single entity. The system pre-fill some of the questions based on users' activities and RDM workflow. User will be able to assign the changes to existing projects and File Sets or create new ones. Users can also rate their own research productivity for a reference of their progress, what will be summarised in the dash board.

```
#!/bin/bash
#HPC job related environment setups......
#< . . >

# Exrta infomation for iCurate
IC_INPUT_FOLDER="/home/fluentuser/example/10/"
IC_INPUT_SCRIPT="/home/fluentuser/example/10/Script.txt"
IC_INPUT_FILE="/home/fluentuser/example/10/example.cas"
IC_OUTPUT_FILE="/home/fluentuser/example/10/Dexample.cas"
IC_EXEC="/apps/Fluent/bin/fluent 3ddp -g -env -ssh -mpi=openmpi -t$nprocs \
        -cnf=$PBS_NODEFILE -i $IC_INPUT_SCRIPT"

# Execute command to launch Fluent
$IC_EXEC
```

Fig. 4. An example of modified PBS script

4.4   HPC Enabled

The iCurate will integrate with PBS (Portable Batch System). The users can declare their meta-data in their job file, and the system-generated meta-data, including successful completion, running time, time started, and so on, will also be recorded. Since every PBS job has its unique number with its hostname, it will be easier to locate the output. However, it is easier to save the screen output from PBS jobs, file outputs can be archived once it is specified. In iCurate, the output files or folder can be specified. Once the jobs are completed, the files or folders will be 'snapshotted' and stored in a separate section.

Figure 4 shows a PBS script with iCurate enabled. More informations are required to be declared before a job submission (Variables name starts with IC_). Those Variables and the system record on particular successful submission will be used in iCurate core for tracing user activities.

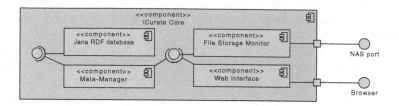

Fig. 5. The components diagram of iCurate

4.5   Implementation

The implementation of iCurate has three parts, core services, NAS, and User Interface, as shown in Fig. 5. The core services can be deployed on the NAS server or an independent server for more stability. Its components include: a Jena RDF (Resource Description Framework) database and reasoning engine, a "Meta-Manager" handling input and output, and a File Storage Monitor extracting user disk activities. The data storage is a modified Linux NAS service, which uses Linux `inotify` to monitor user file system changes. The Web interface is a cross-platform solution for metadata collection. The web interface contains a summary and statistics of user's activities, a score of metadata completion, and metadata questionnaires.

5   Summary

RDM plays a significant part of any research activity and is gaining in importance as a large amount of data is generated from scientific instruments simulations and modelling. In this paper, we have outlined several tools for generation, preservation and curation of data, and identified some of the limitations on the data which are generated in HPC research environments. we have presented out efforts in creating an RDM system, iCurate, which was designed and developed with researchers in mind. This approach allows meta-data be refinedsplncs03 before it is required by formal archiving services. iCurate will provide easy-to-use RDM facility for HPC RDM and it integrates with university repository services. Our future work will involve further development of the system to be deployed not only in the HPC research environment but to be customised for other research domains.

References

1. DataFlow: About datastage. http://www.dataflow.ox.ac.uk/index.php/datastage/ds-about
2. DataFlow: Databank. http://www.dataflow.ox.ac.uk/index.php/databank
3. (DCC), D.C.C.: Dmponline (2015). https://dmponline.dcc.ac.uk
4. European Commission: Results of the consultation on open research data, July 2013. http://ec.europa.eu/digital-agenda/node/67533

5. ISO: Iso 14721:2003 subscribe to updates space data and information transfer systems - open archival information system - reference model (2012). http://www. iso.org/iso/catalogue_detail.htm?csnumber=24683

6. Jones, S.: New checklist for a data management plan July 2013. http://www.dcc. ac.uk/resources/data-management-plans/checklist

7. Allinson, J., François, S., Lewis, S.: SWORD: Simple Web-service Offering Repository Deposit. Ariadne 54, January 2008. http://www.ariadne.ac.uk/issue54/ allinson-et-al/

8. Taverna: Taverna (2014). http://www.taverna.org.uk/

9. The Pennsylvania State University: Data management toolkit, December 2014. https://www.libraries.psu.edu/psul/researchguides/pubcur/datatoolkit.html

10. The Regents of the University of California: Dataup: Describe, manage and share your data. http://dataup.cdlib.org

Multi-agent Systems

Adaptive Model of Multi-objective Agent Behavior in Real-Time Systems

Alexander Alimov[1](✉) and David Moffat[2]

[1] Volgograd State Technical University, Volgograd, Russia
velorth.avelies@gmail.com
[2] Glasgow Caledonian University, Glasgow, Scotland
d.c.moffat@gcu.ac.uk

Abstract. Agents trying to reach their goals in dynamical environments need to be adaptive. Adaptation rules can conflict and their combinations can complicate multi-agent systems development. Agent programs are becoming mostly problem-oriented and high-coupled, and that prevents the reuse of developed programs and their components. High-level generic planning algorithms can be used to adapt agent behavior to environment changes, but they also have high computational complexity, which limits their usefulness in real-time application.

This paper considers a generic agent model that combines planning algorithms with utility theory to reach rational adaptive behavior, with acceptable performance and low component coupling. In the proposed model each agent has a dynamical set of tactical objectives associated with objects in environment. For each kind of objective the agent uses an objective-specific planning algorithm. To handle events from the environment the agent generates new objectives with a decision tree. Each objective has a dynamic priority calculated by heuristic function.

The suggested model may improve performance in real-time environments by decreasing computational complexity. The effectiveness of the model is shown on mini-game example.

Keywords: Multi-agent systems · Artificial intelligence · Adaptive behavior · Real-time planning

1 Introduction

Real-time multi-agent systems (MAS) are effective tools for the modeling of complex processes with large numbers of active autonomous entities. These include urban traffic flow, logistic systems, and simulation of social and economics phenomena. MAS modeling methods are also used for searching and processing information in networks. This general MAS approach is widely used in development of autonomous robot control systems.

One of the most rapidly developing applications of multi-agent systems is game development and virtual reality. In this area multi-agent modeling can be

© Springer International Publishing Switzerland 2015
A. Bikakis and X. Zheng (Eds.): MIWAI 2015, LNAI 9426, pp. 51–60, 2015.
DOI: 10.1007/978-3-319-26181-2_5

used for gathering statistics, or for artificial intelligence implementation of game agents (non-player characters).

Autonomous robots and game characters controlled by AI have to be able to perform in highly dynamical environments. To act effectively and reach their goals they have to adapt their behavior to environment changes. The dynamical nature of the environment dramatically increases the number of factors and cases that the agent must take into account. The complexity of the agent's model directly impacts upon the system's response time and performance.

Implementations of adaptive algorithms and models are also complex and problem-specific program entities. They are hard to reuse even for similar programs and environments; and it is often necessary to modify their source code.

The main objective of this work is the development of a flexible and extensible component-based agent system, that acts effectively in dynamical environments, in real-time.

2 Related Works

In the considered class of multi-agent systems (MASs), agents are able to move in continuous dynamical environments, filled with obstacles, interactive objects and other agents. Agents behave individually and competitively. Cooperation is also possible: but there are no explicit teams or groups. In these conditions the following problems must be solved:

- navigation and movement including path-finding and obstacle avoidance;
- decision making and action planning;
- adaptation to environment changes;
- communication between agents.

The path-finding problem is rather effectively solved with the classical A* algorithm [1], or its improved variations such as hierarchical search [2] and jump point search [3].

Path-finding algorithms work well in static environments, but not so well when agents must avoid collisions with dynamical objects, such as other agents. This problem can be solved with *reactive* behavior algorithms such as Craig Reynolds' well known *Flocking* [4] and *Steering* behaviors [5]. There are also effective and wide-spread algorithms based on reciprocal collision velocity [6,7].

Different approaches represent multiple agents as a continuous substrate, whose properties are similar to liquids. This approach is called "continuous crowds" as presented in [8]. Another robust approach defines potential fields that control agents motions using force law [9].

Although these approaches use fundamentally different models, they all solve one problem, and use similar input data. This enables us to unify them within an abstract navigation subsystem interface.

The navigation subsystem is an important component of the agent's behavior; but the decision making problem is really what defines the agent's architecture at high level. Varieties of architecture can be grouped according to the method that underlies the agent's behavior. There are planning agents that use graph search and inference algorithms; reactive agents based on finite state machines and behavior trees; and utility based agents.

Planning agents try to reach some desired state of the environment, by means of their actions. They represent the plan as a graph where each node is an environment state and each transition is an action which the agent can perform on that state [10,11]. These agents are effective when the objective is single, and known in advance. Planning and inference algorithms have quadratic complexity, and higher; so they are of limited use in real-time systems.

The most common agent architecture, especially in game development, is the finite state machine. FSMs are easy to visualize and understand. Their weak point is extensibility. The more actions an agent can perform, and the more factors it takes into account, the more complex the state machine graph becomes. Faced with these issues, FSMs were extended to hierarchical FSM (HFSM) models and concurrent FSMs. The HFSM aggregates nested finite-state machines in a single one of its states. These structures can be visualized the similar way to classical FSM. In terms of transition function, however, the difference between classical and hierarchical FSM is only a concept of graphical representation. To implement a hierarchical FSM has the same complexity, because transitions can pass through hierarchical layers to any state. The paper [13] considers usage of hierarchical FSM as a behavior model.

Agents based on behavior trees are similar to those based on FSMs. They are wide-spread in game development because they have high performance [12]. However, they have low adaptation capabilities and their implementations are typically highly coupled.

The utility-based agent is the least formalized model. To decide what action should be performed, the agent estimates the utilities of actions. Utility is a scalar value that defines how an action changes the agent's "happiness" value. These agents have high adaptation capabilities, and high performance; but they have limited usage for complex tasks.

The proposed model combines some aspects of the approaches considered above, to provide high adaptation capabilities, acceptable performance and reusable implementations.

3   Hybrid Agent Model

3.1   Generic Model Definition

At each moment of simulation time, the agent's state can be represented with following tuple:

$$S(t) = (R(t), E(t)) \tag{1}$$

where
$R(t) = [r_1(t), r_2(t), ...]$: vector of agent's own variables;
$E(t) = [e_1(t), e_2(t), ...]$: vector of perceived environment variables.

At each simulation moment we can estimate the value of the agent's "Happiness" with utility function $U(t) = f_U(S(t))$. In terms of the utilitarian approach the problem of agent control is reduced to maximization of total utility Y that is gained by agent during the whole simulation period, up to time T.

$$Y(T) = \int_0^T U(t)dt \tag{2}$$

Unfortunately it is impossible to optimize the Y function in its generic form directly. Instead, we can assume that there are some states such that when agent enters one of these it significantly increases its utility. Let us call these "objective states". The objective states are a dynamical set that is represented by the function $Q(t) = \{q_1, q_2, ..., q_n\}$.

Each objective state can be represented with following tuple:

$$q_i(s) = (O_i, P_i(s), c_i(s), A_i)$$

where
O_i : a target object which the objective is associated with;
$P_i(s)$: a function of objective priority;
$c_i(s)$: a predicate that checks if agent state s matches objective conditions;
A_i : an algorithm that leads agents to desired state to complete objective q_i.

When agent enters the state that matches the completion predicate of objective q_i it can be assumed that it gains the relevant utility $u(q_i)$. The utility for states that have not been entered is assumed to remain zero. Thereby, the aggregate utility value during the simulation time can be represented as a sum of all completed objective utilities:

$$Y \simeq \sum_{i=0}^n u(q_i) \tag{3}$$

With this assumption the problem of optimization is replaced with the objective generation problem, and that can be solved with simpler approximation methods.

The architecture scheme of the agent based on the proposed model is presented in Fig. 1, where rectangles represent agent's components, and arrows represent data and control (for actuator) flow. The agent model contains the following components:

- sensors – perceive the environment;
- event handler – handles all perceived events and feedback from actuator;
- memory – stores variables;
- objective manager – stores and process objectives;
- action queue – controls execution of actions, sends commands to actuator and navigator;
- navigator – plans routes in the environment and move agent;
- actuator – interacts with interactive objects and other agents.

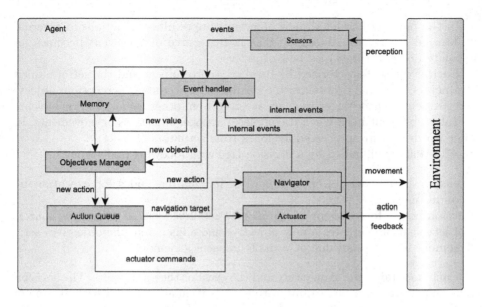

Fig. 1. Proposal adaptive agent scheme

3.2   Objective Management and Actions Queue

The main task of the objective manager is to store the objective set and keep it actual. Objective is actual if it is still not complete and the target state is attainable. Different objectives can be implemented with different classes but the manager doesn't need to know about their internal implementation. Each step of the simulation, the agent must recalculate the priorities of all objectives, and remove the non-actual ones. The agent must select the single objective with the highest priority. It will then start (or resume) the algorithm of that selected objective.

The objective algorithm produces a queue of actions that must be executed step by step. The agent can perform only one action at time. It's action is a program that is executed with the actuator. All actions are targeted on some object in the environment. Typically it is the event source or the objective target. The action execution process consists of the following states: movement, initialization, execution, finalization. In the movement stage the agent must reduce distance to the action target in order to be able to manipulate. The execution stage is the following stage, which depends on the action class. The initialization and finalization stages are instant stages.

3.3   Event Handling

Objectives are removed when they become non-actual, or complete.

When all objectives are achieved, the agent does nothing more, until events may occur that create new objectives.

In the proposed model all new objectives are results of reactions to external or internal events. An object O_i that is the source of the event becomes an associated object (target) for new objective.

Each event is characterized by its class, source object and the set of named parameters. External events are the results of environment perception: objects that are seen and messages that are received from other agents. Internal events are raised by agent itself when it completes actions, objectives or movements. In other words they are processed feedback from actuators.

All events are handled by a decision tree with following node types:

- selector node – selects one or more of its children nodes for execution by condition value;
- action node – forces agent to start a new action associated with event source;
- objective node – creates a new objective and adds it into the objectives set;
- memory node – sets value of variable in agent's memory.

Each node takes the agent state and the event as the parameters. The decision tree can perform multiple actions at time. In particular, it can create two or more conflicted objectives both of which are intended. In that case, the agent will try to complete the objective with the higher utility and priority. In the future, this might be used to create dynamical alliances between agents.

Actions created by the decision tree always have higher priority than actions created by the objective manager. This is required in order to implement instant reactions to some events.

3.4 Objectives Priority Factors

At each simulation step, the objective manager selects the most prioritized objective. These priority calculations are critical for a real-time system, so the priority function must be effective but simple to calculate. It is assumed that the priority function is a convolution of several factors based on objective properties. The following factors are proposed:

- the distance to the navigation target, d;
- the estimated cost of resources that must be spent, c;
- the marginal utility of the objective success, u;
- the probability of the objective success, p.

Two variants of convolution have been tested: weighted additive and multiplicative. Additive convolution led to unstable behavior, where the agent was thrashing between two or more objectives with similar priorities. Multiplicative convolution is calculated by formula 4. Each multiplier is in the range $[0..1]$.

$$P_q(s) = d_q(s) \cdot c_q(s) \cdot u_q(s) \cdot p_q(s) \tag{4}$$

Multiplicative convolution is more stable, but was still observed to thrash in some situations. To prevent this thrashing the relative attention threshold θ was introduced. Once committed to an objective, the agent will switch to another one only if its priority is significantly higher than that of the current one, as in Eq. 5.

$$P_{new}(s) > P_{current}(s)(1 + \theta) \tag{5}$$

Most multipliers in Eq. 4 depend on their objective class; but some generalizations about distance and marginal utility factors are possible. It is often possible to use a heuristic that the agent should select objectives associated with the *nearest* target. In the prototype implementation, the $d_q(s)$ multiplier takes the following form:

$$d_q(s) = \frac{1}{1 + k \cdot |\overrightarrow{p_t} - \overrightarrow{p_a}|} \tag{6}$$

where
$\overrightarrow{p_t}$: target position;
$\overrightarrow{p_a}$: agent position;
k : scale factor.

4   Proof of Concept

Consider the example of a typical shooter game, in which a character's behavior can be implemented with a planning algorithm and the proposed model.

Characters in the game have the following attributes: hit points, ammo count, world position, and rotation angle. Characters can move in the environment, pick up items, use med-kits, and attack other characters, reducing their hit points. For each action the character will gain a score reward. The aim is to maximize the final score.

The basic agent behavior can be represented as a finite state machine with following states: picking item, attacking enemy, healing, evasion. These states will be common for all test implementations.

To simplify the planning model we assume that the agent can move in four directions. The A* search algorithm was used to implement planning. Each graph node represents a state of the agent itself and states of all other objects. Each transition represents the agent's action, and its cost aggregates the acting time, the estimated damage cost, and the spent items cost. In the target nodes all items on the map must be collected and all other agents must be eliminated. Transitions that lead to negative value for the agent's hit points are disabled.

The estimated damage cost linearly depends on count of enemy agents within their attack range.

A naive planning implementation is too heavy in its computational demands. In a dynamical environment long plans cannot be completed and executed, because they must be corrected at each simulation step, or even rebuilt. It is therefore not appropriate to build a complete plan, so we can make the following optimization of the planning model:

1. Introduce heuristic constraints for action transitions: do not reload weapon or use med-kits when there is no need for them;
2. Exclude locomotions from high-level planning. This can decrease accuracy but not significantly;
3. Reduce target nodes requirements: only nearest objectives must be achieved.

The optimized planning model was used in the real-time simulation that is presented in Table 1. The **reduced** planning algorithm requires all objectives within a range to be completed. It works in simple cases, but there are cases where the search didn't complete in time (200000 iterations). The **greedy** planning algorithm requires at least one objective to be achieved. This algorithm would be proven to be not optimal in the long term. In the simulation the agent almost always chose to pick up items as the cheapest action, even if it didn't need them. Finally, the **proposed** model (hybrid) algorithm showed approximately the same performance as the greedy planning did; but the agent chose actions that are more optimal in the long run.

For the proposed model there are the following types of objectives: pick up item, keep hit points level, keep ammo level, attack target, and evade target. The probability of achieving objectives depends on their estimated damage value. An objective's utility is calculated based on the agent's hit points (cost of damage) and the score points for the objective.

Table 1. Performance comparison of models

Test case	Execution time, iterations		
	Reduced planning	Greedy planning	Proposed model
Pick items (locomotion)	Limit exceeded	210	1067
Pick items (approximate)	530	40	24
Single enemy Infinite ammo	188	67	45
Single enemy Limited ammo	198	36	26
Multiple enemies Infinite ammo	631	41	34
Multiple enemies Limited ammo	Limit exceeded	41	38

It can be noticed that there are cases where reduced planning cannot be completed in acceptable time. There are also cases where a solution that completes all the objectives cannot be found at all.

Reduced and greedy planning require to take into account all actions that the agent can perform; but the proposed model excludes actions that do not contribute to the completion of the highest priority objective. In the large scale that can reduce significantly the search space. For simple objectives, the planning algorithm can be replaced with single predefined action or if-then rule set.

In real-time simulation agents that plan their behavior based on utility-based priority function were able to combine objectives in short chains (about three elements) in order to complete more difficult objectives.

5   Conclusion

The key differences between the proposed and classical models is a dynamical objective management system. This system allows to group state and memory variables by the objectives that they are related to. It simplifies the reuse of the objective classes that have been already implemented. The dynamic priority system adapts the agent's behavior to different situations in the environment, and their combinations.

Objective achievement algorithms are based on A* planning that makes the action sequence rational and semi-optimal. But at the same time the planning algorithm takes into account only the data associated with the current objective. This significantly reduces planning time especially for a naive planning implementation.

References

1. Dechter, R., Pearl, J.: Generalized best-first search strategies and the optimality of A*. J. ACM **39**(3), 505–536 (1985)
2. Botea, A., Müller, M., Schaeffer, J.: Near optimal hierarchical path-finding. J. Game Dev. **1**, 7–28 (2004)
3. Daniel, H., Alban, G.: The JPS pathfinding system. In: Proceedings of the Fifth Annual Symposium on Combinatorial Search (2012)
4. Reynolds, C.W.: Flocks, herds and schools: a distributed behavioral model. Comput. Graph. **21**(4), 25–34 (1987)
5. Reynolds, C.W.: Steering behavior for autonomous characters. In: Game Developers Conference (1999)
6. Alonso-Mora, J., Breitenmoser, A., Rufli, M., Beardsley, P., Siegwart, R.: Optimal reciprocal collision avoidance for multiple non-holonomic robots. In: Martinoli, A., Mondada, F., Correll, N., Mermoud, G., Egerstedt, M., Hsieh, M.A., Parker, L.E., Støy, K. (eds.) Distributed Autonomous Robotic Systems. STAR, vol. 83, pp. 203–216. Springer, Heidelberg (2013)
7. Van den Berg, J., Lin, M.C., Manocha, D.: Reciprocal velocity obstacles for real-time multi-agent navigation. In: IEEE International Conference on Robotics and Automation. IEEE (2008)
8. Narain, R., Golas, A., Curtis, S., Lin, M.C.: Aggregate dynamics for dense crowd simulation. ACM Trans. Graph. **28**(5), 1–8 (2009)
9. Reif, J.H., Wang, H.: Social potential fields: a distributed behavioral control for automated robots (1995)
10. Orkin, J., Young, R.M., Laird, J.E.: Agent architecture considerations for real-time planning in games. In: AIIDE. AAAI Press (2005)

11. Orkin, J.: Symbolic representation of game world state: Toward real-time planning in games. In: AAAI Workshop on Challenges in Game Artificial Intelligence. AAAI Press (2005)
12. Cavazza, M.: AI in computer games: survey and perspectives. Virtual Reality **5**, 223–235 (2000). Springer
13. Patel, H.D.: Behavioral hierarchy with hierarchical FSMs (HFSMs). In: Ingredients for Successful System Level Design Methodology, pp. 47–75 (2008)

Boolean Games with Norms

Xin Sun[(⊠)]

Faculty of Science, Technology and Communication, University of Luxembourg,
Luxembourg City, Luxembourg
xin.sun@uni.lu

Abstract. In the present paper we overlay boolean game with norms.
Norms distinguish illegal strategies from legal strategies. Two types of
legal strategy and legal Nash equilibrium are defined. These two equilibrium are viewed as solution concepts for law abiding agents in norm augmented boolean games. Our formal model is a combination of boolean
games and so called input/output logic. We study various complexity
issues related to legal strategy and legal Nash equilibrium.

Keywords: Boolean game · Norm · Input/output logic

1 Introduction

The study of the interplay of games and norms can be divided into two main
branches: the first, mostly originating from economics and game theory [9,19,20],
treats norms as mechanisms that enforce desirable properties of social interactions; the second, that has its roots in social sciences and evolutionary game
theory [10,31] views norms as (Nash or correlated) equilibrium that result from
the interaction of rational agents. A survey of the interaction between games and
norms can be found in Grossi *et al.* [15]. This paper belongs to the first branch.

In this paper we study the combination of boolean games and norms. Boolean
game is a class of games based on propositional logic. It was firstly introduced by
Harrenstein et al. [17] and further developed by several researchers [6,8,11,16,
23,25]. In a boolean game, each agent i is assumed to have a goal, represented
by a propositional formula ϕ_i over some set of propositional variables $\mathbb{P}$. Each
agent i is associated with some subset $\mathbb{P}_i$ of the variables, which are under the
unique control of agent i. The choices, or strategies, available to i correspond to
all the possible assignment of truth or falsity to the variables in $\mathbb{P}_i$. An agent
will try to choose an assignment so as to satisfy his goal ϕ_i. Strategic concerns
arise because whether i's goal is in fact satisfied will depend on the choices made
by other agents.

Norms regulate agents' behaviors in boolean games. Shoham and
Tennenholtz's early work on behavior change under norms [27,28] has considered
only a relatively simple view of norms, where some actions or states are desig-
nated as violations. Alechina *et al.* [1] studies how conditional norms regulate
agents' behaviors, but permissive norms plays no role in their framework. In this

© Springer International Publishing Switzerland 2015
A. Bikakis and X. Zheng (Eds.): MIWAI 2015, LNAI 9426, pp. 61–71, 2015.
DOI: 10.1007/978-3-319-26181-2_6

paper we study how agents' behavior are changed by permissive and obligatory conditional norms. Norms distinguish illegal strategies from legal strategies. By designing norms appropriately, non-optimal equilibrium might be avoided. To represent norms in boolean games we need a logic of norms, which has been extensively studied in the deontic logic community.

Various deontic logic have been developed since von Wright's first paper [32] in this area. In the first volume of the handbook of deontic logic [12], input/output logic [21,22] appears as one of the new achievements in deontic logic in this century. Input/output logic takes its origin in the study of conditional norms. The basic idea is: norms are conceived as a deductive machine, like a black box which produces normative statements as output, when we feed it factual statements as input.

In this paper we use input/output logic as the logic of norms. Given a normative multi-agent system, which contains a boolean game, a set of norms and certain environment. Every strategy of every agent is classified as legal or illegal. Notions like legal Nash equilibrium are then naturally defined.

The structure of this paper is the following: We present some background knowledge, including boolean game, input/output logic and complexity theory in Sect. 2. Normative multi-agent system are introduced and its complexity issues are discussed in Sect. 3. We conclude this paper in Sect. 4.

2 Background

2.1 Propositional Logic

Let $\mathbb{P} = \{p_0, p_1, \ldots\}$ be a finite set of propositional variables and let $L_\mathbb{P}$ be the propositional language built from $\mathbb{P}$ and boolean constants $\top$ (true) and $\bot$ (false) with the usual connectives $\neg, \vee, \wedge, \rightarrow$ and $\leftrightarrow$. Formulas of $L_\mathbb{P}$ are denoted by ϕ, ψ etc. A literal is a variable $p \in \mathbb{P}$ or its negation. $2^\mathbb{P}$ is the set of the valuations for $\mathbb{P}$, with the usual convention that for $V \in 2^\mathbb{P}$ and $p \in V$, V gives the value true to p if $p \in V$ and false otherwise. $\vDash$ denotes the classical logical consequence relation.

Let $X \subseteq \mathbb{P}$, 2^X is the set of X-valuations. A partial valuation (for $\mathbb{P}$) is an X-valuation for some $X \subseteq \mathbb{P}$. Partial valuations are denoted by listing all variables of X, with a "+" symbol when the variable is set to be true and a "−" symbol when the variable is set to be false: for instance, let $X = \{p, q, r\}$, then the X-valuation $V = \{p, r\}$ is denoted $\{+p, -q, +r\}$. If $\{\mathbb{P}_1, \ldots, \mathbb{P}_n\}$ is a partition of $\mathbb{P}$ and $V_1, \ldots, V_n$ are partial valuations, where $V_i \in 2^{\mathbb{P}_i}$, $(V_1, \ldots, V_n)$ denotes the valuation $V_1 \cup \ldots \cup V_n$.

2.2 Boolean Game

Boolean games introduced by Harrenstein et al. [17] are zero-sum games with two players, where the strategies available to each player consist in assigning a truth value to each variable in a given subset of $\mathbb{P}$. Bonzon et al. [7] give a more general definition of a boolean game with any number of players and not necessarily zero-sum. In this paper we further generalizes boolean games such that the utility of each agent is not necessarily in $\{0, 1\}$. Such generalization is reached by representing the goals of each agent as a set of weighted formulas. The idea of using weighted formulas to define utility can be found in many work among which we mention satisfiability games [4] and weighted boolean formula games [23].

Definition 1 (boolean game). *A boolean game is a 4-tuple* $(Agent, \mathbb{P}, \pi, Goal)$, *where*

1. *Agent* $= \{1, \ldots, n\}$ *is a set of agents.*
2. $\mathbb{P}$ *is a finite set of propositional variables.*
3. $\pi : Agent \mapsto 2^{\mathbb{P}}$ *is a control assignment function such that* $\{\pi(1), \ldots, \pi(n)\}$ *forms a partition of* $\mathbb{P}$. *For each agent* i, $2^{\pi(i)}$ *is the strategy space of* i.
4. *Goal* $= \{Goal_1, \ldots, Goal_n\}$ *is a set of weighted formulas of* $L_{\mathbb{P}}$. *That is, each* $Goal_i$ *is a finite set* $\{\langle \phi_1, m_1 \rangle, \ldots, \langle \phi_k, m_k \rangle\}$ *where* $\phi_j \in L_{\mathbb{P}}$ *and* m_j *is a real number.*

A strategy for agent i is a partial valuation for all the variables i controls. Note that since $\{\pi(1), \ldots, \pi(n)\}$ forms a partition of $\mathbb{P}$, a strategy profile S is a valuation for $\mathbb{P}$. In the rest of the paper we make use of the following notation, which is standard in game theory. Let $G = (Agent, \mathbb{P}, \pi, Goal)$ be a boolean game with $Agent = \{1, \ldots, n\}$, $S = (s_1, \ldots, s_n)$ be a strategy profile. s_{-i} denotes the projection of S on $Agent - \{i\}$: $s_{-i} = (s_1, \ldots, s_{i-1}, s_{i+1}, \ldots, s_n)$.

Agents' utilities in boolean games are induced by their goals. For every agent i and every strategy profiles S, $u_i(S) = \Sigma\{m_j : \langle \phi_j, m_j \rangle \in Goal_i, S \models \phi_j\}$. Dominating strategies and pure-strategy Nash equilibria are defined as usual in game theory [24].

Example 1. *Let* $G = (Agent, \mathbb{P}, \pi, Goal)$ *where* $Agent = \{1, 2\}$, $\mathbb{P} = \{p, q, s\}$, $\pi(1) = \{p\}$, $\pi(2) = \{q, s\}$, $Goal_1 = \{\langle p \leftrightarrow q, 1 \rangle, \langle s, 2 \rangle\}$, $Goal_2 = \{\langle p \wedge q, 2 \rangle, \langle \neg s, 1 \rangle\}$. *This boolean game is depicted as follows:*

	$+q, +s$	$+q, -s$	$-q, +s$	$-q, -s$
$+p$	$(3, 2)$	$(1, 3)$	$(2, 0)$	$(0, 1)$
$-p$	$(2, 0)$	$(0, 1)$	$(3, 0)$	$(1, 1)$

2.3   Input/Output Logic

In input/output logic, a norm is an ordered pair of formulas $(\phi, \psi) \in L_{\mathbb{P}} \times L_{\mathbb{P}}$. There are two types of norms which are used in input/output logic, obligatory norms and permissive norms. Let $N = O \cup P$ be a set of obligatory and permissive norms. A pair $(\phi, \psi) \in O$, call it an obligatory norm, is read as "given

ϕ, it is obligatory to be ψ". A pair $(\phi, \psi) \in P$, call it a permissive norm, is read as "given ϕ, it is permitted to be ψ".

Obligatory norms O can be viewed as a function from $2^{L_\mathbb{P}}$ to $2^{L_\mathbb{P}}$ such that for a set Φ of formulas, $O(\Phi) = \{\psi \in L_\mathbb{P} : (\phi, \psi) \in O \text{ for some } \phi \in \Phi\}$.

Definition 2 (Semantics of input/output logic [21]). *Given a finite set of obligatory norms O and a finite set of formulas Φ, $out(O, \Phi) = Cn(O(Cn(\Phi)))$, where Cn is the consequence relation of propositional logic.*[1]

Intuitively, the procedure of the semantics is as following: We first have in hand a set of formulas Φ (call it the input) as a description of the current state. We then close it by logical consequence $Cn(\Phi)$. The set of norms, like a deductive machine, accepts this logically closed set and produces a set of formulas $O(Cn(\Phi))$. We finally get the output $Cn(O(Cn(\Phi)))$ by applying the logical closure again. $\psi \in out(O, \phi)$ is understood as "ψ is obligatory given facts Φ and norms O".

Example 2. *Let p, q, r are propositional variables. Let $O = \{(p, q), (p \vee q, r), (r, p)\}$. Then $out(O, \{p\}) = Cn(O(Cn(\{p\}))) = Cn(\{q, r\})$.* □

Input/output logic is given a proof theoretic characterization. We say that an ordered pair of formulas is derivable from a set O iff (a, x) is in the least set that extends $O \cup \{(\top, \top)\}$ and is closed under a number of derivation rules. The following are the rules we need:

– SI (strengthening the input): from (ϕ, ψ) to (χ, ψ) whenever $\chi \vDash \phi$.
– WO (weakening the output): from (ϕ, ψ) to (ϕ, χ) whenever $\psi \vDash \chi$.
– AND (conjunction of output): from (ϕ, ψ) and (ϕ, χ) to $(\phi, \psi \wedge \chi)$.

The derivation system based on the rules SI, WO and AND is denoted as $deriv(O)$.

Example 3. *Let $O = \{(p \vee q, r), (q, r \to s)\}$, then $(q, s) \in deriv(O)$ because we have the following derivation*

1. $(p \vee q, r)$	*Assumption*
2. (q, r)	*1, SI*
3. $(q, r \to s)$	*Assumption*
4. $(q, r \wedge (r \to s))$	*2,3, AND*
5. (q, s)	*4, WO*

In Makinson and van der Torre [21], the following soundness and completeness theorem is proved:

Theorem 1 ([21]). *Given a set of obligatory norms O,*

$$\psi \in out(O, \{\phi\}) \text{ iff } (\phi, \psi) \in deriv(O).$$

---

[1] In Makinson and van der Torre [21], this logic is called simple-minded input/output logic. Different input/output logics are developed in Makinson and van der Torre [21] as well. A technical introduction of input/output logic can be found in Sun [30].

Permission in Input/Output Logic. Philosophically, it is common to distinguish between two kinds of permission: negative permission and positive permission. Negative permission is straightforward to describe: something is negatively permitted according to certain norms iff it is not prohibited by those norms. That is, iff there is no obligation to the contrary. Positive permission is more elusive. Makinson and van der Torre [22] distinguish two types of positive permission: static and dynamic permission. For the sake of simplicity, in this paper when discuss positive permission we only mean static permission.

Definition 3 (negative permission [22]). *Given a finite set of norms* $N = O \cup P$ *and a finite set of formulas* Φ, $NegPerm(N, \Phi) = \{\psi \in L_{\mathbb{P}} : \neg \psi \notin out(O, \Phi)\}$.

Intuitively, ϕ is negatively permitted iff ϕ is not forbidden. Since a formula is forbidden iff its negation is obligatory, ϕ is not forbidden is equivalent to $\neg \phi$ is not obligatory. Permissive norms plays no role in negative permission.

Definition 4 (positive permission [22]). *Given a finite set of formulas* Φ, *a finite set of norms* $N = O \cup P$ *where* O *is a set of obligatory norms and* P *is a set of permissive norms.*

- *If* $P \neq \emptyset$, *then* $PosPerm(N, \Phi) = \{\psi \in L_{\mathbb{P}} : \psi \in out(O \cup \{(\phi', \psi')\}, \Phi),$
  for some $(\phi', \psi') \in P\}\}$.
- *If* $P = \emptyset$, *then* $PosPerm(N, \Phi) - out(O, \Phi)$.

Intuitively, permissive norms are treated like weak obligatory norms, the basic difference is that while the latter may be used jointly, the former may only be applied one by one. As an illustration of such difference, image a situation in which a man is permitted to date either one of two girls, but not both of them. Alternative definitions of positive permission can be found in Makinson and van der Torre [22], Stolpe [29] and Governatori [14].

2.4 Complexity Theory

Complexity theory is the theory to investigate the time, memory, or other resources required for solving computational problems. In this subsection we briefly review those concepts and results from complexity theory which will be used in this paper. More comprehensive introduction of complexity theory can be found in [3].

We assume the readers are familiar with notions like Turing machine and the complexity class P, NP and coNP. Oracle Turing machine and two complexity classes related to oracle Turing machine will be used in this paper.

Definition 5 (oracle Turing machine [3]). *An oracle for a language* L *is a device that is capable of reporting whether any string* w *is a member of* L. *An oracle Truing machine* M^L *is a modified Turing machine that has the additional capability of querying an oracle. Whenever* M^L *writes a string on a special oracle tape it is informed whether that string is a member of* L, *in a single computation step.*

P^{NP} is the class of problems solvable by a deterministic polynomial time Turing machine with an NP oracle. NP^{NP} is the class of problems solvable by a non-deterministic polynomial time Turing machine with an NP oracle. Another name for the class NP^{NP} is Σ_2^p. Σ_{i+1}^p is the class of problems solvable by a non-deterministic polynomial time Turing machine with a Σ_i^p oracle. Π_i^p is the class of problems of which the complement is in Σ_i^p.

3   From Boolean Game to Normative Multi-agent System

In recent years, normative multi-agent system [2,5] arises as a new interdisciplinary academic area bringing together researchers from multi-agent system [26,33,34], deontic logic [12] and normative system [1,13,18]. By combining boolean games and norms, we here develop a new approach for normative multi-agent system.

Definition 6 (normative multi-agent system). *A normative multi-agent system is a triple (G, N, E) where*

- *$G = (Agent, \mathbb{P}, \pi, Goal)$ is a boolean game.*
- *$N = O \cup P \subseteq L_{\mathbb{P}} \times L_{\mathbb{P}}$ is a finite set of obligatory and permissive norms.*
- *$E \subseteq L_{\mathbb{P}}$ is a finite set of formulas representing the environment.*

3.1   Legal Strategy

In a normative multi-agent system, agent's strategies are classified as either legal or illegal. The basic idea is viewing strategies as formulas and using the mechanism of input/output logic to decide whether a formula is permitted.

Definition 7 (legal strategy). *Given a normative multi-agent system (G, N, E) where $N = O \cup P$, for each agent i, a strategy $(+p_1, \ldots, +p_m, -q_1, \ldots, -q_n)$ is negatively legal if*

$$p_1 \wedge \ldots \wedge p_m \wedge \neg q_1 \wedge \ldots \wedge \neg q_n \in NegPerm(N, E).$$

The strategy is positively legal if

$$p_1 \wedge \ldots \wedge p_m \wedge \neg q_1 \wedge \ldots \wedge \neg q_n \in PosPerm(N, E).$$

Example 4. *Consider the prisoner's dilemma augmented with norms, where the two prisoners are brothers who are morally required to protect each other. Let (G, N, E) be a normative multi-agent system as following:*

- *$G = (Agent, \mathbb{P}, \pi, Goal)$ is a boolean game with*
  - *$Agent = \{1, 2\}$,*
  - *$\mathbb{P} = \{p, q\}$,*
  - *$\pi(1) = \{p\}$, $\pi(2) = \{q\}$,*
  - *$Goal_1 = \{\langle p, 2 \rangle, \langle \neg q, 3 \rangle\}$, $Goal_2 = \{\langle q, 2 \rangle, \langle \neg p, 3 \rangle\}$.*

- $N = O \cup P$ where $O = \{(\top, \neg p)\}$, $P = \{(\top, \neg q)\}$.
- $E = \emptyset$.

	$+q$	$-q$
$+p$	$(2,2)$	$(5,0)$
$-p$	$(0,5)$	$(3,3)$

Then $out(O, E) = Cn(\{\neg p\})$, $\{\neg p, q, \neg q\} \subseteq NegPerm(N, E)$. Therefore $\{-p\}$, $\{+q\}$, $\{-q\}$ are negatively legal while $\{+p\}$ is not. Moreover we have $PosPerm(N, E) = out(O \cup P, E) = Cn(\{\neg p, \neg q\})$, Therefore $\{-p\}$ and $\{-q\}$ are positively legal while neither $\{+p\}$ nor $\{+q\}$ is.   ⊣

Having defined notions of legal strategy. A natural question to ask is how complex is it to decide whether a strategy is legal. The following theorems give a first answer to this question.

Theorem 2. *Given a normative multi-agent system (G, N, E) and a strategy $(+p_1, \ldots, +p_m, -q_1, \ldots, -q_n)$, deciding whether this strategy is negatively legal is NP complete.*

Proof. Concerning the NP hardness, we prove by reducing the satisfiability problem of propositional logic to our problem: Let $\phi \in L_{\mathbb{P}}$ be a formula. Let $N = \{(\neg \phi, \neg p)\}$, $E = \emptyset$. Then $p \in NegPerm(N, E)$ iff $\neg p \notin out(N, E) = Cn(N(Cn(E))) = Cn(N(Cn(\top)))$ iff $\not\vdash \neg \phi$ iff ϕ is satisfiable.

Now we prove the NP membership. We provide the following non-deterministic Turing machine to solve our problem. Let $N = \{(\phi_1, \psi_1), \ldots, (\phi_n, \psi_n)\}$, E be a finite set of formulas and $p_1 \wedge \ldots \wedge p_m \wedge \neg q_1 \wedge \ldots \wedge \neg q_k$ be a formula.

1. Guess a sequence of valuation $V_1, \ldots, V_n, V'$ on the propositional letters appears in $E \cup \{\phi_1, \ldots, \phi_n\} \cup \{\psi_1, \ldots, \psi_n\} \cup \{p_1 \wedge \ldots \wedge p_m \wedge \neg q_1 \wedge \ldots \wedge \neg q_k\}$.
2. Let $N' \subseteq N$ be the set of obligatory norms which contains all (ϕ_i, ψ_i) such that $V_i(E) = 1$ and $V_i(\phi_i) = 0$.
3. Let $\Psi = \{\psi : (\phi, \psi) \in N - N'\}$.
4. If $V'(\Psi) = 1$ and $V'(p_1 \wedge \ldots \wedge p_m \wedge \neg q_1 \wedge \ldots \wedge \neg q_k) = 0$. Then return "accept" on this branch. Otherwise return "reject" on this branch.

It can be verified that $p_1 \wedge \ldots \wedge p_m \wedge \neg q_1 \wedge \ldots \wedge \neg q_k \notin Cn(N(Cn(E)))$ iff the algorithm returns "accept" on some branch and the time complexity of the non-deterministic Turing machine is polynomial.   ⊣

Theorem 3. *Given a normative multi-agent system (G, N, E) and a strategy $(+p_1, \ldots, +p_m, -q_1, \ldots, -q_n)$, deciding whether this strategy is positively legal is coNP complete.*

Proof. The coNP hardness can be proved by a reduction from the tautology problem of propositional logic. Here we omit the details.

Concerning the coNP membership, note that $PosPerm(N, E) = out(O \cup \{(\phi_1, \psi_1)\}, E) \cup \ldots \cup out(O \cup \{(\phi_m, \psi_m)\}, E)$, where $N = O \cup P$, $P = \{(\phi_1, \psi_1), \ldots, (\phi_m, \psi_m)\}$. The NP membership follows from the fact that the NP class is closed under union.   ⊣

3.2   Legal Nash Equilibrium

A (pure-strategy) legal Nash equilibrium is a strategy profile which contains only legal strategies and no agent can improve his utility by choosing another legal strategy, given others do not change their strategies.

Definition 8 (Legal Nash equilibrium). *Given a normative multi-agent system (G, N, E), A strategy profile $S = (s_1, \ldots, s_n)$ is a negatively legal Nash equilibrium if*

– for every agent i, s_i is a negatively legal strategy
– for every agent i, for every negatively legal strategy $s'_i \in S_i$, $u_i(S) \geq u_i(s'_i, s_{-i})$.

Positively legal Nash equilibrium is defined analogously.

Example 5. *In the normative multi-agent system presented in Example 4, there is no negatively legal Nash equilibrium and $(-p, -q)$ is the unique positively legal Nash equilibrium.*

Example 6. *Let (G, N, E) be a normative system as following:*

– $G = (Agent, \mathbb{P}, \pi, Goal)$ is a boolean game with
  • $Agent = \{1, 2\}$,
  • $\mathbb{P} = \{p, q\}$,
  • $\pi(1) = \{p\}$, $\pi(2) = \{q\}$,
  • $Goal_1 = Goal_2 = \{\langle p \wedge q, 2 \rangle, \langle \neg p \wedge \neg q, 3 \rangle\}$.
– $N = O \cup P$ where $O = \{(\top, \neg p), (\top, \neg q)\}$, $P = \emptyset$.
– $E = \emptyset$.

	$+q$	$-q$
$+p$	$(2, 2)$	$(0, 0)$
$-p$	$(0, 0)$	$(3, 3)$

Without norms there are two Nash equilibria: $(+p, +q)$ and $(-p, -q)$. There is only one negatively/positively legal Nash equilibrium: $(-p, -q)$. From the perspective of social welfare, $(+p, +q)$ is not an optimal equilibrium because its social welfare is $2 + 2 = 4$, while the social welfare of $(-p, -q)$ is $3 + 3 = 6$. Therefore this example shows that by designing norms appropriately, non-optimal equilibrium might be avoided.

Theorem 4. *Given a normative multi-agent system* (G, N, E) *and a strategy profile* $S = (s_1, \ldots, s_n)$. *Deciding whether* S *is a negatively legal Nash equilibrium is NP hard and in* $coNP^{NP}$.

Proof. (sketch) It is NP hard because deciding whether a single strategy is legal is already NP hard.

Concerning the $coNP^{NP}$ membership, we prove by giving the following algorithm on a non-deterministic Turing machine with oracle SAT to solve the complement of this problem.

1. Test if S is a negatively legal strategy profile. If no, then return "accept". Otherwise continue.
2. Guess a strategy profile S'.
3. Test if S' is a legal strategy profile. If yes, continue. Otherwise return "reject" on this branch.
4. Test if $u_i(S) < u_i(S')$ for some i. If yes, return "accept" on this branch. Otherwise return "reject" on this branch.

It can be verified that S is NOT a negatively legal Nash equilibrium iff the non-deterministic Turing returns "accept" on some branches. Therefore deciding whether S is a negatively legal Nash equilibrium is in $coNP^{NP}$.   ⊣

Theorem 5. *Given a normative multi-agent system* (G, N, E) *and a strategy profile* $S = (s_1, \ldots, s_n)$. *Deciding whether* S *is a positively legal Nash equilibrium of* G *is coNP hard and in* $coNP^{NP}$.

Proof. (sketch) Similar to the proof of Theorem 4.   ⊣

Theorem 6. *Given a normative multi-agent system* (G, N, E). *Deciding whether there is a negatively/positively legal Nash equilibrium of* G *is* Σ_2^P *hard and in* Σ_3^P.

Proof. (sketch) The lower bound follows from the fact that deciding Nash equilibrium for boolean games without norms is Σ_2^P complete [7]. Concerning the upper bound, recall that $\Sigma_3^P = NP^{\Sigma_2^P}$. The problem can be solved by a polynomial time non-deterministic Turing machine with an Σ_2^P oracle.   ⊣

4   Conclusion

In the present paper we introduce boolean game with norms. Norms distinguish illegal strategies from legal strategies. Using ideas from input/output logic, two types of legal strategies are discussed, as well as two types of legal Nash equilibrium. After formally presenting the model, we use examples to show that non-optimal Nash equilibrium can be avoided by implementing norms. We study the complexity issues related to legal strategy and legal Nash equilibrium.

References

1. Alechina, N., Dastani, M., Logan, B.: Reasoning about normative update. In: Rossi, F. (ed.) IJCAI 2013, Proceedings of the 23rd International Joint Conference on Artificial Intelligence, Beijing, China, 3–9 August 2013. IJCAI/AAAI (2013)
2. Andrighetto, G., Governatori, G., Noriega, P., van der Torre, L.W.N. (eds.) Normative Multi-agent Systems. Dagstuhl Follow-Ups. vol. 4. Schloss Dagstuhl - Leibniz-Zentrum fuer Informatik (2013)
3. Arora, S., Barak, B.: Computational Complexity: A Modern Approach. Cambridge University Press, New York (2009)
4. Bilò, V.: On satisfiability games and the power of congestion games. In: Kao, M.-Y., Li, X.-Y. (eds.) AAIM 2007. LNCS, vol. 4508, pp. 231–240. Springer, Heidelberg (2007)
5. Boella, G., van der Torre, L., Verhagen, H.: Introduction to the special issue on normative multiagent systems. Auton. Agents Multi-Agent Syst. **17**(1), 1–10 (2008)
6. Bonzon, E., Lagasquie-Schiex, M.-C., Lang, J.: Dependencies between players in boolean games. Int. J. Approx. Reasoning **50**(6), 899–914 (2009)
7. Bonzon, E., Lagasquie-Schiex, M.-C., Lang, J., Zanuttini, B.: Boolean games revisited. In: Brewka, G., Coradeschi, S., Perini, A., Traverso, P. (eds.) ECAI 2006, 17th European Conference on Artificial Intelligence, Riva del Garda, Italy, August 29–September 1 2006, Proceedings of Including Prestigious Applications of Intelligent Systems (PAIS 2006). Frontiers in Artificial Intelligence and Applications, vol. 141, pp. 265–269. IOS Press (2006)
8. Bonzon, E., Lagasquie-Schiex, M.-C., Lang, J., Zanuttini, B.: Compact preference representation and boolean games. Auton. Agents Multi-Agent Syst. **18**(1), 1–35 (2009)
9. Coase, R.: The problem of social cost. J. Law Econ. **11**(1), 67–73 (1960)
10. Coleman, J.: Foundations of Social Theory. Belnap Press, Cambridge (1998)
11. Dunne, P.E., van der Hoek, W., Kraus, S., Wooldridge, M.: Cooperative boolean games. In: Padgham, L., Parkes, D.C., Müller, J.P., Parsons, S. (eds.) 7th International Joint Conference on Autonomous Agents and Multiagent Systems (AAMAS 2008), Estoril, Portugal, 12–16 May 2008, vol. 2, pp. 1015–1022. IFAAMAS (2008)
12. Gabbay, D., Horty, J., Parent, X., van der Meyden, R., van der Torre, L. (eds.): Handbook of Deontic Logic and Normative Systems. College Publications, London (2014)
13. Ågotnes, T., van der Hoek, W., Rodríguez-Aguilar, J.A., Sierra, C., Wooldridge, M.: On the logic of normative systems. In: Veloso, M.M. (ed.) IJCAI 2007, Proceedings of the 20th International Joint Conference on Artificial Intelligence, Hyderabad, India, 6–12 January 2007, pp. 1175–1180 (2007)
14. Governatori, G., Olivieri, F., Rotolo, A., Scannapieco, S.: Computing strong and weak permissions in defeasible logic. J. Philos. Log. **42**(6), 799–829 (2013)
15. Grossi, D., Tummolini, L., Turrini, P.: Norms in game theory. In: Handbook of Agreement Technologies (2012)
16. Harrenstein, P.: Logic in conflict. Ph.D. thesis, Utrecht University (2004)
17. Harrenstein, P., van der Hoek, W., Meyer, J.-J., Witteveen, C.: Boolean games. In: Proceedings of the 8th Conference on Theoretical Aspects of Rationality and Knowledge, TARK 2001, pp. 287–298. Morgan Kaufmann Publishers Inc., San Francisco (2001)

18. Herzig, A., Lorini, E., Moisan, F., Troquard, N.: A dynamic logic of normative systems. In: Walsh, T. (ed.) IJCAI 2011, Proceedings of the 22nd International Joint Conference on Artificial Intelligence, Barcelona, Catalonia, Spain, 16–22 July 2011, pp. 228–233. IJCAI/AAAI (2011)

19. Hurwicz, L.: Institutions as families of game forms. Jpn. Econ. Rev. **47**(2), 113–132 (1996)

20. Hurwicz, L.: But who will guard the guardians? Am. Econ. Rev. **98**(3), 577–585 (2008)

21. Makinson, D., van der Torre, L.: Input-output logics. J. Philos. Log. **29**, 383–408 (2000)

22. Makinson, D., van der Torre, L.: Permission from an input/output perspective. J. Philos. Log. **32**, 391–416 (2003)

23. Mavronicolas, M., Monien, B., Wagner, K.W.: Weighted boolean formula games. In: Deng, X., Graham, F.C. (eds.) WINE 2007. LNCS, vol. 4858, pp. 469–481. Springer, Heidelberg (2007)

24. Osborne, M., Rubinstein, A.: A Course in Game Theory. MIT Press, Cambridge (1994)

25. Sauro, L., Villata, S.: Dependency in cooperative boolean games. J. Log. Comput. **23**(2), 425–444 (2013)

26. Shoham, Y., Leyton-Brown, K.: Multiagent Systems - Algorithmic, Game-Theoretic, and Logical Foundations. Cambridge University Press, New York (2009)

27. Shoham, Y., Tennenholtz, M.: On the synthesis of useful social laws for artificial agent societies (preliminary report). In: Swartout, W.R. (ed.) Proceedings of the 10th National Conference on Artificial Intelligence, San Jose. CA, 12 16 July 1992, pp. 276 281. AAAI Press/MIT Press (1992)

28. Shoham, Y., Tennenholtz, M.: On social laws for artificial agent societies: off-line design. Artif. Intell. **73**(1–2), 231–252 (1996)

29. Stolpe, A.: A theory of permission based on the notion of derogation. J. Appl. Log. **8**(1), 97–113 (2010)

30. Sun, X.: How to build input/output logic. In: Bulling, N., van der Torre, L., Villata, S., Jamroga, W., Vasconcelos, W. (eds.) CLIMA 2014. LNCS, vol. 8624, pp. 123–137. Springer, Heidelberg (2014)

31. Ulmann-Margalit, E.: The Emergence of Norms. Clarendon Press, Oxford (1977)

32. von Wright, G.: Deontic logic. Mind **60**, 1–15 (1952)

33. Weiss, G. (ed.): Multiagent Systems, 2nd edn. MIT Press, Cambridge (2013)

34. Wooldridge, M.: An Introduction to MultiAgent Systems). Wiley, Chichester (2009)

Computational Complexity
of Input/Output Logic

Xin Sun[1](✉) and Diego Agustín Ambrossio[1,2]

[1] Faculty of Science, Technology and Communication, University of Luxembourg,
Luxembourg, Luxembourg
{xin.sun,diego.ambrossio}@uni.lu
[2] ICR, SnT, Luxembourg, Luxembourg

Abstract. Input/output logics are abstract structures designed to represent conditional norms. The complexity of input/output logic has been sparsely developed. In this paper we study the complexity of input/output logics. We show that the lower bound of the complexity of the fulfillment problem of 4 input/output logics is coNP, while the upper bound is either coNP or P^{NP}. (This paper is an extension of a short paper [20] by the same authors.)

Keywords: Input/output logic · Norm · Complexity

1 Introduction

In recent years, normative multi-agent system [3,7] arises as a new interdisciplinary academic area bringing together researchers from multi-agent system [17,21,22], deontic logic [9] and normative system [1,2,11]. Norms play an important role in normative multi-agent systems. They are heavily used in agent cooperation and coordination, group decision making, multi-agent organizations, electronic institutions, and so on.

In the first volume of the handbook of deontic logic and normative systems [9], input/output logic [12–15] appears as one of the new achievements in deontic logic of this century. Input/output logic takes its origin in the study of conditional norms. Unlike the modal logic framework, which usually uses possible world semantics, input/output logic adopts mainly operational semantics: a normative system is conceived in input/output logic as a deductive machine, like a black box which produces normative statements as output, when we feed it descriptive statements as input.

Boella and van der Torre [6] extends input/output logic to reason about constitutive norms. Tosatto *et al.* [8] adapts it to represent and reason about abstract normative systems. For a comprehensive introduction to input/output logic, see Parent and van der Torre [15]. A technical toolbox to build input/output logic can be found in Sun [19].

While the semantics and application of input/output logic has been well developed in recent years, the complexity of input/output logic has not been

© Springer International Publishing Switzerland 2015
A. Bikakis and X. Zheng (Eds.): MIWAI 2015, LNAI 9426, pp. 72–79, 2015.
DOI: 10.1007/978-3-319-26181-2_7

studied yet. In this paper we fill this gap. We show that the lower bound of the complexity of the fulfillment problem of 4 input/output logics are coNP, while the upper bound is either coNP or P^{NP}.

The structure of this paper is as follows we present a summary of basic concepts and results in input/output logic and some notes in complexity theory, in Sect. 2. In Sect. 3 we study the complexity of input/output logic. We point out some directions for future work and conclude this paper in Sect. 4.

2 Background

2.1 Input/output Logic

Makinson and van der Torre introduce input/output logic as a general framework for reasoning about the detachment of obligations, permissions and institutional facts from conditional norms. Strictly speaking input/output logic is not a single logic but a family of logics, just like modal logic is a family of logics containing systems K, KD, S4, S5, ... We refer to the family as the input/output framework. The proposed framework has been applied to domains other than normative reasoning, for example causal reasoning, argumentation, logic programming and non-monotonic logic, see Bochman [5].

Let $\mathbb{P} = \{p_0, p_1, \ldots\}$ be a countable set of propositional letters and $L_\mathbb{P}$ be the propositional language built upon $\mathbb{P}$. Let $N \subseteq L_\mathbb{P} \times L_\mathbb{P}$ be a set of ordered pairs of formulas of $L_\mathbb{P}$. We call N a normative system. A pair $(a, x) \in N$, call it a *norm*, is read as "given a, it ought to be x". N can be viewed as a function from $2^{L_\mathbb{P}}$ to $2^{L_\mathbb{P}}$ such that for a set of formulas $A \subseteq L_\mathbb{P}$, $N(A) = \{x \in L_\mathbb{P} : (a, x) \in N$ for some $a \in A\}$. Intuitively, N can be interpreted as a *normative code* composed of conditional norms and the set A serves as explicit input.

Makison and van der Torre [12] define the semantics of input/output logics from O_1 to O_4 as follows:

- $O_1(N, A) = Cn(N(Cn(A)))$.
- $O_2(N, A) = \bigcap\{Cn(N(V)) : A \subseteq V, V$ is complete$\}$.
- $O_3(N, A) = \bigcap\{Cn(N(B)) : A \subseteq B = Cn(B) \supseteq N(B)\}$.
- $O_4(N, A) = \bigcap\{Cn(N(V) : A \subseteq V \supseteq N(V)), V$ is complete$\}$.

Here Cn is the classical consequence operator of propositional logic, and a set of formulas is *complete* if it is either *maximal consistent* or equal to $L_\mathbb{P}$. These four operators are called *simple-minded output*, *basic output*, *simple-minded reusable output* and *basic reusable output* respectively. For each of these four operators, a *throughput* version that allows inputs to reappear as outputs, defined as $O_i^+(N, A) = O_i(N_{id}, A)$, where $N_{id} = N \cup \{(a, a) \mid a \in L_\mathbb{P}\}$. When A is a singleton, we write $O_i(N, a)$ for $O_i(N, \{a\})$.

Input/output logics are given a proof theoretic characterization. We say that an ordered pair of formulas is derivable from a set N iff (a, x) is in the smallest set that extends N and is closed under a number of derivation rules. The following are the rules we need to define O_1 to O_4^+:

- SI (strengthening the input): from (a, x) to (b, x) whenever $b \vdash a$. Here $\vdash$ is the classical entailment relation of propositional logic.
- OR (disjunction of input): from (a, x) and (b, x) to $(a \vee b, x)$.
- WO (weakening the output): from (a, x) to (a, y) whenever $x \vdash y$.
- AND (conjunction of output): from (a, x) and (a, y) to $(a, x \wedge y)$.
- CT (cumulative transitivity): from (a, x) and $(a \wedge x, y)$ to (a, y).
- ID (identity): from nothing to (a, a).

The derivation system based on the rules SI, WO and AND is called D_1. Adding OR to D_1 gives D_2. Adding CT to D_1 gives D_3. The five rules together give D_4. Adding ID to D_i gives D_i^+ for $i \in \{1, 2, 3, 4\}$. $(a, x) \in D_i(N)$ is used to denote the norms (a, x) is derivable from N using rules of derivation system D_i. In Makinson and van der Torre [12], the following soundness and completeness theorems are given:

Theorem 1 ([12]). *Given an arbitrary normative system N and formula a,*

- $x \in O_i(N, a)$ *iff* $(a, x) \in D_i(N)$, *for* $i \in \{1, 2, 3, 4\}$.
- $x \in O_i^+(N, a)$ *iff* $(a, x) \in D_i^+(N)$, *for* $i \in \{1, 2, 3, 4\}$.

2.2   Complexity Theory

Complexity theory is the theory to investigate the time, memory, or other resources required for solving computational problems. In this subsection we briefly review those concepts and results from complexity theory which will be used in this paper. More comprehensive introduction of complexity theory can be found in [4, 18]

We assume the readers are familiar with notions like Turing machine and the complexity class P, NP and coNP. Oracle Turing machine and a complexity class related to oracle Turing machine will be used in this paper.

Definition 1 (oracle Turing machine). *An oracle for a language L is device that is capable of reporting whether any string w is a member of L. An (resp. non-deterministic) oracle Turing machine M^L is a modified (resp. non-deterministic) Turing machine that has the additional capability of querying an oracle. Whenever M^L writes a string on a special oracle tape it is informed whether that string is a member of L, in a single computation step.*

Definition 2. P^{NP} *is the class of languages decidable with a polynomial time oracle Turing machine that uses oracle $L \in NP$.*

3   Complexity of Input/output Logic

The complexity of input/output logic has been sparsely studied in the past. Although the reversibility of derivations rules as a proof re-writing mechanism has been studied for input/output logic framework [12], the length or complexity of such proofs have not been developed. We approach the complexity of input/output logic from a semantic point of view. We focus on the following fulfillment problem:

Given a finite set of norms N, a finite set of formulas A and a formula x, is

$$x \in O(N, A)?$$

The aim of the fulfillment problem is to check whether the formula x appears among the obligations detached from the normative system N and facts A.

3.1 Simple-Minded O_1

Theorem 2. *The fulfillment problem of simple-minded input/output logic is coNP-complete.*

Proof: Concerning the coNP hardness, we prove by reducing the validity problem of propositional logic to the fulfillment problem of simple-minded input/output logic: given an arbitrary $x \in L_{\mathbb{P}}$, $\vdash x$ iff $x \in Cn(\top)$ iff $x \in Cn(N(Cn(A)))$ where $N = \emptyset$ iff $x \in O_1(N, A)$ where $N = \emptyset$.

Now we prove the coNP membership. We provide the following non-deterministic Turing machine to solve the complement of our problem. Let $N = \{(a_1, x_1), \ldots, (a_n, x_n)\}$, A be a finite set of formulas and x be a formula.

1. Guess a sequence of valuations $V_1, \ldots, V_n$ and V' on the propositional letters appears in $A \cup \{a_1, \ldots, a_n\} \cup \{x_1, \ldots, x_n\} \cup \{x\}$.
2. Let $N' \subseteq N$ be the set of norms which contains all (a_i, x_i) such that $V_i(A) = 1$ and $V_i(u_i) = 0$.
3. Let $X = \{x : (a, x) \in N - N'\}$.
4. If $V'(X) = 1$ and $V'(x) = 0$. Then return "accept" on this branch. Otherwise return "reject" on this branch.

The main intuition of the proof is: N' collects all norms which *cannot* be triggered by A.[1] On some branches we must have that N' contains exactly those norms which are not triggered by A. In those lucky branches X is the same as $N(Cn(A))$. If there is a valuation V' such that $V'(X) = 1$ and $V'(x) = 0$, then we know $x \notin Cn(X) = Cn(N(Cn(A)))$.

It can be verified that $x \notin Cn(N(Cn(A)))$ iff the algorithm returns "accept" on some branches and the time complexity of the non-deterministic Turing machine is polynomial. $\dashv$

3.2 Simple-Minded Throughput O_1^+

Lemma 1. $O_1^+(N, A) = Cn(A \cup N(Cn(A)))$.

Proof: The proof is routine and left to the readers. $\dashv$

Theorem 3. *The fulfillment problem of simple-minded throughput input/output logic is coNP-complete.*

---

[1] We say a norm (a, x) is triggered by A if $a \in Cn(A)$.

Proof: Concerning the lower bound, we prove by a reduction from the validity problem of propositional logic: given arbitrary $x \in L_{\mathbb{P}}$, $\vdash x$ iff $x \in Cn(\top)$ iff $x \in Cn(A \cup N(Cn(A)))$ where $N = \emptyset = A$ iff $x \in O_1^+(N, A)$ where $N = \emptyset = A$.

Concerning the upper bound, we prove by giving a non-deterministic Turing machine similar to the one in the proof of Theorem 2. The only change is now in step 4 we test if $V'(A \cup X) = 1$ and $V'(x) = 0$. It can be verified that $x \notin Cn(A \cup N(Cn(A)))$ iff the non-deterministic Turing machine returns "accept" on some branch. By Lemma 1 we know this Turing machine solves our problem. ⊣

3.3   Simple-Minded Reusable O_3

Given a set N of norms and a set A of formulas, we define a function $f_A^N : 2^{L_{\mathbb{P}}} \to 2^{L_{\mathbb{P}}}$ such that $f_A^N(X) = Cn(A \cup N(X))$. It can be proved that f_A^N is monotonic with respect to the set theoretical $\subseteq$ relation, and $(2^{L_{\mathbb{P}}}, \subseteq)$ is a complete lattice. Then by Tarski's fixed point theorem there exists a least fixed point of f_A^N. The following proposition shows that the least fixed point can be constructed in an inductive manner.

Proposition 1 ([19]). *Let B_A^N be the least fixed point of the function f_A^N. Then $B_A^N = \bigcup_{i=0}^{\infty} B_{A,i}^N$, where $B_{A,0}^N = Cn(A)$, $B_{A,i+1}^N = Cn(A \cup N(B_{A,i}^N))$.*

Using the least fixed point, a more constructive semantics of O_3 and O_3^+ are stated as follows, such semantics gives us insights to develop algorithms to solve the fulfillment problem of reusable input/output logic:

Theorem 4 ([19]). *For a set of norms N and a formula a,*

1. *$(a, x) \in D_3(N)$ iff $x \in Cn(N(B_{\{a\}}^N))$.*
2. *$(a, x) \in D_3^+(N)$ iff $x \in Cn(N_{id}(B_{\{a\}}^{N_{id}}))$.*

Theorem 5. *The fulfillment problem of simple-minded reusable input/output logic is between coNP and P^{NP}.*

Proof: The lower bound is easy, here we omit it.

Concerning the upper bound, we provide the following algorithm on a oracle Turing machine with oracle SAT.

Let $N = \{(a_1, x_1), \ldots, (a_n, x_n)\}$, A be a finite set of formulas and x be a formula.

1. Let $X = A, Y = Z = N, U = \emptyset$.
2. for each $(a_i, x_i) \in Y$, ask the oracle if $\neg(\bigwedge X \to a_i)$ is satisfiable.
   (a) If "no", then let $X = X \cup \{x_i\}$, $Z = Z - \{(a_i, x_i)\}$.
   (b) Otherwise do nothing.
3. If $Y == Z$, goto 4. Otherwise let $Y = Z$, goto step 2.
4. for each $(a_i, x_i) \in N$, ask the oracle if $\neg(\bigwedge X \to a_i)$ is satisfiable.
   (a) If "no", then let $U = U \cup \{x_i\}$.

(b) Otherwise do nothing
5. Ask the oracle if $\neg(\bigwedge U \to x)$ is satisfiable.
   (a) If "no", then return "accept".
   (b) Otherwise return "reject".

The correctness of the above algorithm is routine to prove and we leave it to the readers. Concerning the time complexity, the times of loop in step 2 is at most n. Each loop can be finished in polynomial time. Therefore all the loops in step 2 can be done in polynomial time. Step 3 call for step 2 for at most n times. Therefore it can still be done in polynomial time. The times of loop in step 4 is exactly n. Each loop can be finished in polynomial time. Therefore all the loops in step 4 can be done in polynomial time. Step 5 can be done in polynomial time. Therefore the algorithm is polynomial.   ⊣

3.4   Simple-Minded Reusable Throughput O_3^+

Theorem 6. *The fulfillment problem of simple-minded reusable throughput input/output logic is between coNP and P^{NP}.*

Proof: The lower bound is easy, here we omit it.

Concerning the upper bound, we prove by giving an algorithm similar to the one in the proof of Theorem 5. We make the following change:

- In step 2 and 4 we ask the oracle if $\neg(\bigwedge A \wedge \bigwedge X \to a_i)$ is satisfiable.
- In step 5 we ask the oracle if $\neg(\bigwedge A \wedge \bigwedge U \to a_i)$ is satisfiable.   ⊣

4   Conclusion and Future Work

In this paper we develop complexity results of input/output logic. We show that four input/output logics have lower bound coNP and upper bound either coNP or P^{NP}. There are several natural directions for future work:

1. What is the tight complexity results of reusable input/output logic, as well as other input/output logics?
2. What is the complexity of constraint input/output logic? Constraint input/output logic [13] is developed to deal with the inconsistency of output. The semantics of constraint input/output logic is more complicated than those input/output logic discussed in this paper. This might increase the complexity of the fulfillment problem. Constraint input/output logic based on O_3^+ has close relation with Reiter's default logic [16]. Gottlob [10] presents some complexity results of Reiter's default logic, which will give us insights on the complexity of constraint input/output logic.
3. What is the complexity of different types of permission? Three different of permissions are introduced in Makinson and van der Torre [14]. The semantics of these three logics are different, which suggests different complexity for the problems related to permissions.

References

1. Ågotnes, T., van der Hoek, W., Rodríguez-Aguilar, J.A., Sierra, C., Wooldridge, M.: On the logic of normative systems. In: Veloso, M.M. (ed.) Proceedings of the 20th International Joint Conference on Artificial Intelligence, pp. 1175–1180, Hyderabad, India, January 6–12, 2007. http://dli.iiit.ac.in/ijcai/IJCAI-2007/PDF/IJCAI07-190.pdf
2. Alechina, N., Dastani, M., Logan, B.: Reasoning about normative update. In: Rossi, F. (ed.) Proceedings of the 23rd International Joint Conference on Artificial Intelligence, IJCAI/AAAI, Beijing, China, August 3–9, 2013. http://www.aaai.org/ocs/index.php/IJCAI/IJCAI13/paper/view/6884
3. Andrighetto, G., Governatori, G., Noriega, P., van der Torre, L.W.N. (eds.): Normative Multi-Agent Systems, Dagstuhl Follow-Ups, vol. 4. Schloss Dagstuhl - Leibniz-Zentrum fuer Informatik (2013). http://drops.dagstuhl.de/opus/portals/dfu/index.php?semnr=13003
4. Arora, S., Barak, B.: Computational Complexity: A Modern Approach. Cambridge University Press, New York (2009)
5. Bochman, A.: A causal approach to nonmonotonic reasoning. Artif. Intell. **160**(1–2), 105–143 (2004)
6. Boella, G., van der Torre, L.W.N.: A logical architecture of a normative system. In: Goble, L., Meyer, J.-J.C. (eds.) DEON 2006. LNCS (LNAI), vol. 4048, pp. 24–35. Springer, Heidelberg (2006)
7. Boella, G., van der Torre, L., Verhagen, H.: Introduction to the special issue on normative multiagent systems. Auton. Agent. Multi-Agent Syst. **17**(1), 1–10 (2008)
8. Tosatto, C.S., Boella, G., van der Torre, L., Villata, S.: Abstract normative systems: Semantics and proof theory. In: Proceedings of the Thirteenth International Conference on Principles of Knowledge Representation and Reasoning, pp. 358–368 (2012)
9. Gabbay, D., Horty, J., Parent, X., van der Meyden, R., van der Torre, L. (eds.): Handbook of Deontic Logic and Normative Systems. College Publications, London (2013)
10. Gottlob, G.: Complexity results for nonmonotonic logics. J. Log. Comput. **2**(3), 397–425 (1992). http://dx.doi.org/10.1093/logcom/2.3.397
11. Herzig, A., Lorini, E., Moisan, F., Troquard, N.: A dynamic logic of normative systems. In: Walsh, T. (ed.) Proceedings of the 22nd International Joint Conference on Artificial Intelligence, IJCAI/AAAI, pp. 228–233, Barcelona, Catalonia, Spain, July 16–22, 2011. http://ijcai.org/papers11/Papers/IJCAI11-049.pdf
12. Makinson, D., van der Torre, L.: Input-output logics. J. Philos. Logic **29**, 383–408 (2000)
13. Makinson, D., van der Torre, L.: Constraints for input/output logics. J. Philos. Logic **30**(2), 155–185 (2001)
14. Makinson, D., van der Torre, L.: Permission from an input/output perspective. J. Philos. Logic **32**, 391–416 (2003)
15. Parent, X., van der Torre, L.: I/O logic. In: Horty, J., Gabbay, D., Parent, X., van der Meyden, R., van der Torre, L. (eds.) Handbook of Deontic Logic and Normative Systems. College Publications, London (2013)
16. Reiter, R.: A logic for default reasoning. Artif. Intell. **13**(1–2), 81–132 (1980)
17. Shoham, Y., Leyton-Brown, K.: Multiagent Systems - Algorithmic, Game-Theoretic, and Logical Foundations. Cambridge University Press, Cambridge (2009)

18. Sipser, M.: Introduction to the Theory of Computation, 3rd edn. Cengage Learning, Boston (2012)
19. Sun, X.: How to build input/output logic. In: Bulling, N., van der Torre, L., Villata, S., Jamroga, W., Vasconcelos, W. (eds.) CLIMA 2014. LNCS, vol. 8624, pp. 123–137. Springer, Heidelberg (2014)
20. Sun, X., Ambrossio, D.A.: On the complexity of input/output logic. In: the Proceedings of the Fifth International Conference on Logic, Rationality and Interaction (LORI) (2015)
21. Weiss, G. (ed.): Multiagent Systems, 2nd edn. MIT press, Cambridge (2013)
22. Wooldridge, M.J.: An Introduction to Multiagent Systems, 2nd edn. Wiley, Chichester (2009)

Nested Monte-Carlo Search of Multi-agent Coalitions Mechanism with Constraints

Souhila Arib[1]($\boxtimes$), Samir Aknine[2], and Tristan Cazenave[3]

[1] EISTI-Ecole Internationale des sciences du traitement de l'information,
Avenue du Parc, 95000 Cergy, France
souhila.arib@eisti.fr
[2] LIRIS-Université Claude Bernard Lyon 1 - UCBL,
69622 Villeurbanne Cedex, France
samir.aknine@univ-lyon1.fr
[3] LAMSADE-Université Paris Dauphine Place du Mal de Lattre de Tassigny,
75775 Paris Cedex 16, France
cazenave@lamsade.dauphine.fr

Abstract. This paper develops and evaluates a coalition mechanism that enables agents to participate in concurrent tasks achievement in competitive situations in which agents have several constraints. Here we focus on situations in which the agents are self-interested and have not a priori knowledge about the preferences of their opponents, and they have to cooperate in order to reach their goals. All the agents have their specific constraints and this information is private. The agents negotiate for coalition formation (CF) over these constraints, that may be relaxed during negotiations. They start by exchanging their constraints and making proposals, which represent their acceptable solutions, until either an agreement is reached, or the negotiation terminates. We explore two techniques that ease the search of suitable coalitions: we use a constraint-based model and a heuristic search method. We describe a procedure that transforms these constraints into a structured graph on which the agents rely during their negotiations to generate a graph of feasible coalitions. This graph is therefore explored by a Nested Monte-Carlo search algorithm to generate the best coalitions and to minimize the negotiation time.

Keywords: Multi-agent systems · Coalition formation · Coordination · Negotiation

1 Introduction

Forming coalitions of agents which are able to effectively perform tasks is a key issue for many practical application contexts. This paper mainly focuses on self-interested agents which aim to form coalitions with other agents as they cannot reach their objectives individually. Several methods have been developed to control the behaviors of the agents involved in such process [12]. However few CF

© Springer International Publishing Switzerland 2015
A. Bikakis and X. Zheng (Eds.): MIWAI 2015, LNAI 9426, pp. 80–88, 2015.
DOI: 10.1007/978-3-319-26181-2_8

mechanisms cope with the dynamic of the constraints of the agents in such contexts. Indeed, these constraints can gradually be revealed, and relaxed by the agents at different moments of the negotiation in order to meet the requirements of their opponents and thus to ease the convergence. Some coalition methods have been developed to determine formerly to the negotiation the optimal coalitions and take into account the constraints of the agents involved in the coalition process. These methods have addressed important issues such as computational complexity and heuristics approaches for the optimal coalition structure generation, [7,10,13]. In this paper, we focus on contexts where agents neither have the same utility functions, nor they reveal these functions. Thus, it is infeasible to precisely estimate a priori the corresponding utility of each agent for each feasible proposal of solution with current optimal coalitions search algorithms. The issues with processing the constraints of the agents in the negotiation phase for the coalition formation deserve a particular attention and a deep study. Yet, only few works proposes a mechanism to deal with the dynamic of such constraints while agents negotiate them. Note that, since we consider that the agents are self-interested and do not share their information and computations, our aim is not to identify the optimal solution of the coalitions, but to ease the convergence to an agreed common solution for these agents. Our main contribution is a new mechanism that enables agents to negotiate and form coalitions. This mechanism is based on three main abstractions: a constraint graph, a coalition graph and Nested Monte-Carlo search method. First, we develop a constraints based graph which handles the revealed constraints of the agents. This graph of constraints can be used to specify different types of constraints relations, such as a constraints ordering over potential decision outcomes. Building upon this, we transform this representation into a flat representation of coalitions in the graph of coalitions. Each level of this graph allows generating a set of possible coalitions and in this set the agent selects the best coalitions that can be accepted. This graphical representation of constraints and coalitions specifies constraints relations in a relatively compact, intuitive, and structured manner. To explore this graph of coalitions, we first define the problem and link it to other existing problems, so that approximate solution techniques and anytime heuristics that provide increasingly better solutions if given more time can be re-used. We advise new solutions that allow agents use a nested Monte-Carlo search algorithm [1] which finds the best coalition that maximizes the utility of each agent. Nested Monte-Carlo search methods address the problem of guiding the search towards better states when there is no available heuristic. These methods use nested levels of random games in order to guide the search of coalitions. These algorithms have been studied theoretically on simple abstract problems and applied successfully to several games [4]. Specifically, this paper advances the state of the art in the following ways. We advise new anytime heuristics to find approximate solutions fast, we empirically evaluate our algorithm and show that it computes (in less than 600 milliseconds) 689 proposals of solutions for non-trivial problems involving up to 30 agents and 50 tasks. Thus, our work encompasses essential aspects of the coalition formation, from the coalition model, negotiation, and an anytime heuristic. The reminder of the paper is organized as follows. Section 2 briefly describes the related works. Section 3 introduces some preliminaries and the case study. Section 4 presents the coalition formation mechanism, and a final section will conclude the work with a summary of the contributions.

2   Related Work

In game-theoretic perspective, coalitional games with constraints have been addressed by a number of works. However, none of these structures is able to model agents' negotiations for reaching joint agreements. [3] proposes a game-theoretical study and focuses on strategic, core-related issues rather than computational analysis of the coalition formation. This work is more close to [7] where authors propose a constrained coalition formation model and an algorithm for optimal coalition structure generation. They develop a procedure that transforms the specified set of constraints, making it possible to identify all the feasible coalitions. Building upon this, they provide an algorithm for optimal coalition structure generation. [13] address the problem of coalition formation with sparse synergies where the set of feasible coalitions is constrained by the edges of a graph. Their aim is to check whether knowledge of the topology of an underlying social or organizational context graph could be used to speed up coalition enumeration and structure generation. [10] define the problem of allocating coalitions of agents to spatially distributed tasks with workloads and deadlines so as to maximize the total number of tasks completed over time. Nevertheless, these works have not deeply addressed the constraints of the agents in the proposed models or specify how agents negotiate over them to reach agreements. Constraints on coalition sizes have been considered for coalition value calculation [8,9,11]. However, the semantics of these constraints has not been used on the same level as it is done in this paper. [2,5] develop succinct and expressive representations for coalitional games. Such formalism could be used to encode the constraints, but this is not the main concern of the constrained CF mechanism considered in this paper.

3   Preliminaries and Case Study

To illustrate the coalition formation mechanism we propose, let us consider a carpooling example, where some travellers want to move from a city to another, and they want to share their means of transportation. Each traveller formulates to his agent the goals to be achieved. For example "I want to go from NY to Boston", his constraints as departure time, duration of the travel, and unit price of seat. To solve this problem, the agents have to deal with all the constraints and preferences over those of their associated travellers in order to enable them to share transportation. Agents negotiate for the coalitions to form to decrease the unit price of seat, increase the number of passengers, etc. They can step aside in favor of other agents, if an agreement can be found. More formally, consider a set of agents $\mathcal{N} = \{a_1, a_2, ..., a_n\}$, a set of actions $\mathcal{A} = \{b_1, b_2, ..., b_m\}$

and a set of constraints $C_t = \{c_{t1}, c_{t2}, ..., c_{tk}\}$. The agents of $\mathcal{N}$ need to execute the actions of $\mathcal{A}$ by satisfying the constraints in C_t. The constraints are defined as intervals, for instance: departure time: $D \in [10a.m., 12a.m.]$, travel duration in hours: $T \in [1H, 2H]$ and price: $P \in [20, 25]$. The agents' preferences are represented using a preference relation $\succ$ for those they want to share a car with, for instance $a_x \succ_i a_y$ (for agent a_i, a_x is preferred to a_y). We consider a coalition c as a nonempty subset of $\mathcal{N}$ ($c \subseteq \mathcal{N}$). We define $\mathcal{C}$ as the set of all possible coalitions. For a coalition c to be formed, each agent a_i in c should get a certain satisfaction. This satisfaction is defined by a utility function $u_i : \mathcal{C} \mapsto \mathbb{R}$. Note that a coalition is acceptable for agent a_i if it is preferred over, or equivalent to a reference coalition, $u_i(ref)$, which corresponds to the minimal guaranteed gain of the agent during the negotiation. A solution of the negotiation for each agent a_i introduces a coalition structure denoted CS_i which is defined on $\mathcal{N}$ with its associated utility $u_i(CS_i)$. CS_i contains a set of coalitions $\{c_1, c_2, .., c_q\}$ to be formed for the set of actions $\mathcal{A}_i \subseteq \mathcal{A}$ where a_i is involved. Furthermore, for every $q' \in [1, q], c_{q'} \subseteq \mathcal{N}$ and $c_{q'}$ performs a set of actions $\mathcal{A}_{c_{q'}} \subseteq \mathcal{A}$ and $\forall (x, y) \in [1, q]^2, x \neq y, \mathcal{A}_{c_x} \cap \mathcal{A}_{c_y} = \varnothing, \bigcup_{q'=1,...,q} \mathcal{A}_{c_{q'}} \subseteq \mathcal{A}$ and $\bigcup_{q'=1,...,q} c_{q'} \subseteq \mathcal{N}$. The set of all coalition structures is denoted $\mathcal{S}$.

4 Coalition Formation Mechanism (CFM)

In order to satisfy the goals they have to achieve, the agents perform negotiations on the coalitions they want to form. So, the CFM requires an analysis step of constraints that agents exchange in order to guide the choice of the coalitions and a step of generating coalition structures from these constraints. Constraint analysis relies on constructing a graph of constraints and coalition generation is based on the mapping of the constraints to possible coalitions in a coalition graph. Exploring the search graph of coalitions toward better states is based on a Nested Monte-Carlo algorithm.

4.1 Constraint Graph

An effective technique for solving a coalition formation problem is a heuristic search through abstract problem spaces. The first problem space can be represented by a directed connected graph, where nodes correspond to constraint sets and edges correspond to actions (cf. Fig. 1). The constraint graph may include many paths from the start to any node. Since the agents are self-interested, to search among the constraints to deal with in the coalitions, every agent constructs its own graph of constraints based on its own constraints and those revealed by other agents during this negotiation. Given a set of constraints that must be satisfied by an agent to execute a set of actions and starting from the source node labeled with $\{b: \varnothing, c_t: \varnothing\}$, initially there are not constraints and actions associated with this source node, let us define a graph denoted $G(c_t, b)$ as follows.

Definition 1. *Given a node labeled* $\{b: \varnothing, c_t: \varnothing\}$ *the constraint graph* $G(c_t, b)$ *is a directed connected graph, containing all possible nodes of constraints repre-sented by intervals, labeled* $\{X_1, ..., X'_k\}$ *for each action* $b_i, 1 \le i \le m$, *that has to be executed by the agent. Each node has a utility labeled* u, *and directed edges from this node are labeled* $\{b_j, ..., b_k\}$ *where* $1 \le j \,..\, k \le m$.

A constraint graph gathers, the most preferred constraints' intervals in its nodes. At the root node, no action and constraint are added. Each node generates a finite set of child nodes which correspond to the accepted sets of constraints, where the first node of the graph is an outgoing node and the last nodes are incoming nodes. This constraint graph is built following a preference rate on the intervals of constraints.

Let us consider two agents, a_i which has its own interval X and receives from a_j an interval Y. The agent a_i wants to create a new interval Z that meets its constraints and those of a_j, $\{Z \models a_i\}$, by merging its interval and the one received from a_j. We will adopt the convention of the left and right endpoints of an interval X by $\underline{X}$ and $\overline{X}$, respectively [6]. First, a_i tests if $X \subseteq Y$. Thus, if $\underline{Y} \le \underline{X}$ and $\overline{X} \le \overline{Y}$, it will get $X \subseteq Y$ and $Z = X$. Else the agent tests if $X \cap Y$ and calculates the new interval Z. If $X \cap Y = \varnothing$ there are no points in common with a_j. Otherwise, $Z = \{max\{\underline{X}, \underline{Y}\}, min\{\overline{X}, \overline{Y}\}\}$ and tests if it complies with its actions. If a_i does not choose this interval, it calculates $Z = X \sqcup Y$ which means the union X and Y and tests if it complies with its actions. For more details about the operations over the intervals see [6]. Based on this graph, constraint analysis consists for an agent of comparing and grouping its constraints and those received from others. A natural constraint graph analysis involves constructing and linking optimal nodes. Constraints are gathered based on their relations into sets represented in the nodes of this graph. Each level of the graph of constraints refers to an action to be performed by a coalition. The advantage of the suggested method consists in directing the search of the solutions of coalitions towards primary constraints, i.e., important constraints to satisfy, thus, reducing search complexity. To move from one node of this graph to another, an action is added to the graph. The utility of a move, which labels the corresponding edge in the search space, is the utility of the action when it is added and performed by the coalition. A solution path represents a particular succession order of the added actions, and the width of that order is the sum of the edge utilities on the solution path.

Let us consider constructing the constraint graph by the agent a_1 on our previous example using. First, assume that agent a_1 started a negotiation with agents a_2 to a_5 and in which each of these five agents revealed certain of its constraints. The actions that have to be executed are: b_1, b_2, b_3 which correspond respectively to: go from NY to Amherst, find a hotel room in Amherst, and go from Amherst to Boston. The constraints identified by a_1 for the action b_1 are: $D \in [10a.m., 01p.m.]$, $T \in [1H, 2H]$ and $P \in [20, 25]$ and for b_2 are: $D \in [01p.m., 02p.m.]$, $T \in [1H, 2H]$ and $P \in [20, 25]$. The nodes in the first level of the graph assemble possible sets of constraints concerning the action b_1. a_1

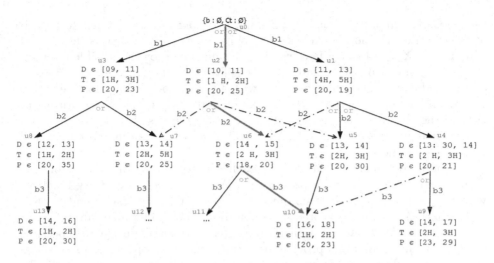

Fig. 1. An example of a graph of constraints against different actions of the agent a_1. Each node is labeled with its associated utility.

compares its own constraints and those received from these agents and creates new intervals of constraints X_{kc_t}.

Let us consider again the agent a_1 who received these intervals of constraints from the agent a_2 concerning the action b_1: $D \in [10a.m., 11a.m.]$, $T \in [1H, 3H]$ and $P \in [20, 35]$. So, $D \in [10a.m., 11a.m.] \subseteq [10a.m., 01p.m.]$ so $Z = [10a.m., 11a.m.] \models (a_1, a_2)$. Thus, a_1 selects the interval Z. For $T \in [1H, 2H] \subseteq [1H, 3H]$, $Z = [1H, 2H] \models (a_1, a_2)$, so a_1 chooses $T \in [1H, 2H]$. For $P \in [20, 25] \subseteq [20, 35]$, $Z = [20, 25] \models (a_1, a_2)$. These results are resumed in the Fig. 1. On the left of this figure, agent a_1 represents the first node created for the action b_1 by the ordered set of intervals $D \in [09a.m., 11a.m.]$, $T \in [1H, 3H]$ and $P \in [20, 23]$. We notice in this node that a_1 chooses the interval $[09a.m., 11a.m.]$ even if its departure time is not completely included in $D \in [10a.m., 01p.m.]$ because it has a good proposal of the seat price. $D \in [10a.m., 11a.m.]$, $T \in [1H, 2H]$ and $P \in [20, 25]$ are associated with the second one and $D \in [11a.m., 01p.m.]$, $T \in [4H, 5H]$ and $P \in [20, 19]$ with the last child. So, for each action, a_1 generates the different possible intervals of constraints that satisfy its action b_1. We observe that the constraints in each child node are created taking into account the end of execution of the antecedent action. So, to generate intervals of constraints that satisfy the action b_{i+1}, the agent takes into account the end time of b_i. This allows the agent to manage the relations between the actions that have to be executed. In this example, in the second level of the graph the agent a_1 identified for the action b_2 these intervals: $D \in [01p.m., 02p.m.]$, $T \in [2H, 5H]$ and $P \in [20, 25]$. The beginning of b_2 is in $[01p.m., 02p.m.]$ because b_1 ends at the latest at $01p.m.$. The dashed arcs show that nodes can share the same child nodes and the red and bold ones show the most preferred path from the root node to the last one, they result from the

Monte-Carlo exploration (detailed below). Every agent a_i which has to negotiate to execute an action b_i while satisfying its constraints c_t, chooses intervals of constraints: $X_{kc_t}(b_i) \models a_i$. It then creates child nodes for the feasible intervals that satisfy b_i. Each node created can be split under appropriate restrictions to other child nodes. The agent a_i starts with a node, labeled X_{kc_t}, for the action b_i. For each action b_{i+1} that must be executed after b_i and need negotiation, a_i creates the new intervals for $b_{i+1} : (X'_{1c_t}, ..., X'_{k'c_t})$, and splits the node b_i, X_{kc_t} to the child nodes $b_{i+1} : (X'_{1c_t}, ..., X'_{k'c_t})$. The agent a_i uses this procedure until no action need negotiation.

A notable detail of the constraints search space construction is that a solution is measured by its maximum path utility. We use an additive utility function, where a path is evaluated by summing its edge utilities. For each iteration, feasible solutions are only explored if their utility is not under a certain reference situation $u_i(ref)$. If an iteration is completed without finding a new possible solution, then all solutions provide less utility than that of the reference situation, $u_i(ref)$, thus, $u_i(ref)$ may be decreased and the search is repeated. The section below explains how a Nested Monte-Carlo algorithm explores a graph of constraints.

4.2   Constraint Graph Exploration with Nested Monte-Carlo Algorithm (NMC)

To optimize the search time for the new coalitions to propose or to accept, agents use the Nested Monte-Carlo algorithm. Nested Monte-Carlo Search is used for problems that do not have good heuristics. It was shown that memorizing the best sequence improves the mean result of the search. Experiments on different games gave very good results, finding a new world record of 80 moves at Morpion Solitaire, improving on previous algorithms at SameGame, and being more than 200,000 times faster than depth-first search at 16x16 Sudoku modeled as a Constraint Satisfaction Problem [1]. In the first step of the mechanism, the NMC explores each level of the graph of constraints and stores the best path of constraints that satisfies the agent a_i. The idea of NMC is to use lower sequences of simulations in order to decide the utility that an agent gets from a path at the current sequence. This step is necessary because agents do not have a priori a knowledge about the utility functions of the others. When all simulations of the underlying sequences have been performed, the agent utility is memorized in the best sequence; it means that it is possible to get the best path. The solution improves monotonically since our algorithm keeps track of the best proposal of solution found so far. Nested Monte-Carlo search combines nested calls with randomness in the playouts and memorization of the best sequence of moves. In nested rollouts, the rollouts are based on a heuristic. It implies that nested rollouts always improve on rollouts and on simply following the heuristic. When the base level does not use a heuristic but random moves, it is possible that a nested search gives worse results than a lower level search. It is useful to memorize the best sequence found so far in order to follow it when the randomized searches give worse results than the best sequence. The basic *sample* function

---

Algorithm 1. Nested play

```
1  begin
2  |   Require:nested (Position, level in the graph);
3  |   best utility = u(ref);
4  |   while not end of the graph do
5  |   |   if level is 1 then
6  |   |   |   move = argmax_move(sample (Play(Position, move)));
7  |   |   end
8  |   |   else
9  |   |   |   move = argmax_move(nested (Play(Position, move), level -1))
10 |   |   end
11 |   |   if utility of a move > best utility then
12 |   |   |   best utility = utility of the move;
13 |   |   |   bestSequence = Sequence after move;
14 |   |   |   bestMove = move of the sequence ;
15 |   |   end
16 |   |   Position = Play(Position, bestMove);
17 |   end
18 |   return best utility
19 end
```

(cf. Algorithm 1) just explores a graph randomly from a given position in the graph, agents use the function $play(position, move)$ which plays the move in the position and returns the resulting position. If none of the moves improves on the best sequence, the move of the best sequence is played, otherwise the best sequence is updated with the newly found sequence and the best move is played.

5 Conclusion

This paper has introduced a new coalition formation mechanism enriched with several principles to deal with the constraints of the agents and a Nested Monte-Carlo based search algorithm. We have detailed how the constraints are modeled as a graph and how this graph is explored using the Nested Monte-Carlo search. From the graph of constraints, each agent gets its most preferred path of constraints and constructs a coalition graph that is used to generate the coalitions to negotiate.

References

1. Cazenave, T.: Nested monte-carlo search. In: IJCAI, pp. 456–461 (2009)
2. Conitzer, V., Sandholm, T.: Complexity of constructing solutions in the core based on synergies among coalitions. Artif. Intell. **170**, 607–619 (2006)
3. Demange, G.: The strategy structure of some coalition formation games. Games Econ. Behav. **64**, 83–104 (2009)
4. Gelly, S., Silver, D.: Combining online and offline knowledge in UCT. In: ICML 2007, pp. 273–280 (2007)
5. Ieong, S., Shoham, Y.: Marginal contribution nets: A compact representation scheme for coalitional games. In: Proceedings of the 6th ACM Conference on Electronic Commerce, pp. 193–202. ACM (2005)

6. Moore, R.E., Kearfott, R.B., Cloud, M.J.: Introduction to Interval Analysis. SIAM, Philadelphia (2009)
7. Rahwan, T., Michalak, T., Elkind, E., Faliszewski, P., Sroka, J., Wooldridge, M., Jennings, N.R.: Constrained coalition formation. In: AAAI (2011)
8. Rahwan, T., Ramchurn, S.D., Dang, V.D., Jennings, N.R.: Near-optimal anytime coalition structure generation. In: IJCAI, pp. 2365–2371 (2007)
9. Rahwan, T., Ramchurn, S.D., Dang, V.D., Jennings, N.R., Giovannucci, A.: An anytime algorithm for optimal coalition structure generation. J. Artif. Int. Res. **34**(1), 521–567 (2009)
10. Ramchurn, S.D., Polukarov, M., Farinelli, A., Truong, C., Jennings, N.R.: Coalition formation with spatial and temporal constraints. In: AAMAS, pp. 1181–1188 (2010)
11. Sandholm, T., Larson, K., Andersson, M., Shehory, O., Tohme, F.: Coalition structure generation with worst case guarantees. Artif. Intell. **111**, 209–238 (1999)
12. Shehory, O., Kraus, S.: Methods for task allocation via agent coalition formation. Artif. Intell. **101**, 165–200 (1998)
13. Voice, T., Ramchurn, S., Jennings, N.: On coalition formation with sparse synergies. In: AAMAS, pp. 223–230 (2012)

Data Mining and Machine Learning

Robust Feature Extraction Based on Teager-Entropy and Half Power Spectrum Estimation for Speech Recognition

Jing Dong[✉], Dongsheng Zhou, and Qiang Zhang

Key Laboratory of Advanced Design and Intelligent Computing,
(Dalian University), Ministry of Education, Dalian 116622, China
wileyzhuanyong@163.com

Abstract. In this paper, we present a robust feature extraction scheme for speech recognition. Compared to standard mel-frequency cepstral coefficients (MFCC), it incorporates perceptual information into half parameter spectrum not into the whole classical spectrum, and combines with Teager-Entropy to construct a new feature vector. Its performance is compared with several techniques, and detailed comparative performance analysis with various types of noise and a wide range of SNR values is presented. The results suggest that our feature achieves superior robustness with HMM-based recognizer on an English digit task. The 8.87 % reduction of average error rate is obtained in comparison to ordinary MFCC. Furthermore, the results also uncover that the half power spectrum-based method leads to superior performance over the whole power spectrum-based method in most given environment.

Keywords: Entropy · MFCC · Spectral analysis · Speech recognition

1 Introduction

Speech recognition has become a very popular subject in both research and flourishing industry domains. Now, it is being applied to the real-world applications from the laboratory research theory, for example, Windows Speech Recognition on Windows Vista and Voice Command on Windows Mobile, these speech interfaces can make many devices easier to operate. However, in a real-life situation, there is bound to be a mismatch between training and testing caused by background noise. The performance of system deteriorates severely, which is the most major obstacle to the commercial use of speech recognition technology. Various methods have been studied, which can be broadly classified into 3 categories – speech enhancement in signal space [1–6], speech model compensation in model space [7–11] and robust feature extraction in feature space [12–21]. In the first category, the enhanced speech signal may lose some important information for recognition, which can degrade the performance of ASR system. In the model-based method, the training model is dynamically modified in an attempt to close to the testing environment and to avoid errors caused by the influence of channel changes. But these model-based methods only suits for relatively stable state. The feature-based method aims at extracting concise, discriminative, noise-insensitive

© Springer International Publishing Switzerland 2015
A. Bikakis and X. Zheng (Eds.): MIWAI 2015, LNAI 9426, pp. 91–101, 2015.
DOI: 10.1007/978-3-319-26181-2_9

feature from noisy speech utterance to make discriminating analysis. Therefore, this method is more suitable for practical operating environment than others. Nowadays it is believed that auditory perception-based features can achieve a high level of success with reasonable robustness among other features. For example, mel-frequency cepstral coefficients (MFCC) have already been well accepted as a good choice and many advanced techniques have been developed based on it [12–14].

In this paper, we focus on the interference of additive noise, and propose a new technology for robust speech recognition. Our approach is closely similar to [14] in a sense that it combines the Teager-Entropy with MFCC derived from half power spectrum estimated by AR model parameters of noisy signal. The experimental results show that significant noise robustness can be achieved in most conditions as compared to MFCC.

The organization of this paper is as follows. In Sect. 2, some related works are explained. In Sect. 3, based on the study of MFCC and Teager-Entropy, the detailed algorithm of the proposed robust feature is presented. The quantitative evaluations are given in Sect. 4. Finally, conclusions and future works are given in Sect. 5.

2 Related Works

2.1 Model-Based Power Spectrum Estimation

Speech is a kind of power signal and its basic characteristic can be represented by power spectrum in frequency domain. There is a widespread cognition that the power spectrum estimation based on parameter model eliminates the influence of window function and the assumption of autocorrelation sequence being zero, it can avoid the spectral leak and has better frequency resolution than FFT-based estimation method, especially for speech signal, the short time stationary stochastic process.

The main idea of parameter model-based power spectrum estimation is that the speech signal is viewed as the output of linear system, where

$$H(z) = \frac{1}{1 + \sum\limits_{k=1}^{p} a_k z^{-k}} \tag{1}$$

$$x(n) = -\sum\limits_{k=1}^{p} a_k x(n-k) + \omega(n) \tag{2}$$

$\omega(n)$ is the white noise sequence of zero mean, σ^2 variance. Here, $x(n)$ is pth order autoregressive (AR) process, namely, the present output is the weighted sum of present input and past inputs. The signal power spectrum can be succinctly indicated as

$$p(f) = \sigma^2 |H(f)|^2 = \frac{\sigma^2}{|A(f)|^2} \tag{3}$$

$$\hat{\sigma}^2 = r_{xx}(0) \prod_{i=1}^{p} \left[1 - |\hat{a}_i(i)|^2 \right] \tag{4}$$

So, this power spectrum estimation includes two steps: for given sequence $x(n)$, first to estimate the model parameter, then compute the power spectrum by above equation.

2.2 Teager-Entropy

The Teager energy operator reflects not only the magnitude change but also the frequency change. The faster the magnitude or frequency changes, the larger the operator is. And the Teager operator can reflect different characteristics aiming at different class signals.

$$TE = |x(n)|^2 \sin^2[f(n)] \approx |x(n)|^2 f^2(n) \tag{5}$$

Entropy is robust to the noise, especially for white noise with uniform distribution spectrum. Given the probability density of frequency spectrum

$$p_i = \frac{|X_w(f_i)|^2}{\sum_{i=0}^{F/2-1} |X_w(f_i)|^2} \tag{6}$$

The negative spectrum entropy is defined as

$$H = \sum_{i=0}^{F/2-1} P_i \log P_i \tag{7}$$

When the noise spectrum is uniform, its entropy is far smaller than that of signal.

3   TE-MHFCC Feature Extraction

MFCC is the most effective feature with reasonable success in many speech recognition systems. In order to avoid some bias and variance of the cepstral coefficients, we consider incorporating the half power spectrum based on parameter model into MFCC to improve its robustness. It is one thing to decrease variance to augment inter-class separability by this new extraction; it is another to reduce calculation quantities by using half power spectrum instead of the whole spectrum.

Formant is one important acoustic characteristic of speech signal to distinguish different vowels, and estimating the spectrum envelop is the key step of extracting formant parameters. Based on minimum mean-square error criterion, the matching effect of formant components is much better than the effect among harmonics for voiced speech spectrum under background noise condition. Different vowels correspond to one group of different formant parameters theoretically, but only the first three formants are enough to describe the corresponding speech signal in the real situation, and they mostly distribute in the first half part of spectrum. In the meantime, the majority part of the speech signals' energy also concentrates in this low frequency region.

It is, therefore, obvious that a representation of the first half part envelope is a desirable characteristic in terms of noise robustness because the low frequency portion of spectrum is relatively less affected by the additive noise than the whole spectrum. With the noise enhances gradually, the deviation of spectrum estimation in the high frequency area increases clearly, but the envelope trends of the first two formants in the low frequency area are still correct, and can approximately describe the signal property in sever noise conditions.

As mentioned above, Teager-Entropy can extract efficient speech information from complex background noise and a successful application has been obtained in the endpoint detection. In this paper, we introduce it into the speech recognition system and combine it with the proposed improved MFCC based on Half power spectrum (MHFCC) to obtain a new robust feature, TE-MHFCC. Firstly, take preprocess for the input speech signal: preemphasis, adding windows and segmenting signal into frames. Based on a standard 12th-order AR model, we estimate the power spectrum by the use of AR parameters, only consider the first half region of the power spectrum to Mel filter. Then, take the DCT transformation and cepstrum mean normalization to obtain the MHFCC. Compute the Teager -Entropy of the speech signal. Finally, integrate TE with MHFCC to form a new feature vector, compute its first derivative, the static and dynamic parameters are combined together for the use of training and testing. Figure 1 shows the schematic diagram of feature extraction.

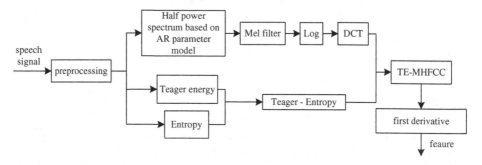

Fig. 1. The flowchart of TE-MHFCC feature extraction

4   Experiment Results

In this section, based on a discrete Hidden Markov Model (HMM) system, the performance of the proposed method was compared with several related techniques: the standard MFCC, the whole power spectrum-based MFCC (MWFCC) and MHFCC. Its performance in noisy speech recognition was evaluated based on two tasks. Speech task A comprised the recordings of 30 male and 30 female adults speaking isolated

English digits sampled at a rate of 10 kHz, each of ten digits was spoken once by every speaker, resulting in a total of 600 files [24]. Speech task B was the monosyllable subset of CASIA Chinese monosyllable isolate word speech corpus provided by institute of automation, Chinese academy of sciences, sampled at 16 kHz and quantized to 16 bits per sample. It comprised 140 syllables, which proportioned all Chinese phonemes [25]. The used window length was 25.6 ms with 8 ms segmentation step. To analyze the influence of different possible situations, six noise circumstances taken from noisex92 database were considered: buccaneer noise, white noise, hfchannel noise, pink noise, m109 noise and f16 noise. They were artificially added to clean speech at a desired SNR, ranging from -5 dB to 20 dB, to generate noisy test set. The performances were compared in terms of spectrum estimation accuracy as well as recognition robustness to noise.

4.1 RMSE

Preliminary test concentrated on the estimation accuracy of half power spectrum measured by Root-mean-square-error (RMSE). Under white noise condition, the RMSE of ten digits' FFT-based power spectrum estimation and their parameter-based half power spectrum estimation are presented in Table 1. As is revealed in Table 1, averaged over different noise levels, the RMSE of half power spectrum estimation is approximately smaller 4 than that of FFT-based power spectrum estimation. The performance for voiced speech signal is much better, such as five, nine and ten, this is due to the fact that the useful formants of voiced speech often lie in the first half region, while the unvoiced speech's energy often distributes in high frequency.

Table 1. The RMSE of whole power spectrum estimation and half power spectrum estimation of ten digits under white noise.

SNR Num	20dB		15 dB		10 dB		5 dB		Mean	
	W	H	W	H	W	H	W	H	W	H
1	9.6	6.0	12.6	8.6	16.0	11.7	19.9	15.2	14.5	10.4
2	9.3	6.0	12.2	8.5	15.6	11.5	19.4	15.0	14.1	10.3
3	9.2	6.5	12.1	9.2	15.5	12.3	19.3	16.0	14.0	11.0
4	10.9	6.7	14.1	9.4	17.7	12.5	21.7	16.0	16.1	11.2
5	8.3	3.1	10.9	4.7	14.1	6.8	17.7	9.6	12.8	6.1
6	10.5	7.3	13.6	10.5	17.1	14.1	21.1	18.1	15.6	12.5
7	9.0	5.4	11.8	7.9	15.2	10.9	19.0	14.2	13.8	9.6
8	13.1	9.2	16.4	12.4	20.1	16.1	24.1	20.1	18.4	14.5
9	9.1	5.1	11.9	7.4	15.3	10.1	19.0	13.3	13.8	9.0
10	9.0	5.4	11.8	7.8	15.1	10.7	18.9	14.1	13.7	9.5
Mean	9.8	6.1	12.7	8.6	16.2	11.7	20.0	15.2	14.7	10.4

Figure 2 (a) and (b) show the RMSE of Chinese monosyllabic isolate words under white noise and pink noise for task B.

It is observable that the half power spectrum-based method provides improvements over the FFT-based method for all SNR levels. For white noise, the RMSE of the

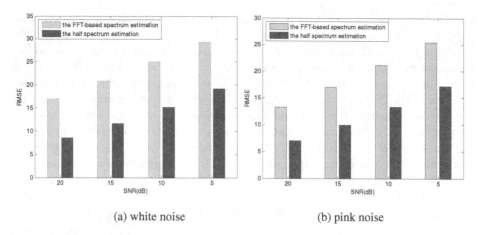

(a) white noise   (b) pink noise

Fig. 2. The RMSE of Chinese monosyllabic isolate words under different noise

proposed method at 10 dB is approximately equal the RMSE of conventional method at 20 dB, similarly for 5 dB with the proposed method and 15 dB with the conventional method. And for Pink noise, the RMSE of the proposed method at 15 dB is even smaller than that of conventional method at 20 dB, similarly for 10 dB with the proposed method and 15 dB with the conventional method, 5 dB with the proposed method and 10 dB with the conventional method.

4.2   Robustness

For comparison, we evaluated the performance of four methods on task A over six different noise conditions to gain some insight into the merit of the TE-MHFCC in the presence of additive noise. The training set consisted of the clean recordings of 15 male and 15 female speakers randomly chosen from the entire database, and the rest were prepared for test. The performance of each group was the average of experimental results obtained from 50 times' stochastic process.

Figure 3 presents the recognition performance of four feature schemes with six different noise types artificially added at SNRs from -5 to 20 dB. As well known the performance deteriorates for decreasing SNR, but it is observable that TE-MHFCC provides improvements for most SNR levels over other three methods except for hfchannel noise -5 dB and 0 dB. The recognition rates of TE-MHFCC representation outperform considerably. The 8.87 % reduction of average error rate is obtained in comparison to ordinary MFCC. It is also clear from Fig. 3 that the best improvements for severe noisy conditions are obtained during 5 to 10 dB. This shows that the proposed new feature is much more robust to noise than other methods. Moreover, the convenience of this new feature is due to the fact that the power spectrum were estimated by using AR parameters only for p = 12, which is far smaller than the order of capon spectrum estimation, accordingly, the proposed method only needs comparatively less amount of calculation.

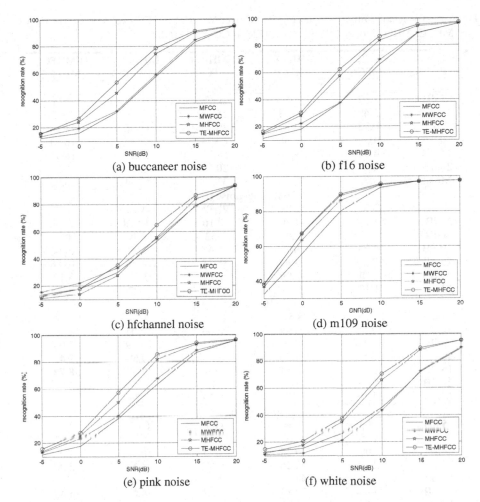

Fig. 3. The performance comparison of four feature schemes under six noise conditions

Based on the above results, in order to determine if the differences in performance between four feature extraction schemes are statistically significant, a relative improvement can be stated for the proposals.

In Table 2, the first three columns show the relative improvements of the other three feature extraction schemes over MFCC, the fourth column is the relative improvement of MHFCC over MWFCC and the last column states the relative improvement of TE-MHFCC over MHFCC.

As presented in Table 2, the improvements are the most obvious during 5 dB to 15 dB. MWFCC improves the results under many noise conditions, especially for m109 noise, the recognition rate is relatively increased 30 % at 5 dB, yet for white noise, there is a little deterioration. On this basis, we only choose the low frequency region to extract MHFCC, the melioration is more notable except hfchannel noise, and this will be explained in the next. Furthermore, by using Teager -Entropy, we can see

Table 2. The relative improvements of MWFC over MFCC, MHFCC over MFCC, TEMHFCC over MFCC, MHFCC over MWFCC and TEMHFCC over MHFC

Type SNR	MWFCC -MFCC	MHFCC -MFCC	TEMHFCC -MFCC	MHFCC -MWFCC	TEMHFCC -MHFCC
F16 noise (%)					
-5 dB	3.52	3.93	5.60	0.42	1.74
0 dB	5.42	12.48	15.08	7.46	2.97
5 dB	0.72	31.85	39.96	31.35	11.90
10 dB	10.44	52.68	60.54	47.16	16.63
15 dB	4.94	49.10	57.72	46.46	16.93
20 dB	-2.71	5.96	20.33	8.44	15.27
Mean	3.72	26.00	33.20	23.54	10.90
Hfchannel noise (%)					
-5 dB	2.31	-2.86	-1.32	-5.29	1.50
0 dB	5.14	-4.70	-0.16	-10.38	4.34
5 dB	4.67	-3.05	7.52	-8.10	10.26
10 dB	4.06	6.07	25.47	2.09	20.66
15 dB	-2.29	23.47	36.38	25.19	16.86
20 dB	-7.09	-2.99	5.35	3.82	8.10
Mean	1.13	2.65	12.20	1.22	10.28
M109 noise (%)					
-5 dB	6.16	8.58	8.01	2.57	-0.62
0 dB	17.68	25.94	26.80	10.04	1.15
5 dB	30.56	45.69	49.52	21.79	7.06
10 dB	21.64	21.33	28.75	-0.39	9.43
15 dB	-0.67	-6.69	2.68	-5.98	8.78
20 dB	-3.78	-6.30	0.42	-2.43	6.32
Mean	11.93	14.75	19.36	4.26	5.35
Pink noise (%)					
-5 dB	2.03	2.64	4.83	0.62	2.25
0 dB	6.95	9.70	12.11	2.95	2.67
5 dB	3.09	19.14	31.13	16.56	14.83
10 dB	11.91	50.26	60.63	43.54	20.85
15 dB	12.06	47.77	54.84	40.61	13.53
20 dB	-1.43	12.83	17.58	14.05	5.45
Mean	5.76	23.72	30.18	19.72	9.93
White noise (%)					
-5 dB	-2.27	-1.05	2.26	1.19	3.28
0 dB	-4.48	2.58	6.25	6.76	3.77
5 dB	-7.03	11.47	14.89	17.28	3.87
10 dB	-3.88	37.09	45.58	39.44	13.50
15 dB	2.16	58.85	63.09	57.94	10.32
20 dB	3.96	53.53	56.08	51.62	5.48
Mean	-1.92	27.07	31.35	29.03	6.70

that recognition performance is mostly improved. In white noise condition, the performance improves relatively 63 % at 15 dB, and in pink noise condition, it improves 60 % at 10 dB with respect to MFCC. Simultaneously, we note the relative improvements are negative at -5 dB and 0 dB under hfchannel noise, they are -1.321 % and -0.16 % respectively, which means the new feature is not suitable for these conditions, but if we consider its relative improvements at 5-20 dB, 5 %-36 %, the proposed method is still considered to be effective.

According to the results in the fourth column of the table, we observe that the feature based on half power spectrum can reduce the recognition rate with respect to the feature based on whole power spectrum with the interference of m109 and hfchannel noise at some SNR periods. This is due to the fact that majorities of energy of these noises exist in low frequency band, when we only choose the first half region to extract MHFCC, the interference degree increases obviously. But for the noise with high frequency component, such as white noise, its spectrum is uniform distribution, discarding the later part can effectively weaken its effect on speech. From the last column in Table 2, we also observe that the usage of TE contributes to the increase of recognition performance for stationary noise, e.g. the white noise, the improvement is apparent. Although there is a slight decrease of MHFCC performance in hfchannel condition, the introduction of TE can effectively improve it. As can be seen that TE feature relatively improves the results 20 % at 10 dB, and 16 % at 15 dB with respect to MHFCC. Furthermore, it also can be noticed that, averaged over different noise levels, the results are ameliorated most obviously during 5 to 15 dB, in reality, most noise interference is just under this condition, which meets the requirement of practical application quite well. Accordingly, the experimental results show that significant noise robustness can be achieved by the use of the proposed feature in most cases.

5   Conclusions

In this paper, a robust feature extraction technique for speech recognition is presented. The perceptual information was directly incorporated with half power spectrum estimated from parameter model. Based on this incorporating, the improvements on both robustness and computational efficiency are obtained. In addition, by introducing Teager-Entropy, the robustness of recognition systems in stationary noise environments further increases. Experiments clearly showed that the proposed new feature yields better recognition performance in sever noisy conditions. One limitation of this algorithm is that it performs better for stationary noise, for the future research, a study on non-stationary background noise will be considered.

Acknowledgements. This work is supported by the National Natural Science Foundation of China (No. 61370141, 61300015), Natural Science Foundation of Liaoning Province (No. 2013020007), the Scientific Research Fund of Liaoning Provincial Education Department (No. L2013459, No. L2015015), the Program for Science and Technology Research in New Jinzhou District (No. 2013-GX1-015, KJCX-ZTPY-2014-0012).

References

1. Gupta, M., Douglas, S.C.: A spatio–temporal speech enhancement technique based on generalized eigenvalue decomposition. IEEE Trans. Audio Speech Lang. Process. **17**(4), 830–839 (2009)
2. Baby, D., Virtanen, T., Gemmeke, J.F., Van Hamme, H.: Coupled dictionaries for exemplar-based speech enhancement and automatic speech recognition. IEEE/ACM Trans. Audio Speech Lang. Process. **23**(11), 1788–1799 (2015)

3. Shao, Y., Chang, C.-H.: Bayesian separation with sparsity promotion in perceptual wavelet domain for speech enhancement and hybrid speech recognition. IEEE Trans. Syst. Man Cybern. Part A Syst. Hum. **41**(2), 284–293 (2011)
4. Chan, K.Y., Yiu, C.K.F., Dillon, T.S., Nordholm, S., Ling, S.H.: Enhancement of speech recognitions for control automation using an intelligent particle swarm optimization. IEEE Trans. Industr. Inf. **8**(4), 869–879 (2012)
5. Rajakumar, P.S., Ravi, S., Suresh, R.M.: Speech enhancement models suited for speech recognition using composite source and wavelet decomposition model. In: 2010 International Conference on Signal and Image Processing (ICSIP), pp. 511–514 (2010)
6. Langarani, M.S.E., Veisi, H., Sameti, H.: The effect of phase information in speech enhancement and speech recognition. In: 2012 11th International Conference on Information Science, Signal Processing and their Applications (ISSPA), pp. 1446–1447 (2012)
7. Leutnant, V., Krueger, A., Haeb-Umbach, R.: A new observation model in the logarithmic mel power spectral domain for the automatic recognition of noisy reverberant speech. IEEE/ACM Trans. Audio Speech. Lang. Process. **22**(1), 95–109 (2014)
8. Hong, W.-T.: HCRF-based model compensation for noisy speech recognition. In: 2013 IEEE 17th International Symposium on Consumer Electronics (ISCE), pp. 277–278 (2013)
9. Nguyen, D.H.H., Xiao, X., Chng, E.S., Li, H.: An analysis of vector taylor series model compensation for non-stationary noise in speech recognition. In: 2012 8th International Symposium on Chinese Spoken Language Processing (ISCSLP), pp. 131–135 (2012)
10. Zhao, Y., Juang, B.-H.: Nonlinear compensation using the gauss-newton method for noise-robust speech recognition. IEEE Trans. Audio Speech Lang. Process. **20**(8), 2191–2206 (2012)
11. Mushtaq, A., Hui-Lee ,C.: An integrated approach to feature compensation combining particle filters and hidden markov models for robust speech recognition. In: 2012 IEEE International Conference on Acoustics, Speech and Signal Processing (ICASSP), pp. 4757–4760 (2012)
12. Shannon, B.J., Paliwal, K.K.: Effect of speech and noise cross correlation on amfcc speech recognition features. In: IEEE International Conference on Acoustics, Speech and Signal Processing, vol. 4, pp. IV-1033–IV-1036 (2007)
13. Rajnoha, J., Pollak, P.: Modified Feature Extraction Methods in Robust Speech Recognition. In: 17th International Conference on Radioelektronika, pp. 1–4 (2007)
14. Dharanipragada, S., Yapanel, U.H., Rao, B.D.: Robust feature extraction for continuous speech recognition using the MVDR spectrum estimation method. IEEE Trans. Audio Speech Lang. Process. **15**(1), 224–234 (2007)
15. Sharma, U., Maheshkar, S., Mishra, A.N.: Study of robust feature extraction techniques for speech recognition system. In: 2015 International Conference on Futuristic Trends on Computational Analysis and Knowledge Management (ABLAZE), pp. 654–658 (2015)
16. Gerazov, B., Ivanovski, Z.: Kernel power flow orientation coefficients for noise-robust speech recognition. IEEE/ACM Trans. Audio Speech Lang. Process. **23**(2), 407–419 (2015)
17. Alam, M.J., Kenny, P., Dumouchel, P., O'Shaughnessy, D.: Robust feature extractors for continuous speech recognition. In: 2014 Proceedings of the 22nd European Signal Processing Conference (EUSIPCO), pp. 944–948 (2014)
18. Mitra, V., Franco, H., Graciarena, M., Vergyri, D.: Medium-duration modulation cepstral feature for robust speech recognition. In: 2014 IEEE International Conference on Acoustics, Speech and Signal Processing (ICASSP), pp. 1749–1753 (2014)
19. Du, J., Huo, Q.: A feature compensation approach using high-order vector taylor series approximation of an explicit distortion model for noisy speech recognition. IEEE Trans. Audio Speech Lang. Process. **19**(8), 2285–2293 (2011)

20. Milner, B., Darch, J.: Robust acoustic speech feature prediction from noisy mel-frequency cepstral coefficients. IEEE Trans. Audio Speech Lang. Process. **19**(2), 338–347 (2011)
21. Kim, W., Hansen, J.H.L.: An advanced feature compensation method employing acoustic model with phonetically constrained structure. In: 2013 IEEE International Conference on Acoustics, Speech and Signal Processing (ICASSP), pp. 7083–7086 (2013)
22. Proakis, J.G., Manolakis, D.G.: Digital Signal Processing: Principles, Algorithms, and Applications, ch. 12. Prentice-Hall Inc., New Jersy (2004)
23. Han, J. Q., Zhang, L., Zheng, T.R.: Speech Signal Processing. Tsinghua Publishing House, ch. 2, Beijing (2007)
24. Childers, D.G.: Speech Processing and Synthesis Toolboxes (Authorized Published). Wiley, Tsinghua Publishing House, New York (2004)
25. Chinese Linguistic Data Consortium. http://www.chineseldc.org/

On-line Monitoring and Fault Diagnosis of PV Array Based on BP Neural Network Optimized by Genetic Algorithm

Hanwei Lin, Zhicong Chen[✉], Lijun Wu, Peijie Lin,
and Shuying Cheng[✉]

Qi Shan Campus of Fuzhou University, 2 Xue Yuan Road,
University Town, Fuzhou 350108, Fujian, People's Republic of China
{zhicong.chen, sycheng}@fzu.edu.cn

Abstract. The vast majority of photo voltaic (PV) arrays often work in harsh outdoor environment, and undergo various fault, such as local material aging, shading, open circuit, short circuit and so on. The generation of these fault will reduce the power generation efficiency, and even lead to fire disaster which threaten the safety of social property. In this paper, an on-line distributed monitoring system based on ZigBee wireless sensors network is designed to monitor the output current, voltage and irradiate of each PV module, and the temperature and the irradiate of the environment. A simulation PV module model is established, based on which some common faults are simulated and fault training samples are obtained. Finally, a genetic algorithm optimized Back Propagation (BP) neural network fault diagnosis model is built and trained by the fault samples data. Experiment result shows that the system can detect the common faults of PV array with high accuracy.

Keywords: PV module · BP neural network · Genetic algorithm · Zigbee

1 Introduction

Because of the increasing depletion of fossil energy resources and the increasing environmental pollution, many countries in the world are actively seeking alternative renewable clean energy [1]. Solar PV power generation has the advantages of clean, no pollution, sustainability and broad [2]. Therefore, the use of solar power has been widely valued by many countries [3]. However, the PV array work in the complex outdoor environment. It is easy to appear the local material aging, open circuit or short circuit and other problems. These problems greatly reduce the battery life. At the same time, the failure of the PV array will reduce the power generation efficiency, even the fire will happen when they are serious enough, endangering the safety of social property. It can be seen that the fault detection of PV power generation is very necessary.

The routine maintenance of PV power generation system mainly rely on artificial periodic cleaning solar panel, PV module connection and between the component and the support, and the inverter voltage and current monitoring data manual judgment whether the normal power generation [4]. This method which is time-consuming, lack of real time, dependent on the subjective experience of maintenance personnel is

© Springer International Publishing Switzerland 2015
A. Bikakis and X. Zheng (Eds.): MIWAI 2015, LNAI 9426, pp. 102–112, 2015.
DOI: 10.1007/978-3-319-26181-2_10

difficult to find and diagnose faults quickly and accurately. In order to overcome the complexity and subjectivity of artificial detection of PV power generation system, the monitoring technology of PV power generation system is concerned by many domestic and foreign scholars and related organizations. Off line monitoring system mainly includes the infrared image analysis method, the ground capacitance measurement method, based on the parameter model of the fault diagnosis method. Online monitoring system mainly includes the satellite detection method, the fault diagnosis method based on the working voltage window, the time domain analysis method and so on. Drews uses the satellite to observe the light intensity data in the area of the PV array, and inputs it to the model of the photo voltaic array to get the power output of the simulation. We determine whether exits fault by comparing simulation powers obtained with actually powers detected [5]. A fault detection method based on power loss analysis is proposed by Chouder [6]. The parameters of the electrical model of the array are fitted by using the measured I–V curve of the PV array. The expected output power is obtained by combining with the measured ambient light intensity and component temperature. Comparing the expected power with the measured power, it is used to determine whether the array is in fault. Gokmen proposed a fault diagnosis method based on the working voltage window. The number of open circuit and short circuit faults can be detected by the range of the series voltage. Considering the cost and the complexity of the system, the monitoring system is mostly realized by the array level/component level monitoring. These systems can not accurately obtain the location of faulty PV modules.

PV module fault has the characteristics of fuzzy and multilevel. In order to distinguish the fault types, the intelligent algorithm is introduced to the fault diagnosis. The neural network method proposed by Syafaruddin. By establishing and training several neural network structure [8], he judge whether there is a short circuit fault, An online diagnostic method based on BP neural network is proposed by Wang Yuanzhang. In order to distinguish three states of the components, including normal, short and abnormal aging, it use the open circuit voltage of photovoltaic module, short-circuit current, maximum power point voltage current as the inputs of the trained neural network. In these methods, the number of feature variables used in fault diagnosis is limited, and the characteristic variables are difficult to measure in practice. Therefore these methods can only distinguish the limited fault causes and influence the accuracy of fault diagnosis.

In order to obtain the location of PV module, the real-time fault monitoring based on component level is a necessity. At the same time, in order to improve the accuracy of fault diagnosis, the correct use of intelligent algorithm is also an inevitable. With the continuous progress of low-cost wireless sensor technology, you can install a wireless sensor on each PV module. In this subject, we use algorithm based on distributed on-line monitoring of photo voltaic array of Zigbee wireless sensor network and BP neural network algorithm based on genetic algorithm optimization to study fault diagnosis. Because the PV system have the unpredictable nature of the various faults, fault types and the fault performance characteristics are not easy to be diagnosed by environmental factors. In this paper, an online fault diagnosis method based on BP neural network is proposed. The model can store the corresponding relationship between the failure state of PV modules and the causes of faults in the structure of the neural network. So the data that input to the trained neural network can judge whether the components are faulty or the type of fault, so as to realize the on-line fault diagnosis of PV modules [5, 11]. The way ensure the output efficiency and normal operation of PV power plant, saving a lot of material resources and manpower.

2   Online Monitoring System

The on-line monitoring system that is designed by this paper is mainly composed of the PV module wireless acquisition module, data gateway and remote data management center. The zigbee network nodes that are embedded in the PV cell module are the monitoring unit of the whole system [5]. Their work is a real time acquisition of the output parameters of PV modules, and the acquisition data is transmitted to the data gateway by the serial port. The data gateway is the buffer, and the data is transmitted to the remote data management center. When the data management center receives this collection of data, it is to carry on a series of processing and storage. So it can get the fault type and location of PV modules.

All the monitoring nodes in the system adopt the same design principle. They are directly mounted on the PV panels. The monitoring node is composed of power supply module, collection module and wireless transceiver module.

2.1   Power Supply Module

The Zigbee node can be powered by a PV cell board through a buck circuit and a voltage regulator circuit. Due to output voltage of PV cells in work is around 18 V. It can have a voltage stabilization using LM2596 in 5 V. And then has a reduction voltage from 5 V to 3.3 V with LM1117. It meet the voltage required by CC2531. The specific circuit is shown in Fig. 1:

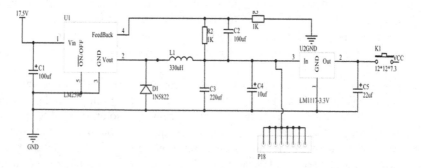

Fig. 1. Power module

2.2   Collection Module

This system mainly monitors the voltage, current, temperature and light intensity of the PV module. CC2531 chip has 12 input configurable 8 bit Analog/Digital. The reference voltage of 2.5 V of CC2531 interface is provided with 8 A/D by using TI LM4040. PV module output the maximum voltage of about 21.5 V. Therefore, the voltage first to carry out partial pressure, and then Analog Digital sampling before the voltage is sampled. The partial pressure ratio is 20. Specific circuit as shown in Fig. 2:

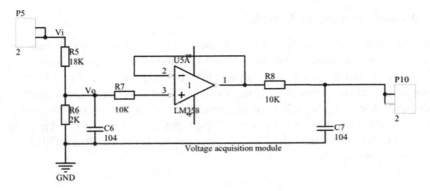

Fig. 2. Voltage acquisition module

R5 is 18 K, and R6 is 2 K. According to partial pressure formula $V_0 = \frac{R6}{R5+R6} Vi$. We can get the $V_0 = \frac{1}{20} Vi$. It through a voltage following circuit that plays a protective isolation role in the circuit. This ensures that the sampling voltage is within the reference range of the CC2531.

For current sampling, the current is converted to voltage when it is sampled. This topic selected MAX4080 to complete the collection of current. MAX4080 is a current sense amplifier. The specific circuit is shown in Fig. 3:

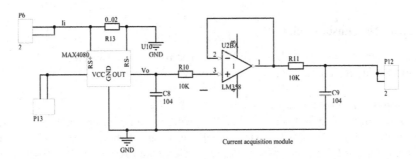

Fig. 3. Current acquisition module

Sampling resistor R13 select precision resistor 0.02 Ω. According to the MAX4080 data sheet, we can be aware of $Vo = 20Ii * R13$. As the reference voltage of the AD CC2531 sampling interface is 2.5 V, and the maximum current of PV modules is about 6 A. It can be calculated into the formula, and the sampling resistance is most appropriate for 0.02 Euro. The voltage Vo is used for the AD interface of the CC2531 after the voltage following circuit.

DS18B20 temperature sensor and Hl750FVI illumination sensor were used to collect temperature and light intensity. According to the data of the two manuals can be known, their measurement range and measurement accuracy are very good to meet the system requirements.

2.3   Wireless Transceiver Module

Wireless communication module is the important data transmission channel in the system. The data transmission and networking function between the nodes and the sink node, the node and the node are realized. In this design, each component string is composed of a cluster, and the sensing node of a certain component is used as the cluster head. A star topology is used to cluster the first and the cluster members, and cluster head and sink node are also used in the topology of the network. Each cluster head sends data to the sink node. The sink node forwards the data to the data gateway by the serial port.

3   Fault Diagnosis

Because PV power generation system works in a relatively harsh environment, such as deserts, mountains or roofs. So it is easy to have a more common failure. In this paper, four kinds of fault are studied, such as short circuit, open circuit, shadow and abnormal aging. The output of PV modules is influenced by environmental factors. The output characteristic of the components and the environmental factor is the nonlinear relationship. Therefore, the problem of fault diagnosis of PV modules is a complicated nonlinear problem. BP neural network algorithm is introduced to realize the on-line fault diagnosis of components. The algorithm is currently widely used and mature.

3.1   Fault Diagnosis Model

The fault diagnosis model is shown in Fig. 4. The model has 3 layers, which are input layer, hidden layer and output layer respectively.

3.1.1   Input and Output Layer

The feature selection based on BP neural network is very important for the feature selection of PV fault diagnosis model, which can improve the accuracy of fault diagnosis. Considering the actual measurement problems and related theory research, we will be the input variables of the neural network with the maximum power point voltage, the maximum power point current, the ambient light intensity, and the ambient temperature. As the PV module has an open circuit fault, the staff can directly obtain from the measured data. Therefore, in this paper, it is not the fault type of BP neural network. Normal, short circuit, abnormal aging, the shadow will be as neural network output variable.

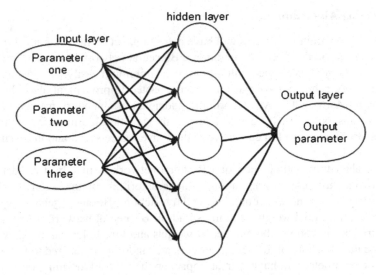

Fig. 4. BP neural network model

3.1.2 Implicit Layer

In the BP neural network, the number of hidden layer nodes is a more important and complex problem. If the election is too small, the neural network cannot produce enough to satisfy the right combination of connected learning neural network to the sample. If the election is too much, it can make the system error smaller, but the network training error time increases at the same time, and easy to fall into local minimum point. In this paper, the common formula $n_1 = \sqrt{(n+m)} + a$ will be used. n1 is the number of the hidden layer units. n is the input unit. M is the output unit A is the constant between [1 and 10].

The BP neural network model that is designed in this paper has 4 input nodes and 4 output nodes. After many tests, the network error is the smallest when the implicit node is 11.

3.1.3 Normalization of Input Data

Because the maximum power point voltage, maximum power point current, light intensity, temperature of the number of units and units are not the same. If the original data directly put into the neural network for training, it will make the network performance and convergence poor. So the input data must be pre processed before training the neural network. There are two main methods of data normalization.

The first method is the maximum and minimum method. Function form is as follows: $X_k = \frac{(X_k - X_{\min})}{(X_{\max} - X_{\min})}$. $X_{\min}$ is the data sequence of the smallest. $X_{\max}$ is the data sequence of the biggest.

The second method is the mean variance method. Function form is as follows: $X_k = \frac{(X_k - X_{mean})}{X_{var}}$. X_{mean} is the data sequence of the mean. X_{var} is the data sequence of the variance.

In this paper, we use the first data normalization method. The normalized function uses the MATLAB own function mapminmax.

3.2   Learning Algorithm

BP neural network training process will have a great impact on the accuracy of the fault diagnosis of PV modules [12]. The traditional BP neural network has a lot of problems. Such as slow convergence speed, falling into local minimum easily, and low accuracy. At present, many scholars have put forward the related improvement algorithm. Such as Levenberg-Marquardt (LM) algorithm, additional momentum method, etc. By comparison, this paper finally decided to use the genetic algorithm to optimize the neural network. The algorithm can improve the convergence speed and the accuracy of network training.

Genetic algorithm optimization of BP neural network is mainly divided into BP neural network structure determination, genetic algorithm optimization weight and threshold, BP neural network training and prediction. Because genetic algorithm parameters are the initial weights and thresholds of BP neural network. As long as the network structure is known, the number of weights and thresholds can be known. The weights and the threshold of the neural network are randomly initialized to [−1,1]. This initialization parameter will have a great impact on the network training, but it is often impossible to obtain accurately [8]. Genetic algorithm is introduced to optimize the initial weights and thresholds. The process of the algorithm is shown in Fig. 5:

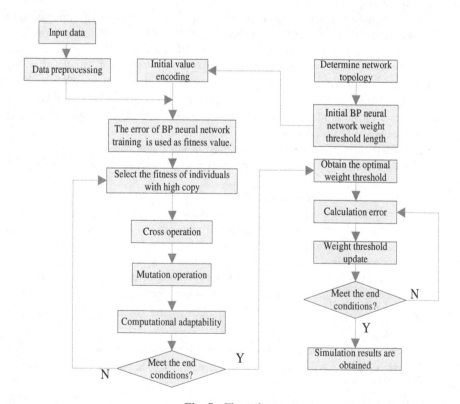

Fig. 5. Flow chart

Genetic algorithm optimizing BP neural network is used to optimize the initial weights and thresholds of the BP neural network. Each individual in the population contains all the weights and thresholds of a network. The fitness function is applied to calculate individual fitness value. The genetic algorithm finds the optimal fitness value corresponding to the individual by choosing, crossing and mutation operation. BP neural network prediction uses genetic algorithm to get the optimal individual to the network initial weights and threshold assignment, and the network is trained to predict the function output. In this paper, the BP neural network structure is set up to 4-11-4. The input layer consists of 4 nodes and the hidden layer consists of 11 nodes and the output layer has 4 nodes. There are 4 * 5 + 5 * 4 = 40 weight and 11 + 4 = 15 threshold in total. So the genetic algorithm individual encoding length is 40 + 15 = 21. At the same time, The absolute value of the prediction error of training data is the individual fitness value, and the smaller the fitness value of the individual is, the more excellent the individual is.

4 Simulation Analysis

The PV array model is built in Mat-lab/Simulation. So as to collect data from different types of fault, it is necessary to classify the PV array model. For abnormal aging fault, we choose the series of external resistance to simulate and the size of resistance represents the aging degree. In the sampling data, the selection of resistance is 1, 3, 7, 10 Ω resistor and get the simulation data of different aging degree. For the shadow failure, this paper uses the light intensity of different gain to obtain different degrees of light intensity. For the short circuit fault, this paper uses the small resistance of 5 Ω in parallel to the output of the PV modules, thus it can simulate the short-circuit fault of PV modules.

In the light intensity (100–1000, step size to 50) and module temperature (25–36, step size to 1), The maximum power point voltage and the maximum power point current are sampled respectively, and the standard simulation data for each fault type is completed. According to the above four different models of the PV model, we can extract four types of corresponding feature data. The extracted data are stored in data1, data2, data3, data4 database files. Each group of data is 5 dimensional. The first dimension is the number of fault types, and the other 4 dimension is the fault type of data. Then the four types of data are combined into a group. In these data, the 1600 groups of data are randomly selected as training data, and the 400 groups of data are used as test data. The training data were normalized. Set the expected output value of each group's fault categories on the basis of the category identification. For example, When the identity of the class is 1, the expected output vector is [1 0 0 0], which means that the desired output is normal type.

4.1 MATLAB Simulation Results

By using genetic algorithm to optimize the BP neural network, we can find out the initial weights and thresholds of the network, and the optimal initial weights and thresholds are assigned to the neural network. Predict the forecast data in the trained BP neural network, the results obtained as shown in Fig. 6:

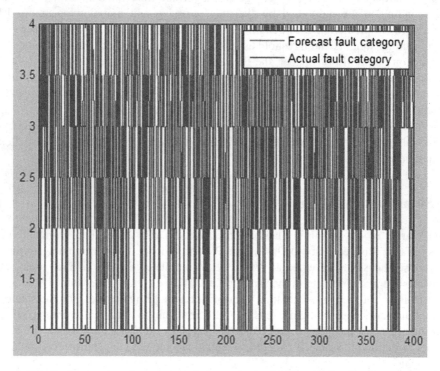

Fig. 6. Comparison between prediction and actual failure (Color figure online)

From Fig. 6 it can be found that the predicted fault types and the actual fault types are relatively close. The red line and the blue line are basically in coincidence. Some red lines will not coincide with the blue line because the model has some diagnostic error rate. From Fig. 7 we can also find that the BP neural network based on genetic algorithm has higher accuracy and it can accurately identify the type of fault. In the random sample of 400 sets of test data, the number of the fault of each class is compared with the total number of the corresponding categories, which can be obtained each class of diagnostic error rate. Thus the accuracy of the algorithm is obtained. Specific simulation results are shown in Table 1. The short circuit fault model is lower than the other three types.I have tried many methods to simulate by changing the implicit node, learning rate, population size and the number of iterations. The results are almost the same as above. The simulation results are obtained by changing the value of the implicit nodes, learning rate, population size and the number of iterations. At this point, The implicit node is 0.1 and the learning rate is 11, and the population size is 15 and the iteration number is 10.

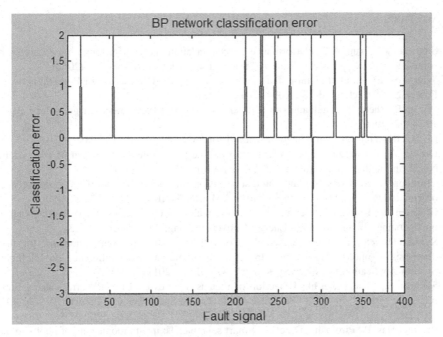

Fig. 7. BP network classification error

Table 1. Diagnostic accuracy

Fault signal category	Normal	Short circuit	Shadow	Aging
Correct recognition rate	1	0.7869	1	0.9795

5 Summary

In this paper, the electrical parameters and environmental parameters of PV modules are acquired by Zigbee sensor node, and the data is transmitted to the gateway through the Zigbee wireless sensor network. The faults are detected by a trained BP neural network. A genetic algorithm is used to optimize the BP neural network fault diagnosis model. The model can detect four types of PV array operating condition, including normal, short circuit, shadow, abnormal aging. Since the open circuit fault can be directly detected from the acquired data, it is not treated as the output fault type of the diagnosis model. Simulation results shows that the fault diagnosis system features high accuracy. Through the designed system, the operators or managers can log in to view the parameters of each PV module, and quickly find out the location of the fault PV module.

Acknowledgment. This research is supported by the grant No. JA14038 and No. JK2014003 from the Educational Department of Fujian Province, the grant No. 2015J05124 from Science and Technology Department of Fujian Province, the grant No. LXKQ201504 from ministry of education of China.

References

1. Sharma, V., Chandel, S.: Performance and degradation analysis for long term reliability of solar PV systems: a review. Renew. Sustain. Energy Rev. **27**, 753–767 (2013)
2. Wang, Y., Li, Z., et al.: Online fault diagnosis of PV module based on BP neural network. Power Netw. Technol. **37**(8), 2094–2100 (2013)
3. Wang, P., Zheng, S.: Fault analysis of solar PV array based on infrared image. Solar J. **31**(2), 197–202 (2010)
4. Li, B.: Research on fault detection method for PV array. Tianjin University (2010)
5. Drews, A., De Keizer, A., et al.: Monitoring and remote failure detection of PV systems based on satellite observations. Sol. Energy **81**(4), 548–564 (2007)
6. Chouder, A., Silvestre, S.: Automatic supervision and fault detection of PV systems based on power losses analysis. Energy Convers. Manag. **51**(10), 1929–1937 (2010)
7. Gokmen, N., Karatepe, E., et al.: An efficient fault diagnosis method for PV systems based on operating voltage-window. Energy Convers. Manag. **73**, 350–360 (2013)
8. Syafaruddin, S., Karatepe, E., et al.: Controlling of artificial neural network for fault diagnosis of photo voltaic array. In: 2011 16th International Conference on Intelligent System Application to Power Systems (ISAP). IEEE (2011)
9. Spataru, S., Sera, D., et al.: Detection of increased series losses in PV arrays using fuzzy inference systems. In: 2012 38th IEEE Photo Voltaic Specialists Conference (PVSC). IEEE (2012)
10. Papageorgas, P., Piromalis, D., et al.: Smart solar panels: in-situ monitoring of photo voltaic panels based on wired and wireless sensor networks. Energy Procedia **36**, 535–545 (2013)
11. Ando, B., Baglio, S., et al.: SENTINELLA: a WSN for a smart monitoring of PV systems at module level. In: 2013 IEEE International Workshop on Measurements and Networking Proceedings (M&N). IEEE (2013)
12. Ducange, P., Fazzolari, M., et al.: An intelligent system for detecting faults in photo voltaic fields. In: 2011 11th International Conference on Intelligent Systems Design and Applications (ISDA). IEEE (2011)

Elitist Quantum-Inspired Differential Evolution Based Wrapper for Feature Subset Selection

Vadlamani Srikrishna[1], Rahul Ghosh[2], Vadlamani Ravi[3(✉)],
and Kalyanmoy Deb[4]

[1] Indian Institute of Technology Bombay, Mumbai 400076, India
vadsrikrishna@gmail.com
[2] Indian Institute of Technology Guwahati, Guwahati 780139, India
rahul.ghosh@iitg.ac.in
[3] Institute for Development and Research in Banking Technology,
Hyderabad 500057, India
rav_padma@yahoo.com
[4] Michigan State University, East Lansing, MI 48824, USA
kdeb@ecs.msu.edu

Abstract. In a Feature Subset Selection (FSS) problem, the objective is to obtain an optimal feature subset on which the learning algorithm can focus and neglect the irrelevant features. A wrapper formulates the FSS as a combinatorial optimization problem. In this paper, we propose an elitist quantum inspired Differential Evolution (QDE) algorithm for FSS. The performance of QDE is found to be significantly better than that of Binary Differential Evolution (BDE) algorithm on three benchmark problems taken from literature. In both cases, logistic regression was chosen as the classifier. Further, QDE outperformed not only the extant algorithms reported in literature but also the t-statistic cum logistic regression based filter method for FSS.

1 Introduction

Feature Subset Selection (FSS) is an important stage in data mining. The objective of FSS is to identify a set of relevant features, while discarding all the irrelevant ones. FSS has important ramifications, namely, shorter computational time, simplification of a model, better human comprehensibility, shorter length 'if-then' rules in decision trees and the possible avoidance of over-fitting. Algorithms to perform FSS can be divided into two main categories: filter and wrapper based.

Wrapper based feature subset selection [1] involves searching for feature subsets through a space of possible feature subsets and scoring each subset based on its fitness value. Each new feature subset generated is used to train a model and the fitness value, which in classification problems is classification accuracy on the test-set, is computed. Obtaining the feature subset with the optimum fitness values is formulated as a combinatorial optimization problem, which is solved using various optimization techniques. In this context, evolutionary optimization techniques can be preferred because of their superior properties. Among them, differential evolution (DE), proposed by

© Springer International Publishing Switzerland 2015
A. Bikakis and X. Zheng (Eds.): MIWAI 2015, LNAI 9426, pp. 113–124, 2015.
DOI: 10.1007/978-3-319-26181-2_11

Storn and Price [2, 3] has been shown to be more robust, faster in convergence and more efficient than conventional Evolutionary Algorithms (EA) [2, 3]. DE explores the search space using a simple mutation and crossover function after which a greedy algorithm is used to optimize the objective function. Proper selection of the DE parameters has a large impact on the optimization performance.

Since DE was proposed to solve optimization problems in continuous space, it was modified so that it could solve combinatorial optimization problems. One such algorithm is Binary Differential Evolution (BDE) [4], which uses the sigmoid function to map the continuous space to the binary space. A Binary Version of DE algorithm (BDE) was also proposed to solve 0-1 Knapsack problem [5].

The use of ideas from Quantum Computing has been proposed to make the algorithms dramatically faster and extremely parallel. These have been shown to produce superior performance compared to their classical counterparts [6, 7]. Thus, Quantum inspired Evolutionary Algorithms (QEA) [8] and Quantum Inspired Particle Swarm Optimization (QPSO) [9] were developed for continuous and binary space.

In quantum inspired evolutionary algorithms, the individuals or solutions are modeled as vectors of quantum bits (Qubits), which are obtained by superposing binary bits. Quantum superposition and Quantum probabilities along with a set of optimized parameter values speed up the search process for an optimal solution. QEA has been used for solving many combinatorial optimization problems like optimal feature selection based on clone genetic strategy for text categorization [10], N-Queens problem [11], while QPSO has been used for feature selection and parameter optimization in the context of evolving spiking neural network [12]. An adaptive quantum inspired Differential Evolution (AQDE) algorithm was proposed by Hota and Pat [13] for solving the 0-1 knapsack problem.

In this paper, an elitist Quantum Inspired Differential Evolution (QDE) Algorithm is proposed to obtain an optimal subset of features for classification tasks using Logistic Regression for fitness function evaluation.

The rest of the paper is structured as follows: Sect. 2 presents a brief review of literature and overview of wrapper based FSS, DE, BDE and QDE. Section 3 presents the proposed algorithm in detail. Section 4 presents the results and discussion, while Sect. 5 concludes the paper.

2   Literature Review

2.1   Wrapper Based Feature Subset Selection

Wrappers for FSS, first proposed by Kohavi & John [1], use a search algorithm that searches for a good subset, whose strength is then evaluated using fitness function. Each state in the model represents a feature subset. The number of bits in each state is same as the number of features and each bit indicates the presence (1) or absence (0) of the feature in that state. New feature subsets are found and evaluated until the convergence is reached (all the solutions have accuracy higher than a certain threshold) or the maximum number of pre-specified iterations is over. Many search algorithms have been used like the hill-climbing search engine and the best-first search [11].

2.2   Differential Evolution

In the family of evolutionary algorithms, DE is a stochastic, population based optimization algorithm which, like any genetic algorithm, consists mainly of three steps: mutation, crossover and selection. DE maintains a population of solution vectors and each vector undergoes all the three steps in an iteration of the algorithm. A mutant vector is generated for each target vector in the population. The mutant vector and the target vector undergo crossover to generate a trial vector. Finally, the selection step chooses the vector for the next generation of population. This process is repeated until convergence is reached (all the solutions have accuracy higher than a certain threshold) or the maximum number of pre-specified iterations is over.

In a d-dimensional search space, a fixed number of vectors corresponding to the population size is initialized. In each generation t, a new mutant vector (M_i^t) is generated for every target vector (X_i^t) using the following rule:

$$M_i^t = X_{i1}^t + F\left(X_{i2}^t - X_{i3}^t\right) \tag{1}$$

where X_{i1}^t, X_{i2}^t and X_{i3}^t are three solution vectors randomly chosen from the current generation all distinct from each other as well as X_i^t. F is a user defined control parameter chosen in the interval [0, 1].

In the crossover step, each target vector undergoes mating with the mutant vector to generate a trial vector (U_i^t). The recombined vector is generated by following simple crossover rule:

$$u_{i,j}^t = \begin{cases} m_{i,j}^t, & if\left(rand_{(0,1)}[j] \leq CR\right) \vee \left(j = I_{rand}\right) \\ x_{i,j}^t, & if\left(rand_{(0,1)}[j] > CR\right) \wedge \left(j \neq I_{rand}\right) \end{cases} \tag{2}$$

where j = 1,2,...d, $u_{i,j}^t$ is the jth bit of U_i^t, $rand_{(0,1)}[j]$ is the jth evaluation of a random number generator in the interval [0, 1] from a uniform distribution. I_{rand} is randomly chosen index from 1,2,... d to ensure that U_i^t and X_i^t are distinct. CR is a control parameter of the DE and typically chosen between 0 and 1.

The last stage is the selection step where a greedy algorithm is followed while selecting a solution vector for the population. The fitness of both the target vector and the trial vector is calculated and the fitter vector is allowed to form the population for the next generation.

2.3   Binary Version of Differential Evolution

Due to the good performance of DE, an algorithm was developed which enabled the use of DE in binary space. It is inspired by the Discrete Version of Particle Swarm Optimization (DPSO). The proposed algorithm maintains a population of d-dimensional binary strings as the solution individuals.

The components of the DE-individuals are initialized as floating-point numbers in the interval [0, 1] following uniform distribution. A sigmoid function is then used to map these floating-point vectors to binary vectors according to Eq. (3). A mutation operator that is similar to that of DE is used. There is a possibility that the resultant mutant vector (M_i^t) is no longer binary. The same sigmoid based discretization process from continuous to binary space is used.

$$m_{i,j}^t = \begin{cases} 1, if \ rand_{(0,1)} \leq sig\left(m_{i,j}^t\right) \\ 0, if \ rand_{(0,1)} > sig\left(m_{i,j}^t\right) \end{cases} \tag{3}$$

where $m_{i,j}^t$ is the jth bit of M^t, rand(0, 1) is a random number uniformly selected in the range [0, 1] and function sig() is the sigmoid mapping function. Then, the same crossover and selection operator as that of the DE are employed. There are no other stand-alone binary evolutionary algorithms applied to this problem in literature.

2.4   Quantum Inspired Differential Evolution

These algorithms are designed to run on classical computers but as the name suggests they draw their ideas from the theory of quantum computing. In a quantum computer smallest bit of information is called Quantum-Bit (Q-bit or qubit) and denoted as $[\alpha \ \beta]^t$ where α and β are complex numbers. $|\alpha^2|$ denotes the probability that the qubit will be in state '0' and $|\beta^2|$ denotes the probability that it will be in state '1'. Since $|\alpha^2| + |\beta^2| = 1$, this represents the equation of a circle and each qubit can be represented by a single variable θ defined between [0, 2π] Representation for an individual s of d-dimensions in QDE will be as follows:

$$s = [\theta_1 \theta_2 \theta_3 \ldots \theta_d] \tag{4}$$

Hence, each individual s in the population is a vector containing d qubits. Each solution in the population is initialized randomly uniform distribution in the interval [0, 2π]. The discretization process used to map to the binary space is as follows:

$$s_{i,j} = \begin{cases} 1, if \ rand_{(0,1)} < sin^2\left(\theta_{i,j}\right) \\ 0, if \ rand_{(0,1)} > sin^2\left(\theta_{i,j}\right) \end{cases} \tag{5}$$

where $s_{i,j}$ is the jth bit of the ith individual in the population and $\theta_{i,j}$ is the corresponding qubit.

A mutation operator, that is same as that of DE, mutates the solution individuals, the only difference being that it now operates on individuals which have qubits as components. The crossover operator operates on the mutated individuals and the population individuals and in the selection process, population is updated by following a greedy method based on the accuracy of the individuals.

3 Proposed Methodology

Proposed wrapper based FSS using QDE is presented here. Figures 1 and 2 depict the pseudo code and the schematic of the QDE respectively. In Fig. 2, first QDE is initialized with a random feature subset, whose fitness (accuracy) is evaluated using logistic regression and each solution is tested for elitism. If the solutions are not elite, then they are subjected to the heuristics of QDE indicated by the dotted box. Otherwise, they are added to the elite population. Selected solutions of a population that underwent the QDE operations are used to update the population. If the maximum number of iterations is reached (indicated by MAXITR) then the algorithm converges. Else, we test the elitism for the population members again and loop continues until convergence.

```
Procedure
t ← 0
initialize the population Q(t)
   make S(t) using (5)
evaluate fitness of S(t)
while t < Max_Iter do
t ← t+1
        use (6) for the mutation step to get M(t)
        use (7) for the crossover step to get Q'(t)
        make S'(t) using (5)
evaluate fitness of S'(t)
update S(t+1) and Q(t+1) using (8)
end while
end Procedure
```

Fig. 1. Pseudo code of the QDE based wrapper for FSS

A. Initialization: Each individual of the population consists of qubits represented by θ which are initialized by a uniform random number in the interval $[0, 2\pi]$ The binary vector corresponding to each individual is derived according to (5).

B. Mutation Step: The mutation step is similar to that of DE and operates on the individuals of the population having qubits as components to generate a population of mutated vectors (M^t). Mutated vector θ_i^m corresponding to the ith target individual (θ_i) in the population is generated according to the following rule:

$$\theta_i^m = \theta_{i1} + F(\theta_{i2} - \theta_{i3}) \tag{6}$$

where the three individuals are chosen randomly from the population such that i_1, i_2, i_3 and i are distinct. F is the mutation parameter chosen between 0 and 1.

C. Crossover Step: In the crossover step, a new trial vector (θ_i^c) is generated from the target individual and the mutated individual according to the Eq. (7):

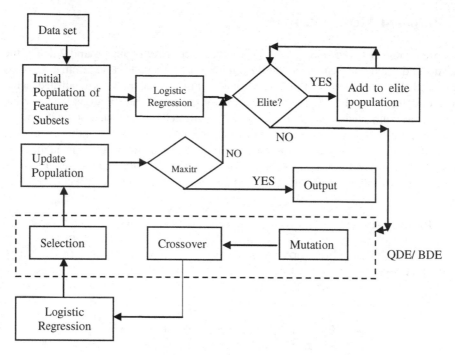

Fig. 2. Schematic of the proposed QDE based wrapper for FSS

$$\theta_{i,j}^c = \begin{cases} \theta_{i,j}^m, if\left(rand_{(0,1)}[j] \le CR\right) \vee (j = I_{rand}) \\ \theta_{i,j}, if\left(rand_{(0,1)}[j] > CR\right) \wedge (j \ne I_{rand}) \end{cases} \tag{7}$$

where $\theta_{i,j}^c$ the jth bit of the ith individual θ_i^c formed after the crossover step, I_{rand} is a random integer chosen from [1,2,3....D], $rand_{(0,1)}[j]$ is the jth random number from uniform distribution [0,1] and CR is the crossover rate chosen from [0,1].

D. Selection Step: The target (θ_i) and the trial (θ_i^c) individual are mapped to the binary space to generate binary vectors (V_i^t and V_i^{ct}) corresponding to each of the individuals using the mapping rule given in (5). The feature subsets corresponding to the binary vectors are obtained and used to train a logistic regression model. The population is updated using a greedy algorithm by observing the classification accuracy. Population is updated according to the Eqs. (8a) and (8b):

$$V_i^{t+1} = \begin{cases} V_i^{ct}, if\ f\left(V_i^{ct}\right) > f\left(V_i^t\right) \\ V_i^t, if\ f\left(V_i^{ct}\right) < f\left(V_i^t\right) \end{cases} \tag{8a}$$

$$\theta_i^{t+1} = \begin{cases} \theta_i^c, if\ f\left(P_i^{ct}\right) > f\left(P_i^t\right) \\ \theta_i, if\ f\left(P_i^{ct}\right) < f\left(P_i^t\right) \end{cases} \tag{8b}$$

Where $f\left(V_i^t\right)$ is the classification accuracy of the feature subset corresponding to the ith individual. θ_i^{t+1} and V_i^{t+1} belongs to the next generation (t + 1).

We introduced the concept of elitism in both QDE and BDE, which ensures that a solution undergoes the process of mutation, cross-over and selection only if its accuracy is less than a certain threshold, which can be set by the user.

4 Experimental Setup and Results

The Wrapper based FSS using QDE as well as BDE were implemented and the results were compared. Both the algorithms, implemented in C++, were tested for their effectiveness on three different benchmark data sets namely (i) Parkinson Data Set [14] having 197 instances and 23 features (ii) Breast Cancer Wisconsin (Diagnostic) Data Set [15] having 569 instances and 32 features and (iii) Ionosphere Data Set [15] having 351 instances and 34 features.

We followed stratified random sampling and tested the algorithms using the 10-fold cross validation method. Further, both the algorithms were run 30 times. The population size is fixed at 30 for all the data sets. Thus, we conducted 300 experiments for each dataset resulting in the evaluation of fitness values of 9,000 individuals for each data set. We also used the elitism concept in both algorithms, according to which a solution undergoes the mutation, crossover and selection steps only if its accuracy is less than 90 %. The same parameter values, after testing several combinations, were used for QDE and BDE, which are as follows: For Parkinson Data Set and Breast Cancer Wisconsin (Diagnostic) Dataset, F and CR are fixed at 0.4 and 0.6 respectively. For the Ionosphere Dataset, F and CR are fixed at 0.7 and 0.6 respectively. The maximum number of iterations is fixed at 200 for both QDE and BDE across all data sets.

4.1 Feature Subsets and Accuracy

In this subsection, the feature subsets with maximum accuracy are selected and out of those, subsets with the minimal most length are presented in Table 1. It is observed that in all three datasets, QDE yielded optimal feature subsets whose cardinality is less than that obtained by BDE. It is a remarkable achievement of QDE. Further, there are many feature subsets, albeit with some common features, which yielded the highest accuracy for QDE. However, this number is far too less for BDE in the case of Parkinsons data (see Table 1). In the case of the other two datasets, this number is the same for both QDE and BDE. One possible reason could be that the Parkinson's data seems to have multi modal solutions, which is difficult to verify though.

4.2 Convergence and Function Evaluations

The proposed algorithm was also compared to its binary counterpart based on the number of function evaluations. The results presented in Table 4 indicate that the average function evaluations consumed by QDE over all 30 runs is nearly 2/3 times that of BDE in all datasets. This is another interesting achievement of QDE.

Table. 1. Parkinsons dataset

Algorithm	Feature subset	Accuracy
QDE	1,8,9,11	100
	3,13,17,19	100
	3,5,11,15	100
	8,16,17,19	100
	8,11,19,20	100
	5,7,19,20	100
	2,8,19,20	100
	1,3,4,22	100
	12,13,15,16	100
	2,6,8,16	100
	4,9,11,16	100
	6,10,12,22	100
	4,9,14,16	100
	6,11,15,16	100
	1,10,11,19	100
	3,7,19,20	100
	3,7,13,19	100
BDE	1,8,13,16,19	100
	2,3,10,15,21	100
	3,4,7,19,20	100
	1,4,12,16,21	100
	1,2,8,11,16	100
	1,5,10,16,22	100
	2,10,16,17,19	100

Table. 2. Breast cancer wisconsin (diagnostic) data set

Algorithm	Feature subset	Accuracy
QDE	1,3,14,19,21	100
BDE	1,4,7,8,12,16,17,20,24,30	100

Table. 3. Ionosphere dataset

Algorithm	Feature subset	Accuracy
QDE	5,8,12,27,28,29,31	94
BDE	3,5,6,12,13,15,20,21,24,27,30,33,34	88

4.3   Repeatability

In this subsection, all the feature subsets with the highest accuracy are considered, the frequencies of occurrence of a feature in those subsets are calculated and the features are sorted accordingly. In order to assess the performance of QDE over that of BDE on

Table. 4. Average function evaluations

Data set	QDE	BDE
Parkinson	203763	317279
Breast cancer	200412	307843
Ionosphere	290086	355442

Table. 5. Repeatability of features of QDE

Data set	Features
Parkinson	1 3 4 5 9 11 13 19 20
Breast Cancer	1 4 7 11 14 21 23
Ionosphere	1 3 4 5 8 11 12 17 27 28 29 31

Table. 6. Repeatability of features of BDE

Data set	Features
Parkinson	1 2 4 7 11
Breast cancer	1 4 7 11 24
Ionosphere	1 4 11 17 22 28 32 34

this metric, we selected those features which had at least 50 % frequency of occurrence for all the data sets. The resulting features are presented in Tables 5 and 6.

Let us define S and S' as the union of all the maximum accuracy achieving subsets obtained on the Parkinson data by QDE and BDE respectively. Now, among all the features in S, the number of features that appear in more than half of the maximum accuracy achieving solutions is 9, as is evident from Table 5. However, from Table 6, we observe that the number of features of S' which appear in more than half of the solutions is 5. In the other datasets too, the number of features presented in Table 5 is greater than that presented in Table 6. This observation can be explained as follows.

If we take a set of solutions obtained by BDE, we will find significant variations between the elements of this set. This is because only a few features are obtained frequently and the others keep changing as we run through the solutions one after the other. As the frequency of appearance of features is less, we will face difficulties in finding out which features are the most discriminating. On the other hand, a set of solutions obtained by QDE is likely to have greater homogeneity, i.e., most of the solutions will have a significant number of features in common. This helps us find out which features are the most useful for classification, thereby helping us achieving the aim of FSS problem.

Moreover, every feature in the subsets presented in Table 1 for the Parkinson data set is repeated in at least 33 % of solutions. Similarly, the percentage turns out to be 17 % and 37 % for the Breast Cancer Wisconsin data set (Table 2) and Ionosphere data set (Table 3) respectively. Meanwhile the figures reported by BDE are 20 %, 25 % and 14 % for Parkinson, Breast Cancer Wisconsin and Ionosphere data set respectively.

We also compared the accuracy of QDE with that of existing methods for the data sets and QDE proved to be the better performer. The result is evident as 97 % accuracy is achieved in Breast Cancer Wisconsin data set by Street et al. [16]. Similarly 95 % accuracy was reported by Tsanas et al. [17] on the Parkinsons data set, while Sigillito et al. [18] obtained 96 % accuracy with a back-propagation trained neural network on the Ionosphere data set. However, all these studies did not employ 10-fold cross validation.

Table 7 presents the results of a filter based feature selection method viz., t-statistic [19] where, again, logistic regression is used as a classifier. We chose the top 5 features for the Parkinson's data and Wisconsin Breast Cancer data, while for the ionosphere data we selected top 7 features, just to be in line with the number of features chosen by QDE and BDE. The t-statistic based filter yielded 100 % accuracy in just one fold in the case of Parkinson's data and Wisconsin Breast Cancer data, while in the case of Ionosphere data it yielded 91.67 % only in just one fold. However, QDE produced many subsets with 100 % accuracy in the case of Parkinson's and Wisconsin Breast Cancer datasets, while it simply outperformed t-statistic based filter in the case of Ionosphere dataset. It is interesting to note that some of the features selected by QDE as the most repeatable features are also among the top 10 features selected by t-statistic.

Table. 7. Features obtained by t-statistic

Data set	Features
Parkinson	9 12 13 19 22
Breast cancer	3 8 21 23 28
Ionosphere	1 4 6 8 22 29 31

In the case of QDE, each individual is made up of N qubits (N is the number of features in our case). Each qubit is a linear combination of the 0 and 1 states. Hence, each individual whose constituent qubits have angles, say, strictly between 0 and $\pi/2$, has some 'information' of all the 2^N solutions in the solution space. Depending on the individual angles, the 'amount' of information of each of those solutions in the individual will differ. Hence, the mutation and crossover operations make use of information from large areas of the search space to produce a crossed-over individual.

Furthermore, it is clear from the Table 4 that the average number of iterations required by QDE to converge to the optimal solutions is much less than that required by the BDE algorithm. Also, the BDE failed to obtain many of the solutions obtained by QDE. These observations can be rationalized as follows. It should be noted that QDE depends on the function Cos^2 (.) for observing, while BDE depends on sigmoid for transforming the real individuals to binary ones. Therefore, QDE has better search capability compared to that of BDE because of the oscillatory nature of Cos^2 (.) as opposed to the monotonically increasing nature of sigmoid in BDE. In other words, massive amount of exploration taking place in QDE resulted in its superior performance and less function evaluations compared to BDE.

5 Conclusions

In this paper, we proposed a novel FSS wrapper based on elitist QDE. The performance of the proposed algorithm was compared with that of the BDE on three datasets taken from literature. Owing to the inherent superior search capability, QDE outperformed BDE on many counts namely, higher accuracy, less iterations, obtaining multi-modal solutions, obtaining feature subsets with minimal cardinality. QDE also outperformed the t-statistic based filter type FSS method. The future directions include incorporating other concepts of quantum computing viz., quantum rotation using gates, entanglement etc. in the present software and investigating its utility for large scale practical problems. Further, it is easily seen that many other combinatorial optimization problems can also be solved by using the proposed method.

References

1. Kohavi, R., John, G.: Wrappers for feature selection. Artif. Intell. **1–2**, 273–324 (1997)
2. Storn, R., Price, K.: Differential evolution-a simple and efficient heuristic for global optimization over continuous spaces. J. Global Optim. **11**(4), 341–359 (1997). Kluwer Academic Publishers
3. Storn, R., Price, K.: Differential Evolution-a Simple and Efficient Adaptive Scheme for Global Optimization over Continuous Spaces. Technical Report TR-95–012, ICSI, ftp.icsi. berkeley.edu (1995)
4. Engelbrecht, A., Pampara, G.: Binary differential evolution strategies. In: Proceedings of the IEEE Congress on Evolutionary Computation, pp. 1942–1947 (2007)
5. Peng, C., Jian, L., Zhiming, L.: Solving 0-1 knapsack problems by a discrete binary version of differential evolution. In: Second International Symposium on Intelligent Information Technology Application, pp. 513–516 (2008)
6. Grover, L.K.: A fast quantum mechanical algorithm for database search. In: Proceedings of the 28th ACM Symposium on Theory of Computing, pp.212–219 (1996)
7. Shor, P.: Polynomial-time algorithms for prime factorization and discrete logarithms on a quantum computer. SIAM J. Comput. **26**, 1484–1509 (1997)
8. Han, K.-H., Kim, J.-H.: Quantum-inspired evolutionary algorithm for a class of combinatorial optimization. IEEE Trans. Evol. Comput. **6**(6), 580–593 (2002)
9. Sun, J., Feng, B., Xu, W.B.: Particle swarm optimization with particles having quantum behavior. In: proceedings of the IEEE Congress on Evolutionary Computation, pp. 325–331 (2004)
10. Chen, H., Zou, B.: Optimal feature selection algorithm based on quantum-inspired clone genetic strategy in text categorization. In: Proceedings of the First ACM/SIGEVO Summit on Genetic and Evolutionary Computation (2009)
11. Draa, A., Meshoul, S., Talbi, H., Batouche, M.: A quantum inspired differential evolution algorithm for solving the N-queens problem. Int. Arab J. Inf. Technol. **7**(1), 21–27 (2010)
12. Hamed, H.N.A., Kasabov, N.K., Shamsuddin, S.M.: Quantum inspired particle swarm optimization for feature selection and parameter optimization in evolving spiking neural networks for classification tasks. In: Evolutionary Algorithms, vol. 1, pp. 132–148 (2011)
13. Hota, A.R., Pat, A.: An adaptive quantum-inspired differential evolution algorithm for 0-1 knapsack problem. Comput. Sci. Neural Evol. Comput. **1**, 2–8 (2011)

14. Little, M.A., McSharry, P.E., Roberts, S.J., Costello, D.A., Moroz, I.M.: Exploiting nonlinear recurrence and fractal scaling properties for voice disorder detection. Biomed. Eng. Online **6**(1), 23 (2007)
15. Lichman, M.: UCI Machine Learning Repository. [http://archive.ics.uci.edu/ml]. Irvine, CA: University of California, School of Information and Computer Science (2013)
16. Street, W.N., Wolberg, W.H., Mangasarian, O.L.: Nuclear feature extraction for breast tumor diagnosis. In: International Symposium on Electronic Imaging: Science and Technology, IS&T/SPIE 1993, vol. 1905, pp. 861–870, San Jose, CA (1993)
17. Tsanas, A., Little, M.A., McSharry, P.E., Ramig, L.O.: Accurate telemonitoring of parkinsons disease progression using non-invasive speech tests. IEEE Trans. Biomed. Eng. **57**, 884–893 (2010)
18. Sigillito, V.G., Wing, S.P., Hutton, L.V., Baker, K.B.: Classification of radar returns from the ionosphere using neural networks. Johns Hopkins APL Tech. Digest **10**, 262–266 (1989)
19. Liu, H., Li, J., Wong, L.: A comparative study on feature selection & classification methods using gene expression profiles and proteomic patterns. Genome Inform. **13**, 51–60 (2002)

Towards Efficiently Mining Frequent Interval-Based Sequential Patterns in Time Series Databases

Phan Thi Bao Tran[✉], Vo Thi Ngoc Chau, and Duong Tuan Anh

Faculty of Computer Science & Engineering,
HCMC University of Technology, Ho Chi Minh, Vietnam
{tranptb, chauvtn, dtanh}@cse.hcmut.edu.vn

Abstract. Nowadays time series mining has been taken into account in many various application domains. One of the most popular mining tasks in the existing works is the frequent pattern mining task on time series databases. Periodic patterns in a time series are often examined in this task. Such patterns help us understand more the corresponding object observed on a regular basis. As we extend our consideration to a group of many different objects to find out their common behaviors repeating over time, we need a pattern type to be more informative and thus a solution to discover the hidden patterns. Therefore, our work aims at so-called interval-based sequential patterns frequently in a time series database. We also provide two different solutions to mining such frequent patterns: the first one based on the existing ARMADA solution with the additional preprocessing and post-processing and the second one based on our new FITSPATS algorithm with the use of stems as suffix expansion and a temporal pattern tree. Experimental results have shown that our solutions are capable of discovering the frequent interval-based sequential patterns in a time series database and the FITSPATS algorithm is more effective and efficient for the task.

Keywords: Interval-based sequential pattern · Frequent pattern mining · Time series mining · Temporal pattern tree · Support

1 Introduction

In 1995, sequential pattern mining has been introduced by Agrawal et al. in [1]. After that, many other sequential pattern mining algorithms were defined. It is realized that most of the works on sequential pattern mining went through sequences of ordered events with no consideration for explicit temporal information of the single events in the database. Recently some of them listed in [5, 7, 13, 15] have focused on more temporal aspects of the resulting patterns. In [7], the authors proposed a sequential pattern mining algorithm for generalized sequential pattern mining with item intervals. In [5], the authors defined a coincidence representation and CTMiner algorithm to mine frequent time-interval based patterns. As for [15], the authors considered to find all interesting periodic patterns with various gaps from matching between a given pattern and a given character sequence. In [13], two algorithms MAPB and PAPD with an incomplete Nettree structure were introduced to discover frequent periodic patterns

© Springer International Publishing Switzerland 2015
A. Bikakis and X. Zheng (Eds.): MIWAI 2015, LNAI 9426, pp. 125–136, 2015.
DOI: 10.1007/978-3-319-26181-2_12

with wildcard gaps. Different from the works on sequence databases, our work has formulated a concept of frequent interval-based sequential pattern on time series databases where our patterns are more informative with various time gaps and various subsequences of different lengths automatically discovered from the database. To the best of our knowledge, the type of frequent interval-based sequential patterns defined in our work has not yet been received much consideration from the existing works. Therefore, we have designed two different approaches with the corresponding frequent pattern mining algorithms on time series databases: the first one based on the existing work in [12] named the ARMADA-based algorithm and the second one based on a temporal pattern tree combined with the idea of using the stems in [12] for a more efficient solution named the FITSPATS algorithm.

As compared to the existing works [2, 4, 9, 10, 16] that performed the frequent pattern mining task on time series to find periodic patterns, our work provides the patterns also more informative. This is because periodic patterns help us understand more the corresponding object observed on a regular basis. As we extend our consideration to a group of many different objects to find out their common behaviors repeating over time, we need a pattern type to be more informative and thus a solution to discover the hidden patterns. Such a need is one of our motivations.

For more consideration on the pattern and rule mining task on time series databases, [3] has recently focused on discovering recent temporal patterns from interval-based sequences of temporal abstractions with two temporal relationships: before and co-occur. Mining recent temporal patterns in [3] is one step in learning a classification model for event detection problems. Different from [3], our work belongs to the frequent pattern mining task. Indeed, we would like to discover more complex frequent temporal patterns in many different time series with temporal relationships. For more applications, such patterns can be used in other time series mining tasks such as clustering, classification, and prediction in time series. Based on the temporal concepts of duration, coincidence, and partial order in interval time series, [8] defined pattern types in a hierarchy with levels corresponding to the temporal concepts duration, coincidence, and partial order. Such a kind of patterns is different from that considered in our work where frequent interval-based sequential patterns are of interest to find the repeating group of subsequences in the time series database. For another form of patterns, [6] aimed to capture the similarities among stock market time series such that their sequence-subsequence relationships are preserved. In particular, [6] identified patterns representing collections of contiguous subsequences which shared the same shape for specific time intervals. Their patterns show pairwise similarities among sequences, called timing patterns using temporal relationships. As compared to [6], our work supports temporal relationships with explicit time. In addition, we discovered those relationships between the subsequences, which repeated in the number of time series in the database that is greater than or equal to the minimum support count. More recently, [14] provided a method to detect all repeated patterns in a time series by using the actual suffix arrays with Moving Longest Expected Repeated Pattern on external media. In contrast to the pattern kind related to a single time series, the frequent patterns in our work are discovered from a set of the time series where these patterns occur frequently in many different time series. As for [11], we figured out that the task in [11] was pattern matching rather than pattern mining because the work in [11] aimed at finding hidden patterns from time series data sets and a set of predetermined patterns. In our work, there was no such a set of predetermined patterns. In [17], a new type of patterns in multivariate time series is defined as temporal associations to capture a wide range of local relations along and across individual time series. As compared to our patterns, a pattern in [17] is a short sequence of consecutive characters along the lifespan of a time series while ours are composite ones with explicit temporal relationships between the elements.

In short, the main concentration of our work is to find an effective and efficient solution to the frequent interval-based sequential pattern mining task on time series databases. Several experiments have been conducted with two aforementioned solutions and their results show that both ARMADA-based algorithm and FITSPATS algorithm are capable of discovering the frequent interval-based sequential patterns from a time series database with respect to a minimum support. In addition, the effectiveness and efficiency of our new FITSPATS algorithm have also been confirmed.

2   A Frequent Pattern Mining Task on Time Series Databases

Before going into detail over a frequent pattern mining task on time series databases in this paper, we provide several fundamental concepts as follows.

2.1   Fundamental Concepts

Definition 1 (*time series database* (**DB**)). A time series database is a collection of time series such that: $DB = \{ts_1 \dots ts_n\}$. Each time series ts_i has the same length $|TS|$. $|DB|$ is the size of the database DB which is the number of time series in DB. As preprocessing for normalization, dimensionality reduction, discretization, etc. is excluded from our work, each time series in DB is given as a symbolic trend-based time series. Without loss of generality, in this work, we use a set of symbols $\Sigma = \{D, I, U\}$ to represent the popular main trends in time series where D is used for a decreasing trend, I for an increasing trend, and U for an unchanged trend.

Example 1: An example of a symbolic trend-based time series database DB is given in Table 1 with $|DB| = 3$ and $|TS| = 10$.

Table 1. A symbolic trend-based time series database DB

Identifier	0	1	2	3	4	5	6	7	8	9
ts_0	U	I	I	D	I	U	I	I	I	D
ts_1	D	I	U	I	D	I	D	U	D	U
ts_2	U	D	I	U	D	U	I	D	I	U

Definition 2 (*k-pattern* (**p**)). A k-pattern p is defined as a collection of k subsequences sequentially connected to each other by a "before" relationship with explicit time gaps. Association information of p includes time series indices that p belongs to and occurrence positions in each of its corresponding time series. In addition, we call each

k-pattern in our work an interval-based sequential *k*-pattern due to the order of its interval-based elements and relationships between the elements.

Example 2: I b4 D (Cnt = 3, ts[0][4], ts[1][1, 3], ts[2][2]) is a 2-pattern that includes two subsequences *I* and *D* showing that *I* occurs before *D* four time units starting at position 4 in time series 0 denoted as *ts[0][4]*, position 1 and 3 in time series 1 as *ts[1]* *[1, 3]*, position 2 in time series 2 as *ts[2][2]*. The number of occurrences of this 2-pattern denoted as *Cnt* is 3.

Definition 3 (*subpattern* (*p*)). A pattern *p* is a *subpattern* of pattern *p'* if *p* can be obtained by removing some subsequences from pattern *p'*.

Definition 4 (count (*cnt*)). A count cnt_p of pattern *p* is the number of time series that contain *p*.

Definition 5 (*support* (*σ*)). A support σ_p of pattern *p* is a percentage of cnt_p over the number of time series in a database sized $|DB|$, i.e. $\sigma_p = cnt_p/|DB|$.

Definition 6 (*frequent pattern* (*p*)). Given a minimum support *minsup*, a pattern *p* is a *frequent* pattern if its support *σ* is greater than or equal to *minsup*.

Definition 7 (*stem*). A *stem* is a subsequence which is added to the end of a *k*-pattern *p* to form a [*k* + 1]-pattern *p'*. That is, [*k* + 1]-pattern *p'* = *k*-pattern *p* + *stem*.

Example 3: Given a 1-pattern *D (Cnt = 3, ts[0][3, 9], ts[1][0, 4, 6, 8], ts[2][1, 4, 7])* and a stem *UI*, we obtain a 2-pattern as: *D + UI = D b1 UI (Cnt = 2, ts[0][3], ts[1][0])*.

2.2   A Frequent Pattern Mining Task Definition on Time Series Databases

Given a symbolic trend-based time series database *DB* and a specific minimum support *minsup*, our work focuses on discovering **all** the frequent sequential patterns whose supports are greater than or equal to *minsup*.

In the following, we highlight the characteristics of our resulting patterns:

- Each element in a resulting pattern which is a subsequence in a time series is of any various length up to the knowledge intrinsic in the database. All the elements of each pattern are automatically discovered.
- Each time gap between any two elements in a pattern is also of any various length. All the time gaps in each pattern are also automatically discovered.
- No constraint on the resulting patterns is required to be predetermined except for their frequent occurrences with respect to a specific minimum support.

Besides, in the frequent pattern mining area on sequence databases, each single event has a unit length and at each point in time, a single event or a set of consecutive events is considered for each sequence while in our work, each subsequence represents a composite event with various lengths and at each point in time, many of such composite events are considered in a time series. Hence, if we transform our task into a frequent sequential pattern mining task on a sequence database, the number of different single events to be considered will increase very much, leading to the combinatorial explosion problem. In particular, each time series TS in the database is |TS| long and each single event in the database is from the set of symbols Σ. The total number of subsequences that might exist for consideration is $|\Sigma| * \left(|\Sigma|^{|TS|} - 1\right)/(|\Sigma| - 1)$. Those subsequences are composite events in our contexts and will be mapped to single events in the task on sequence databases. For example, given $|TS| = 10$ and $|\Sigma| = 3$, the total number of mapped single events in the sequential pattern mining task is 88572 and if $|TS| = 100$ then that number will be $1.5*(3^{100} - 1)$, which is extremely large. In addition, a temporal relationship with an explicit time gap is then discovered for any two consecutive composite events in a resulting pattern in a time series and then such a pattern is examined for its repeating occurrence in many different time series in a time series database. So, our patterns are different from periodic patterns in a single sequence or a single time series in [2, 4, 9, 10, 16].

3 Towards Efficiently Mining Frequent Interval-Based Sequential Patterns in Time Series Databases

In order to form a solution to the aforementioned task on time series databases, we approach the task in two directions: the first one called the ARMADA-based algorithm by adapting an existing solution on temporal databases because time series databases are the specific case of temporal databases and the second one called the FITSPATS algorithm by constructing a new straightforward solution on time series databases. These two approaches enable us to find out an appropriate solution to efficiently mining frequent interval-based sequential patterns in time series databases.

3.1   The ARMADA-Based Algorithm

The ARMADA-based algorithm is a solution based on the proposed work in [12] using an index set – a kind of list – to store potential patterns, their occurrence positions, and stems. It followed a level-wise iterative approach with a target of frequent interval-based sequential patterns with explicit time gaps in temporal relationships. As proposed in [12], the original ARMADA algorithm has an interval-based dataset as its input and returns a set of frequent patterns without explicit time gaps between their elements. Therefore, we need a preprocessing phase to transform our time series database into the original format of the input as required and after the execution of the ARMADA algorithm, a post-processing phase is done to further process the result for the final frequent patterns. The details of the ARMADA-based algorithm are below.

In the preprocessing phase, we convert the symbolic trend-base time series DB to an interval-based database using a sliding window. All frequent 1-patterns are then captured in a tree structure. This phase results in a converted interval-based database. The details of the tree structure will be presented in the next subsection.

In the main phase, we invoke the original ARMADA algorithm in [12]. It is worth noting that the original algorithm used an index set, a kind of list, to store only the first occurrence position of 1-patterns and stems. This leads to a lack of the frequent patterns related to those 1-patterns which have other occurrence positions in the time series. It is from the difference between temporal databases and time series databases where the number of combinations gets higher and the search space gets larger than that for a temporal database or a sequential database.

In the post-processing phase, in order to avoid scanning the database once more, we made use of the final lists of index sets of the frequent patterns that do not contain explicit time gaps. For each frequent pattern, we examine its index set for many various time gaps between the elements and then count its occurrences to check if it is frequent with respect to a given minimum support.

3.2   The FITSPATS Algorithm

The FITSPATS algorithm is a frequent interval-based sequential pattern mining algorithm on time series databases that we develop using a so-called temporal pattern tree (TP-tree) to overcome the limitation of the ARMADA-based algorithm on time series databases. In particular, the FITSPATS algorithm performs only one database scan to capture and organize all frequent 1-patterns and their occurrences in the time series into a tree structure called a temporal pattern tree. This tree enables us to compact the database. Based on these frequent 1-patterns in the tree, FITSPATS discovers the other frequent k-patterns for $k \geq 2$ in a level-wise iterative manner and thus, results in a complete set of the frequent interval-based sequential patterns hidden in the given time .series database. In addition, we speed up the process of FITSPATS by means of the looking-ahead and tree pruning mechanisms to reduce the search space for true candidates of the resulting frequent patterns. These mechanisms help us save the cost of the process. The details of the FITSPATS algorithm are described as follows.

Input: a symbolic trend-based time series database DB, a minimum support $minsup$

Output: a list of all the frequent interval-based sequential patterns

The FITSPATS algorithm: mining the frequent interval-based sequential patterns in a time series database

1: Discovering all frequent 1-patterns and stems
2: Mining the other frequent (k+1)-patterns from the frequent k-patterns

The following delineates the mining approach with the FITSPATS algorithm which includes two main phases: Phase 1. Discovering all frequent 1-patterns and stems used in the next phase, Phase 2. Mining the other frequent ($k + 1$)-patterns from k-patterns. The input of phase 1 is the input of the algorithm which includes a time series database DB and a minimum support $minsup$. The output of phase 1 will be fed into phase 2 as an input which is a list of all frequent 1-patterns and stems. The output of phase 2 is also the output of the algorithm which is a set of all frequent interval-based sequential patterns hidden in the database DB.

Phase 1. Discovering all frequent 1-patterns and stems

In this phase, we use a sliding window to scan *DB* and get all subsequences, called potential 1-patterns, present in *DB*. These subsequences are then stored in TP-tree for finding and counting the frequent patterns afterward. A looking-ahead mechanism is applied in this phase. They are detailed as follows.

A temporal pattern tree (TP-tree). This is a tree rooted by "null". Each node has the following fields.

- *Node label:* a subsequence which is a string of symbols {D, U, I}.
- *Cnt:* the number of time series that contains the corresponding pattern.
- *ListOfTimeStamp:* a list of timestamps to capture the occurrence positions of the pattern. Each timestamp includes two parts: the first part is the time series index to which the pattern belongs and the second one contains occurrence positions of the pattern in the corresponding time series.
- *ListOfChild:* a list of child nodes. Child nodes at the same level have been sorted in lexicographic order from left to right.

A child node has a node label obtained from the label of its parent node and extended with one more symbol. For example, node UI is one of the children of the node U.

A looking-ahead mechanism. A looking-ahead mechanism enables us not to scan the rest of time series in the database and not to extend all other subsequences of a subsequence just examined. It is based on checking an *abandon* condition with the current count *cnt* and the rest of time series in *DB* which are remained unchecked with respect to the minimum support *minsup*: an *abandon* condition is *true* if (*cnt* + the rest of time series) < *minsup**|*DB*|.

Example 4: Given *DB* with 10 time series and *minsup* 0.8. After scanning 4 time series, *cnt* of subsequence DID is 1, the number of the rest time series is 6. The abandon condition: (1 + 6) < 0.8*10 is true for DID. So, DID is unable to be frequent. This implies that we do not need to scan and store its extended subsequences DIDU, DIDUD, etc. This also helps us to reduce time and storage cost accordingly.

Scanning the database for the count of each potential 1-pattern. Using the tree and looking-ahead mechanism described above, we proceed to scan the database by means of a sliding window and then, store all subsequences in TP-tree.

Example 5: Given a database in Table 1, we build its corresponding TP-tree below.

After scanning *DB*, we have 19 subsequences in the following order:

U, ts[0][0] $\longrightarrow$	UI, ts[0][0]	UII, ts[0][0]
UIID, ts[0][0]	UIIDI, ts[0][0]	UIIDIU, ts[0][0]
UIIDIUI, ts[0][0]	UIIDIUII, ts[0][0]	UIIDIUIII, ts[0][0]
UIIDIUIIID, ts[0][0]	I, ts[0][1]	II, ts[0][1]
IID, ts[0][1]	IIDI, ts[0][1]	IIDIU, ts[0][1]
IIDIUI, ts[0][1]	IIDIUII, ts[0][1]	IIDIUIII, ts[0][1]
IIDIUIIID, ts[0][1]		

First, we create the root of the tree and label it "null". U is linked as a child of the root. Then UI is linked to U, UII is linked to UI, and so on. The pattern I is linked as a child of the root and sorted before pattern U from left to right. Then II is linked to I, IID is linked to II, etc. The potential 1-patterns are stored in TP-tree shown in Fig. 1. Notice that each node has its own information for each aforementioned field but due to space limitation, we provide the information of node U and node UI in the tree. Using the looking-ahead mechanism, TP-tree in Fig. 1 has several subtrees rooted by a marked (/) node pruned. For instance with a marked node DID, we do not need to store DID and its extended subsequences DIDU, DIDUD, and so on.

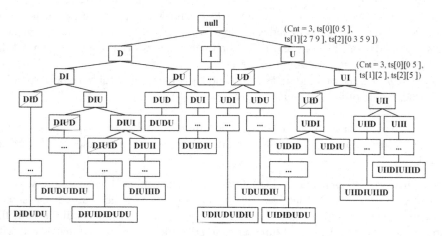

Fig. 1. TP-tree stores all potential 1-patterns and has subtrees rooted by a marked node pruned.

Discovering true frequent interval-based 1-patterns and stems. From the result of the previous step, we remove from the tree all the subsequences which do not satisfy *minsup*. After that, we have a TP-tree storing all frequent 1-patterns. This TP-tree is also a tree of stems used for the next phase. In addition, we extract a list of all frequent 1-patterns from the tree. This list becomes the input of next phase in order to generate frequent $(k + 1)$-patterns for $k \geq 2$. An example of a TP-tree of frequent 1-patterns which is also a tree of stems is given in Fig. 2.

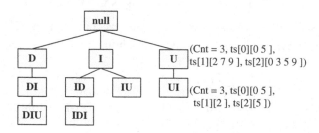

Fig. 2. A tree of frequent 1-patterns which is also a tree of stems

Phase 2. Mining the other frequent $(k + 1)$-patterns from the frequent k-patterns
The main purpose of this phase is to mine the other frequent k-patterns at higher levels starting from the frequent 1-patterns in a level-wise approach for $k \geq 2$.

Firstly, a potential $(k + 1)$-pattern is created by combining a frequent k-pattern p and a stem. Different time gaps between p and a stem create different potential $(k + 1)$-patterns. For example, a frequent 1-pattern D + a stem UI = potential 2-patterns D b1 UI and D b2 UI. The 2-pattern D b1 UI is different from D b2 UI. Such a combination is carried out in a depth-first search strategy over TP-tree of stems. As each stem is in fact a frequent 1-pattern, we expand a frequent k-pattern p by adding a stem to the end of p as p's suffix if there exists at least one time series where a starting point in time of the stem is greater than an ending point in time of p. After that, a resulting potential $(k + 1)$-pattern $p*$ from this expansion has its occurrences checked with respect to the minimum support $minsup$. If satisfying $minsup$, $p*$ is a frequent $(k + 1)$-pattern. Secondly, p is further combined with the child nodes from the current stem until no more combination is made. Otherwise, i.e. if not satisfying $minsup$, we do not need to combine p with all the child nodes in the subtree rooted by the stem. This is our tree pruning mechanism. Finally, we combine p with another stem in a similar way.

Using the combination process, we obtain all frequent $(k + 1)$-patterns from each frequent k-pattern p and the stems. In a recursive manner, we mine the other frequent patterns at the higher levels such as $(k + 2)$, $(k + 3)$, and so on using the discovered frequent patterns and the stems at the levels such as $(k + 1)$, $(k + 2)$, and so on, respectively. The detailed pseudo-code is excluded from our paper due to space limitation.

4 Experimental Results

For further evaluation above approaches, we conducted several experiments by varying the value of a minimum support $minsup$, the number of time series in the database $|DB|$, and the length of time series $|TS|$ in DB. These time series databases are synthetic, symbolic trend-based, and generated randomly with three trends: increasing (I), decreasing (D), and unchanged (U). For real time series databases, we can pre-process them to achieve symbolic trend-based ones using some preprocessing techniques such as normalization, dimensionality reduction, discretization, etc. and then apply the proposed algorithms to discover all frequent interval-based sequential patterns. In each experiment, we record the processing time $Time\#$ in millisecond, the number of candidates $Candidate\#$, and the number of resulting frequent patterns $Pattern\#$. The algorithms were programmed by Visual C# and executed on a Windows 8 PC, Intel core i5, 8 Gb RAM. Experimental results are shown in the Tables 2, 3 and 4.

Through all experimental results, we realize that the number of candidates generated by the ARMADA-based algorithm is very high and especially a few times up to a few hundred times higher than that by the FITSPATS algorithm. This leads to the fact that the ARMADA-based algorithm needs much more processing time than the FITSPATS algorithm does for each experiment. It is interesting to figure out that with less processing time, the FITSPATS algorithm can discover more frequent interval-based sequential patterns that are hidden in any position in the related time series while the

ARMADA-based algorithm cannot. Even though the higher number of the frequent patterns, there are not many corresponding candidates from the FITSPATS algorithm. Thus, the ratio of real frequent patterns to the candidates from the FITSPATS algorithm is much smaller than that from the ARMADA-based one. The reason for such a result is that the FITSPATS algorithm has a tree pruning method and a looking-ahead mechanism more appropriate to process time series databases.

Table 2. Experimental results as $|DB| = 100$, $|TS| = 30$, and varying *minsup* in [0.25, 0.65]

minsup	ARMADA-based algorithm			FITSPATS algorithm		
	Time#	Candidate#	Pattern#	Time#	Candidate#	Pattern#
0.25	3,564,126	102,271,658	173	9,472	62,685	11,758
0.30	3,081,809	69,670,159	91	5,240	40,689	5,933
0.35	2,827,958	55,580,887	55	4,900	34,344	4,592
0.40	2,462,976	41,903,944	40	3,715	29,685	3,572
0.45	2,246,256	31,736,247	39	3,099	25,566	2,662
0.50	2,012,187	24,471,551	39	2,345	21,624	1,828
0.55	1,793,433	19,113,091	37	1,906	18,194	1,167
0.60	1,539,206	13,963,097	35	1,432	14,635	572
0.65	1,402,554	11,369,929	31	1,006	12,794	369

Table 3. Experimental results as $|DB| = 50$, varying $|TS|$ in [5, 35], and *minsup* = 0.5

$|TS|$	ARMADA-based algorithm			FITSPATS algorithm		
	Time#	Candidate#	Pattern#	Time#	Candidate#	Pattern#
5	9	230	4	1	120	4
10	46	4,616	12	10	750	37
15	1,015	46,178	15	31	1,979	88
20	10,141	339,922	28	117	4,258	245
25	87,305	2,375,592	37	384	8,176	771
30	656,508	15,254,313	38	1,010	14,695	1,956
35	4,527,626	94,799,998	39	2,662	23,826	3,807

Regarding the difference in the number of the resulting frequent patterns between the ARMADA-based algorithm and the FITSPATS algorithm, we would like to remind that the original ARMADA algorithm has used an index set to keep the first occurrence position of 1-patterns and stems and thus led to a lack of a capability of discovering all frequent patterns based on those 1-patterns at other occurrence positions in the time series. In contrast, the FITSPATS algorithm is capable of storing all positions of all 1-patterns and stems in the time series and after that, able to mine all desired frequent patterns effectively.

In short, adopting and adapting some existing solution is proved to be possible for mining frequent interval-based sequential patterns in time series databases. However, the existing solution incurs the expense of effort for the preprocessing and post-processing, leading to the expense of time to perform the task. A new direct solution is therefore necessary and confirms the contribution of our work in which appropriate data structures have been defined and algorithms have been designed to efficiently and effectively conduct the task. It is more important when the space of the solution is huge like the frequent interval-based sequential pattern mining task on time series that we discussed in Sect. 2.2.

Table 4. Experimental results as varying $|DB|$ in [20, 90], $|TS| = 30$, and *minsup* = 0.2

$|DB|$	ARMADA-based algorithm			FITSPATS algorithm		
	Time#	*Candidate#*	*Pattern#*	*Time#*	*Candidate#*	*Pattern#*
20	469,696	77,829,347	557	10,569	360,485	78,614
30	724,621	80,417,294	359	8,883	179,692	45,310
40	1,069,812	89,100,539	297	9,073	139,089	36,797
50	1,449,034	98,067,324	268	10,799	122,467	32,402
60	1,797,220	102,812,172	252	13,747	115,644	30,158
70	2,372,990	118,572,757	250	13,259	112,888	29,130
80	2,888,988	128,478,715	251	15,352	109,750	27,655
90	3,482,113	138,954,480	246	16,881	108,799	26,762

5 Conclusions

In this paper, we have considered a frequent interval-based sequential pattern mining task on time series databases. Such patterns help us to get fascinating insights into the common behaviors in many different objects each of which is observed and recorded as a time series. In order to discover all the frequent interval-based sequential patterns hidden in a time series database, we approach the task in two ways: the first one using some existing solution on temporal databases and adapting it to the task on time series databases and the second one defining a new algorithm named FITSPATS as a direct solution. The first approach needs to add the preprocessing and post-processing into the existing solution which is the ARMADA algorithm in [12] while the second one is equipped with the use of stems for extending the frequent patterns at the lower level to those at the higher levels and the invention of a temporal pattern tree to compress the time series database so that the database is scanned once. Experimental results show that both approaches enable us to reach the frequent interval-based sequential patterns in a time series database. Moreover, the experimental results confirm the effectiveness and efficiency of the FITSPATS algorithm in comparison with the ARMADA-based algorithm.

As our future works, extending the FITSPATS algorithm with the Moving Longest Expected Repeated Pattern process for actual suffix arrays in external media [14] is of our interest for longer time series in a larger database. Besides, we consider a new programming model with MapReduce to deal with big data in the future. Discovering association rules and putting those patterns and rules into practice for classification and prediction in an application domain like finance and medicine are also our plan.

Acknowledgments. This paper is funded by Ho Chi Minh City University of Technology, Vietnam National University at Ho Chi Minh City, Vietnam, under the grant number TNCS-2015-KHMT-07.

References

1. Agrawal, R., Srikant, R.: Mining sequential patterns. In: Proceedings of ICDE, pp. 3–14 (1995)
2. Assfalg, J., Bernecker, T., Kriegel, H.-P., Kröger, P., Renz, M.: Periodic pattern analysis in time series databases. In: Zhou, X., Yokota, H., Deng, K., Liu, Q. (eds.) DASFAA 2009. LNCS, vol. 5463, pp. 354–368. Springer, Heidelberg (2009)
3. Batal, I., Fradkin, D., Harrison, J., Mörchen, F., Hauskrecht, M.: Mining recent temporal patterns for event detection in multivariate time series data. In: Proceedings of KDD, pp. 280–288 (2012)
4. Chanda, K., Saha, S., Nishi, M.A., Samiullah, M., Ahmed, C.F.: An efficient approach to mine flexible periodic patterns in time series databases. Eng. Appl. Artif. Intell. **44**, 46–63 (2015)
5. Chen, Y.-C., Jiang, J.-C., Peng, W.-C., Lee, S.-Y.: An efficient algorithm for mining time interval-based patterns in large databases. In: Proceedings of CIKM, pp. 49–58 (2010)
6. Dorr, D.H., Denton, A.M.: Establishing relationships among patterns in stock market data. Data Knowl. Eng. **68**, 318–337 (2009)
7. Hirate, Y., Yamana, H.: Generalized sequential pattern mining with item intervals. J. Comput. **1**(3), 51–60 (2006)
8. Mörchen, F., Ultsch, A.: Efficient mining of understandable patterns from multivariate interval time series. Data Min. Knowl. Disc. **15**, 181–215 (2007)
9. Nishi, M.A., Ahmed, C.F., Samiullah, M., Jeong, B.-S.: Effective periodic pattern mining in time series databases. Expert Syst. Appl. **40**, 3015–3027 (2013)
10. Rashee, F., Alshalalfa, M., Alhajj, R.: Efficient periodicity mining in time series databases using suffix trees. IEEE Trans. Knowl. Data Eng. **23**(1), 79–94 (2011)
11. Shameem, M.R., Naseem, M.R., Subanivedhi, N.K., Sethukkarasi, R.: A dynamic approach for mining generalised sequential patterns in time series clinical data sets. In: Meghanathan, N., Nagamalai, D., Chaki, N. (eds.) ACITY 2012. AISCC, vol. 177, pp. 667–674. Springer, Heidelberg (2013)
12. Winarko, E., Roddick, J.F.: ARMADA – an algorithm for discovering richer relative temporal association rules from interval – based data. Data Knowl. Eng. **63**, 76–90 (2007)
13. Wu, Y., Wang, L., Ren, J., Ding, W., Wu, X.: Mining sequential patterns with periodic wildcard gaps. Appl. Intell. **41**, 99–116 (2014)
14. Xylogiannopoulos, K.F., Karampelas, P., Alhajj, R.: Analyzing very large time series using suffix arrays. Appl. Intell. **41**, 941–955 (2014)
15. Zhang, M., Kao, B., Cheung, D.W., Yip, K.Y.: Mining periodic patterns with gap requirement from sequences. ACM Trans. Knowl. Disc. Data (TKDD) **1**(2), August 2007
16. Zhou, H., Hirasawa, K.: Traffic conduction analysis model with time series rule mining. Expert Syst. Appl. **41**(14), 6524–6535 (2014)
17. Zhuang, D.E.H., Li, G.C.L., Wong, A.K.C.: Discovery of temporal associations in multivariate time series. IEEE Trans. Knowl. Data Eng. **26**(12), 2969–2982 (2014)

Optimizing Ontology Learning Systems that Use Heterogeneous Sources of Evidence

Gerhard Wohlgenannt, Stefan Belk$^{(\boxtimes)}$, and Katharina Rohrer

Vienna University of Economics and Business, Welthandelsplatz 1,
1200 Vienna, Austria
{gerhard.wohlgenannt,stefan.belk,karohrer}@wu.ac.at
http://www.wu.ac.at

Abstract. As the manual construction of ontologies is expensive, many systems to (semi-)automatically generate ontologies from data have been built. More recently, such systems typically integrate multiple and heterogeneous evidence sources. In this paper, we propose a method to optimize ontology learning frameworks by finding near-optimal input weights for the individual evidence sources. The optimization process applies a so-called source impact vector and the Tabu-search heuristic to improve system accuracy. An evaluation in two domains shows that optimization provides gains in accuracy of around 10 %.

Keywords: Heterogeneous evidence sources · Ontology learning · Optimization · Spreading activation

1 Introduction

Ontologies are a cornerstone of the Semantic Web. As the manual construction of ontologies is expensive and cumbersome, systems for (semi-automatic) learning of ontologies have been created, which bootstrap the ontology construction process using data-driven methods. Naturally, as the task at hand is very complex, automatically generated ontologies are (i) typically lightweight (containing few axioms), and (ii) contain correct, but also wrong, constituents. Therefore an obvious goal in ontology learning is improving system accuracy. In contrast to seminal work on ontology learning, which used a single domain text corpus to extract facts, more recently there has been some work which uses multiple and heterogeneous sources (see also next section). Using multiple sources can provide accuracy gains, as it can better exploit redundancy of facts found in different sources. Redundancy of evidence in various sources can be seen as a measure of trust and relevance [5].

In previous work, we studied – in an ontology learning context – how many sources are necessary to benefit from heterogeneous sources, and how much evidence is sufficient per single source [11]. This research was an important step to find guidelines on how to configure an ontology learning system regarding the number of evidence sources, and Wohlgenannt [11] deliberately used the same

© Springer International Publishing Switzerland 2015
A. Bikakis and X. Zheng (Eds.): MIWAI 2015, LNAI 9426, pp. 137–148, 2015.
DOI: 10.1007/978-3-319-26181-2_13

input weights for all sources – in order to isolate the effect of using multiple and heterogeneous (unstructured, semi-structured, and structured) sources.

In this paper, we build on previous work, and aim to further optimize system accuracy by using an optimization algorithm (Tabu search [7]) to find the best combination of input weights for the individual evidence sources.

The research questions are as follows: (i) How well can an ontology learning system be optimized by adapting source input weights? – especially if quality of evidence varies between sources. (ii) What is the influence of the number of sources used in the system on the optimization results? (iii) What other findings and guidelines can be extracted from the data collected in the optimization runs?

To address the research questions, we did two batches of optimization runs. The first one was conducted in 2013 with all 32 evidence sources used in our system, which was not very well tuned at that point. The second set of optimization runs was done in 2015, then with a better tuned system, and a reduced set of evidence sources (according to our findings in our previous work [11] that a limited number of sources is sufficient for high accuracy).

The structure of this paper is as follows: Sect. 2 discusses related work, and Sect. 3 provides an overview of the ontology learning system used, and of the heterogeneous evidence sources. Then we introduce the optimization method (Sect. 4), and evaluate the optimization process with different configuration settings in Sect. 5. Section 6 concludes the paper.

2   Related Work

There has been a lot of initial effort in building ontology learning systems around the year 2000. These systems usually process high quality domain text with statistics- and linguistics-based methods. For example, Text2Onto [3] combines machine learning approaches with linguistic processing to build so called Probabilistic Ontology Models. These models are independent of a concrete target language and attach probabilities to learned ontological structures.

Obviously, there are also more recent efforts to build systems to learn ontologies from text, for example OntoGain [6], which uses a number of algorithms such as formal concept analyses or association rule mining for unsupervised ontology construction. Other systems often try to compute a probability for ontological elements learned, which is used in the selection and integration process. For example, Abeyruwan et al. [1] suggest a method for unsupervised bottom-up ontology generation which selects ontological elements by a respective Bayesian probability. For details and information on other ontology learning systems see the recent survey publication by Wong et al. [14].

There have been few efforts yet to learn ontologies from heterogeneous evidence sources. Manzano-Macho et al. [5] outline some of the potential benefits of using heterogeneous evidence sources: the combination of sources leads to an overlap of the ontological elements suggested, this redundancy can be seen as a measure of relevance and trustiness for a certain domain. And, obviously, some methods will add valuable complementary information that other sources or methods did not detect. In their approach to combine heterogeneous sources of evidence, Manzano-Macho et al. [5] aim at building a taxonomy of concepts with higher accuracy compared to using a single evidence source only.

Another method for the integration of heterogeneous sources has been proposed by Cimiano and Völker [2]. They focus on learning taxonomic relations between concepts combining multiple methods, and then convert the evidence found into first order logic features. Standard classifiers are applied to find useful combinations of evidence sources.

In this paper there is no focus on improving single evidence sources, but rather on the smart combination of heterogeneous evidence sources. Existing systems typically use a domain corpus as input, whereas our framework integrates a wider variety of sources, including evidence from APIs available on the Social Web and a Linked Data source. The sources are heterogeneous regarding a number of aspects: quality, type, number of evidences, etc. (see also Sect. 3.1).

We use spreading activation as our main method to integrate evidence. Spreading activation is a method to search neural and semantic networks, its suitability for information retrieval tasks has been demonstrated e.g. in [4].

3   The Ontology Learning Framework

All experiments presented in the paper are based on an existing ontology learning system which evolved over the years. This section can only give a quick overview of the framework, for details on the workings of the system see Wohlgenannt et al. [12], or Liu et al. [9], who present the original version of the framework. In a nutshell, the system starts from a (typically small) seed ontology, which is extended with additional concepts and relations. So the main tasks are the selection of new concept candidates and the positioning of concepts with regards to the seed ontology.

We compute new ontologies for the domains in question in regular intervals (monthly) to trace the evolution of the domains. Figure 1 shows (parts of) the graphical representation of an example ontology learned in the *climate change* domain on data from January 2014. The concept colors indicate the ontology extension stage, the darker, the earlier the concepts were introduced. Before explaining the system in more detail, we take a look at the evidence sources used in the process.

3.1   The Evidence Sources

As the name suggests, the *evidence sources* provide the data needed to extend (learn) ontologies. In general, the input to evidence acquisition is a term (typically the label of a seed concept), and the result is a list of terms related to the seed term, and optionally significance values. So, for example, the system sends the seed concept label "CO2" to the Flickr API[1], and gets a list of related

---

[1] www.flickr.com/services/api/flickr.tags.getRelated.html.

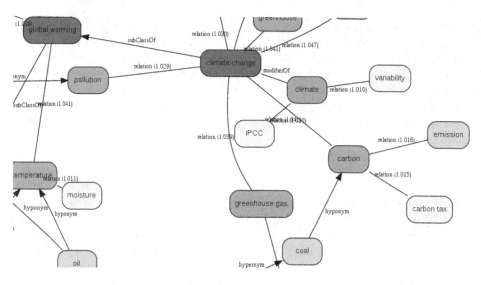

Fig. 1. Part of a sample ontology in the domain of climate change after three stages (levels) of extension

terms – which will then be used in the ontology learning process together with information from all other sources of evidence.

The evidence sources are heterogeneous regarding various aspects, such as (i) the number of evidences returned; (ii) the average quality (i.e. domain relevance) of terms provided; (iii) the update frequency of the source – some sources are dynamic, e.g. social sources and news media, some rather static, e.g. WordNet and DBpedia; (iv) the availability of a significance score for the terms; and (v) the type of underlying data (text, structured, etc.).

By default, the ontology learning system currently uses 32 sources of evidence, which are listed in Tables 1 and 2. The first batch of experiments (see Sect. 5) conducted in year 2014 is based on all 32 sources, the second batch (year 2015) is based on the sources marked with boldface fonts. A major fraction of evidence sources is made up by the 16 sources based on keywords computed with co-occurrence statics on documents published and mirrored in the respective period of time, and filtered with a domain-detection service. The system computes page- and sentence-level keywords for documents collected from: US news media, UK news media, AU/NZ News Media, Websites of NGOs, Fortune 1000 company Websites, Twitter tweets, Youtube postings, Google+ postings, and public Facebook pages and postings. Furthermore, we use Hearst patterns [8] on those corpora, which constitutes further 10 evidence sources. Currently, the system includes two evidence sources based on calls to APIs of Social Media sites (Twitter, Flickr). Structured evidence sources contribute the remaining 4 sources of evidence, i.e. hypernyms, hyponyms and synonyms from WordNet, and related terms from DBpedia. For more details on evidence sources used see Wohlgenannt et al. [13].

Table 1. The first 26 evidence sources are based on domain-specific text collected from the Web.

Data sources	Extraction method		
Domain text from:	Keywords/page	Keywords/sentence	Hearst patterns
US news media	**1**	2	**3**
UK news media	4	**5**	6
AU/NZ news media	7	**8**	9
Other news media	**10**	11	12
Social media – Twitter	13	-	**14**
Social media – Youtube	**15**	-	16
Social media – Facebook	17	-	18
Social media – Google+	19	-	20
NGOs Websites	21	22	**23**
Fortune 1000 Websites	**24**	25	26

Table 2. The other 6 sources of evidence, based on WordNet, DBpedia and Social Media APIs.

Data source:	Method				
	Hypernyms	Hyponyms	Synonyms	API	SPARQL
WordNet	**27**	**28**	29	-	-
DBpedia	-	-	-	-	30
Twitter	-	-		**31**	-
Flickr	-	-	-	**32**	-

3.2 Source Impact

As mentioned, evidence sources are heterogeneous in number and quality of terms provided, we use a so-called Source Impact Vector (SIV) to manage the influence of a particular evidence source on the ontology learning process. Equation 1 demonstrates that the SIV consists of one impact value per evidence source (and point in time). The impact value is in the interval $[0.0, 1.0]$, a value of 1.0 results in high impact in the learning processes, wheres 0.0 in fact omits evidence suggested by the respective source.

$$SIV_t = \begin{bmatrix} I_{es_1,t} & I_{es_2,t} & \cdots & I_{es_n,t} \end{bmatrix} \tag{1}$$

The SIV is used to set the weights in the spreading activation network (see next subsection for details), which selects new concept candidates for the ontology. Initial versions of the system [9,10] applied a manually picked and static source impact, in this paper we propose novel ideas and experiments to optimize the ontology learning system via the SIV. The optimization process aims to find a configuration of the SIV which maximises the ratio of relevant new concept candidates suggested by the system.

3.3 The Ontology Learning Process

Having described the evidence sources and the SIV, we can introduce the basic workflow of the ontology learning system: The ontology learning run starts with a small seed ontology (in the climate change usecase we use two concepts, namely *climate change* and *global warming*). The system collects evidence for the seed concepts from the 32 (or 14) evidence sources. After integrating all this evidence into a semantic network, a transformation process using the SIV converts the semantic network into a spreading activation network. The spreading activation algorithm yields new concept candidates. We currently pick the 25 candidates with the highest level of activation. The concept candidates are evaluated for domain relevance by domain experts, this is the only part of the system that involves human intervention. Finally, the system positions new concepts rated as relevant with regards to the existing ontology. For the next ontology extension step the framework uses the result from the previous iteration as new seed ontology, and further extends it. We typically do three ontology extension iterations. Figure 2 gives on overview of the workflow.

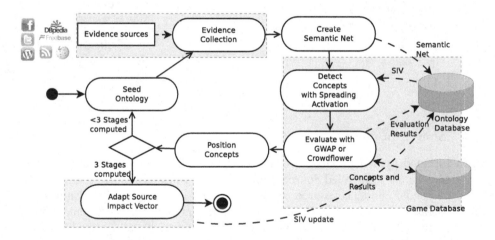

Fig. 2. The ontology learning framework

The goal of this research (and the optimization) is to improve the ratio of relevant to non-relevant concept candidates, i.e. to improve the output of the spreading activation algorithm. The SIV is a key factor in this optimization process, as it determines – in combination with significance scores provided by the evidence sources – the weights in the spreading activation network.

4   The Optimization Process

Although a spreading activation network has the fundamental characteristics of a neural network, we did not find a way to apply classic neural network learning techniques to optimize the output for a number of reasons:

- The spreading activation network doesn't have an explicit output layer, the *results* of the spreading activation algorithm are the activation levels of nodes all over the network.
- We select a preset number of nodes (e.g. 25) with the highest activation level as concept candidates. The use of an error function (as used e.g. in back-propagation) is not straightforward, as we only assess the preset number of nodes with the highest activation, but any other node might be a relevant domain concept as well. So there is no distinct correct output of the spreading activation network that could be used.
- The learning algorithm can not freely optimize the weights in the network, as values of the SIV are only factors in the connection weights. First of all, when multiple SIV factors make up a connection weight, it is not clear which specific SIV factor should be changed. And more importantly, if a specific SIV value is changed for one connection, it needs to be changed simultaneously everywhere in the spreading activation network wherever used, leading to unpredictable effects.

The characteristics described above led us to experiment with heuristics to improve the output of the ontology learning framework based on the modification of the SIV. This includes a baseline with a static SIV (Sect. 4.1), and a model that aims to optimize the SIV (Sect. 4.2).

Overall, the crucial factor which has an impact on the results of the ontology learning process are not so much the absolute values in the SIV, but the differences between evidence sources. Higher source impact for an evidence source results in increased activation levels and therefore a higher chance of being a candidate concept for evidence suggested by the particular source.

4.1   Static Source Impact Values

The simplest way to use the SIV is to have static values for any source, not changing over time or across domains. We use a source impact of 0.2 for all 32 evidence sources. This uniform source impact has been used in the experiments regarding the number and balancing of evidence sources presented by Wohlgenannt [11], and provides good results, which we use as a baseline and starting point of the optimization experiments.

4.2   Optimization

With this strategy, instead of having a single static SIV, the system investigates different SIV settings and their results. In the first batch of experiments, we set the source impact for every evidence source to values in the interval $[0.0, 1.0]$ with a step-size of 0.1, i.e. *eleven* values per source. With 32 evidence sources, this leads to an enormous number (11^{32}) of potential permutations. As a single ontology learning run (depending on settings) takes around four hours of computation time, we decided to use the Tabu Search heuristic [7], and simply optimize every evidence source by itself, with settings for other sources constant. The leads to 352 $(11 * 32)$ ontology learning runs. In the second batch, we used a step-size of 0.2 and 14 evidence sources, resulting in 84 $(6 * 14)$ ontology learning runs.

The following Pseudocode shows the Tabu Search-based optimization strategy:

Algorithm 1. Optimize SIV with Tabu Search

Initial solution ← *Static SIV*
\# do for all 32 evidence sources
for each evidence source e **do**
   \# create neighborhood
   for X in interval $[0.0, 1.0]$ **size** 0.1 **do**
      \# evaluate every neighbor
      $SIV_e \leftarrow X$
      compute ontology (all 3 extension steps) using SIV
      $Q_x \leftarrow$ Evaluate quality of ontology
      Remember result (X, Q_x)
   end for
   \# keep value X with best result – skip the rest
   $SIV_e \leftarrow$ pick best result from neighborhood
   Put all other solutions from neighborhood on Tabu list
end for

Basically, the heuristic looks for the *best* source impact value for a single evidence source, and uses this value when optimizing the other evidence sources. One of the downsides of this method is that the order in which evidence sources are processed obviously has an impact on the result. The system randomizes the order of evidence sources before every optimization run. It will typically not find a global optimum, but hopefully a good solution with a limited number of permutations. Furthermore, the optimization helps to visualize and understand how specific SIV settings contribute to ontology quality.

5   Evaluation

This section summarizes the findings of optimization runs performed to gain insights about the improvements of accuracy which can be reached by optimizing the combination (the impact) of evidence sources.

5.1   Evaluation Setup

In previous experiments conducted in year 2014 (see Sect. 5.3 for results) we used a step-size of 0.1 in a source impact interval of 0.0 to 1.0, and 32 evidence sources. For the recent batch of evaluation experiments we used a more computationally efficient setup, with optimization runs for 14 evidence sources, and a step-size of 0.2. Previous work shows that 10–15 evidence sources are sufficient to have good results [11].

Relevance assessment of concept candidates is being done by domain experts. The accuracy values used in this section are simply the number of concept candidates rated as domain-relevant by the domain experts divided by all concept candidates suggested by the system. We decided to use the ratio of relevant concepts as evaluation metric because (i) the relevance of domain concepts is critical to generating useful domain ontologies, and (ii) relevance assessment for concept candidates is the only part of the system where manual input is applied.

5.2   Recent Optimization Experiments

These experiments were conducted with the latest version of the ontology learning system in the first half of year 2015. We compared the results of using a static SIV (uniform source impact of 0.2 for all evidence sources) to optimizing source impact.

Table 3. System accuracy and gains by optimizing the SIV as compared to a static SIV, in two different domains.

Domain	Static SIV	Optimized	Improvement
Climate Change (en)	67.15 %	76.88 %	9.73 %
Tennis (en)	44.57	54.42 %	9.85 %

Table 3 summarizes the results for two different domains, the domains of *climate change* and of the sport *tennis*. The values in the table represent the average accuracy for ontology generation runs with a static SIV and for the optimization processes. The data indicates a substantial improvement in accuracy of around 10 % which can be reached by optimizing the SIV.

The difference in accuracy between the two domains can be attributed to the following reasons: (i) In the *climate change* domain we are supplied with much bigger and domain-specific corpora, whereas with *tennis* we use general news corpora which are then filtered for tennis-related documents ex-post. Besides corpus-size and quality, (ii) the domain of tennis has a lot of overlap with other sports domains. Concepts such as ball, tournament, etc. attract related but not domain-relevant terms from other sports, whereas climate change seems to be more "closed".

However, the most interesting fact is the improvement in accuracy, which is statistically significant, as confirmed with a binomial test.

5.3 Previous Experiments

This section discusses experiments done in early 2014 with an older version of the system, which wasn't as well tuned, with lower general accuracy. This is reflected by the accuracy values for *Static SIV* in Table 4. We evaluated two settings, where we used either up to 50 (*limit = 50*) evidences per source and concept, or up to 200 (*limit = 200*)² – for details on these settings see Wohlgenannt [11].

Table 4. System accuracy of the previous system version, in the domain of *climate change*, for two settings.

Domain	Static SIV	Optimized	Improvement
Climate Change (en) – *limit = 50*	63.33 %	78.18 %	14.85 %
Climate Change (en) – *limit = 200*	64.13 %	77.33 %	13.20 %

In year 2014 we started from a lower baseline (around 63–64%), and experience improvements from optimization between 13–15%, more than in the recent batch of experiments. Our interpretation of the results is the following:

- The lower baseline leaves more room for improvement.
- In the 2014 experiments we used a *step-size* of 0.1, which resulted in higher computational cost, but also a more fine-tuned optimization.
- The number of evidence sources was much higher (32 sources), therefore the potential for fine-grained optimization of sources was higher.

With regards to the research questions posed in Sect. 1, the evaluation shows that system accuracy can be raised substantially by optimization using the SIV. It helps to have a high number of evidence sources and also a fine-grained step-size, this allows for a more precise optimization process.

5.4 Analysis of Evidence Sources

The evidence sources provide terms and relations of different quality to the learning algorithms. Wohlgenannt [11] discusses the quality and characteristics of evidence sources in some detail. In a nutshell, the number and quality (domain relevance) of evidence is very heterogeneous. Keyword-based sources typically provide a high number of terms, with good quality for the terms with highest co-occurrence significance, but degrading with more terms added having a lower significance. Terms for structured sources such as DBpedia and WordNet generally offer good quality, but low term numbers. In our experiments, APIs of social sources such as Twitter and Flickr yield mostly low quality terms – but we still have them included to (i) benefit from the effect of redundancy between sources, and (ii) as they often provide very recent and complementary terminology.

---

² The more recent evaluations in Sect. 5.2 were conducted with *limit = 50* settings.

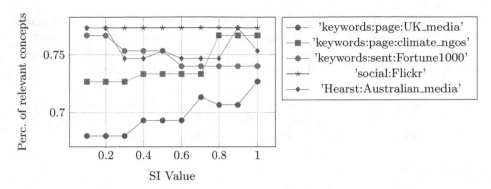

Fig. 3. Influence of source impact settings for a number of selected evidence sources.

Figure 3 visualizes the influence of source impact (SI) settings for some individual sources on system accuracy. The data is taken from an optimization run in the domain of *climate change*, and helps to explain the characteristics and experiences with SIV optimization. Usually evidence sources fall into one of the following categories:

- *Increasing the SI raises accuracy.* These evidence sources obviously yield relevant terms and helpful contributions to the ontology learning system. With higher impact of the source the accuracy goes up. keywords:page:UK_media and keywords:page:climate_ngos in Fig. 3 fall into this category.
- *Increasing the SI lowers accuracy.* This applies to sources which do not contribute much helpful data. For example keywords:sent:Fortune1000.
- *Accuracy independent of SI.* This usually happens when a source provides a very low number of evidences, social:Flickr in our example.
- *Erratic.* As with Hearst:Australian_media, sometimes the effects of the SI are rather erratic. Such cases are the biggest challenge for optimization.
- A mix of the basic categories described above.

Erratic behavior or a mix of the categories described above results from the fact that the system selects the 25 concept candidates with the highest spreading activation level. Raising the influence of a single evidence source gives more importance to all its evidence, relevant or not. The Tabu search heuristic will not find an optimal, but typically good, combination of sources (i.e. the SIV).

6   Conclusions

Ontology learning aims at (semi-)automatically constructing lightweight ontologies from sources of evidence. When using multiple and heterogeneous sources, balancing and optimizing the influence of evidence sources is crucial. In this paper, we introduce and evaluate a strategy for optimizing such ontology learning systems, and see improvements in accuracy (in the concept detection phase) of ca. 10–15%. The contributions are as follows: (i) Presenting a novel method to configure and optimize ontology learning systems using the source impact vector and the Tabu-search heuristic, and (ii) experiments in two domains to estimate the accuracy gains from this optimization technique. Future work includes the repetition of experiments in other domains, also based on corpora in other languages, and the application of alternative optimization strategies.

Acknowledgments. The work presented was developed within uComp, a project which receives the funding support of EPSRC EP/K017896/1, FWF 1097-N23, and ANR-12-CHRI-0003-03, in the framework of the CHIST-ERA ERA-NET.

References

1. Abeyruwan, S., Visser, U., Lemmon, V., Schürer, S.: PrOntoLearn: unsupervised lexico-semantic ontology generation using probabilistic methods. In: Bobillo, F., Costa, P.C.G., d'Amato, C., Fanizzi, N., Laskey, K.B., Laskey, K.J., Lukasiewicz, T., Nickles, M., et al. (eds.) URSW 2008-2010/unidl 2010. LNCS, vol. 7123, pp. 217–236. Springer, Heidelberg (2013)
2. Cimiano, P., Pivk, A., Schmidt-Thieme, L., Staab, S.: Ontology learning from text. In: Learning Taxonomic Relations from Heterogeneous Sources of Evidence, pp. 59–76. IOS Press, Amsterdam (2005)
3. Cimiano, P., Völker, J.: Text2Onto. In: Montoyo, A., Muñoz, R., Métais, E. (eds.) NLDB 2005. LNCS, vol. 3513, pp. 227–238. Springer, Heidelberg (2005)
4. Crestani, F.: Application of spreading activation techniques in information retrieval. Artif. Intell. Rev. **11**(6), 453–482 (1997)
5. Manzano-Macho, D., Gómez-Pérez, A., Borrajo, D.: Unsupervised and domain independent ontology learning: combining heterogeneous sources of evidence. In: Calzolari, N., et al. (eds.) LREC 2008. ELRA, Marrakech, May 2008
6. Drymonas, E., Zervanou, K., Petrakis, E.G.M.: Unsupervised ontology acquisition from plain texts: the *OntoGain* system. In: Hopfe, C.J., Rezgui, Y., Métais, E., Preece, A., Li, H. (eds.) NLDB 2010. LNCS, vol. 6177, pp. 277–287. Springer, Heidelberg (2010)
7. Glover, F., Laguna, M.: Tabu Search. Kluwer, Norwell (1997)
8. Hearst, M.A.: Automatic acquisition of hyponyms from large text corpora. In: Proceedings of COLING 1992, Nantes, France, pp. 539–545 (1992)
9. Liu, W., Weichselbraun, A., Scharl, A., Chang, E.: Semi-automatic ontology extension using spreading activation. J. Univ. Knowl. Manag. **0**(1), 50–58 (2005)
10. Weichselbraun, A., Wohlgenannt, G., Scharl, A.: Refining non-taxonomic relation labels with external structured data to support ontology learning. Data Knowl. Eng. **69**(8), 763–778 (2010)
11. Wohlgenannt, G.: Leveraging and balancing heterogeneous sources of evidence in ontology learning. In: Gandon, F., Sabou, M., Sack, H., d'Amato, C., Cudré-Mauroux, P., Zimmermann, A. (eds.) ESWC 2015. LNCS, vol. 9088, pp. 54–68. Springer, Heidelberg (2015)
12. Wohlgenannt, G., Weichselbraun, A., Scharl, A., Sabou, M.: Confidence management for learning ontologies from dynamic web sources. In: Proceedings of KEOD 2012, Barcelona, Spain, pp. 172–177. SciTePress, October 2012
13. Wohlgenannt, G., Weichselbraun, A., Scharl, A., Sabou, M.: Dynamic integration of multiple evidence sources for ontology learning. J. Inf. Data Manag. (JIDM) **3**(3), 243–254 (2012)
14. Wong, W., Liu, W., Bennamoun, M.: Ontology learning from text: a look back and into the future. ACM Comput. Surv. **44**(4), 20:1–20:36 (2012)

Machine-Learning Based Routing Pre-plan for SDN

Fengqing Chen[1,2(✉)] and Xianghan Zheng[1,2]

[1] College of Mathematics and Computer Science, Fuzhou University,
Fuzhou 350108, China
cfqfzu@163.com
[2] Fujian Key Laboratory of Network Computing and Intelligent Information Processing,
Fuzhou 350108, China

Abstract. In Software Defined Network (SDN) environment, controller has to compute and install routing strategy for each new flow, leading to a lot of computation and communication burden in both controller and data planes. In this background, intelligent routing pre-design mechanism is regarded to be an important approach for routing efficiency enhancement. This paper investigates and proposes efficient SDN routing pre-design solution in three aspects: flow feature extraction, requirement prediction and route selection. First, we analyze and extract data packet and association features from user history data, apply these features into semi-supervised clustering algorithm for efficient data classification, analysis and feature extraction. After that, flow service requirement could be predicted through extraction of user, flow and data plane load features and implementation of supervised classification algorithm. Furthermore, we propose corresponding handling strategies related to data plane topology, flow forwarding and multi-constraint weight assignment, and proposes personalized routing selection mechanism.

Keywords: SDN · Routing pre-design · Semi supervised clustering algorithm

1 Introduction

Software defined network (SDN) is a new network architecture. The core idea is to separate control plane and data plane and programmatically control the Routing and network traffic, build a dynamic, open and controlled network environment [1–3]. But the SDN controllers need to be developed for each new stream and installed routing policy, which caused a huge burden of computing and communications. As we seen, it is not efficient [4]. So, intelligent route planning mechanism is considered to be an effective way to solve this problem. In SDN routing domain, it based on the application of machine learning theory, and gradually penetrate with feature extraction from a stream, demand forecasting, and path planning. SDN route planning studies have academic and applied significance. Existing IP network is a distributed routing control system which is integrated with the control module and the data module. Each node is independently responsible for the synchronization of routing information, and calculates the path over to the other node and controls data where routing interactions are highly complex. Meanwhile, routing calculated only considers destination address information

© Springer International Publishing Switzerland 2015
A. Bikakis and X. Zheng (Eds.): MIWAI 2015, LNAI 9426, pp. 149–159, 2015.
DOI: 10.1007/978-3-319-26181-2_14

regardless of the original address and other network information (for example, latency, overhead, and so on). It is difficult to meet the business needs of the best route. In contrast, software-defined networking mode is more flexible, which divides the network into three tiers (as shown in Fig. 1) [4]. SDN controller is responsible for the maintenance of a global network view, and through South to the interface (for example, Openflow [6], and so on) to updates flow table information in the switch, and finally realizes network central control. Application layer interacts with the control layer through Northern interface and develops corresponding business rules (for example, network configuration, applications, etc.), programmable network control and service. From a theoretical, SDN routing in the network architecture is under the view of global perspective to consider bandwidth, load, quality of service (QoS), security and other factors, which provides optimization methods, and it has the natural advantage of decision.

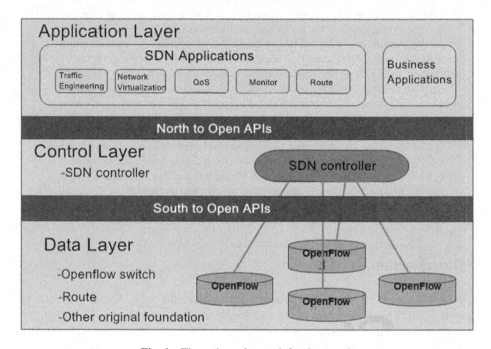

Fig. 1. Three-tier software defined network

However, since software has defined network (SDN) environments, the controller needs to calculate each new flow routing paths in real time, and install routing strategies on each Openflow switch in the path. while a large network environment brings control plane and data plane a huge burden on computing and communications, resulting in higher latency. So it is not efficient. SDN routing optimization method focused on the existing switch software and hardware upgrade (for example, switch hardware upgrade [1], Openflow protocol optimization [6]) and optimization of the distributed architecture of the controller (for example, Hyper Flow [7], DIFANE [8], Onix [9], SOX [10]) aspects, but it is still unable to solve the validity of above defects effectively. In this context, under the environment of SDN "route planning" concept coming into being [4], the core idea is to predict the user (or VM virtual machine) business requirements, plan ahead and set up routing policies, with a view to reduce the effect of the delay. Therefore, pre-planning is an important route to plan supplementary route. It is also an important way to improve the effectiveness of SDN routing.

Nevertheless, current research does not present a feasible method of route pre planning. It mainly faces following 3 aspects challenges: first, data of SDN network environment is existing and processing as the flow form, and the definition of flow of grain degrees was fuzzy, not only the traditional five-tuple (source node IP, source port, destination IP, destination port, and Protocol), also in the view of business applications. It is difficult to accurately identify data flow and business properties (for example, boundary determination, small elephant flow or data flow, real-time or non-real-time data), route feasibility challenges of judgment. Secondly, the SDN business demand forecasting of flow is a prerequisite for route planning strategy. It requires a combination of a user, information flow, data plane load characteristics, and current research on SDN. There is no reference in the field of theoretical and practical basis. In addition, routing path selection of pre-planning would give personalized path calculation method after considering about many factors. It requires a calculation method based on the current path so that optimization studies would be continue. In response to the above technical challenges, we start with the deep data and network characteristics, and follow by conducting three studies: First, the studies of SDN category identification, grain size analysis and feature extraction methods, provide basic premise for route planning. Second, carry out the study of the features of the SDN flow and data plane load to learn and predict SDN flow business requirements, and provide policy guidance for route planning. On this basis, we correspond flow of business process strategy and data plane topology optimization strategies, and introduce multiple constraints, design based on the new calculation method of the path. Finally, optimize choice of implementation. These three studies are layer-by-layer deep, complementary and indispensable.

The develop of machine learning theory as well as its efficient application in the area of data analysis offers a new approach for pre-planning SDN environment. It contains theory of probability theory, statistics, machine learning algorithms, and many other areas of knowledge, based on existing data and past experience, combined with a variety of learning methods, predicting and problem-solving. Judging from the task type, the applications of machine learning research mainly focused on the following three areas [11–13]: (1) unsupervised clustering, classification of similar objects to different group or subset through static methods makes the Members of objects have similar properties in the same set. The category identification of project flow, particle size analysis and feature extraction can be attributed to the field. (2) The supervised classification, based on known sample data and category properties into unknown patterns for analysis and classification. The demand forecasts of SDN flow in this project can be attributed to the field. (3) Problem solving, is that obtain knowledge which can improve the efficiency of problem solving through learning. Multi-objective optimization of path planning in this project can be attributed to the field. In recent years, machine learning has received wide attention from academia and industry because of its excellent ability of problem solving and independent study skills. It has the potential to be a big weapon to solve the problem of SDN route pre-planning.

In the software-defined networking environments, there are still many problems intend be solved. For example, how to identify categories, analysis particle size and extract feature in a large network environment? How to combine with user and data plane load characteristics to predict SDN flow business needs? How to design a network application and mechanism of flow path planning that meet personalized business needs? These issues are worthy of further study, and also a technical difficulty of SDN route pre-planning. This topic is intended for software defined network area, and focus on the effectiveness of the current routing flaw, which emphasize routing programming based on machine learning method and key technologies for route optimization in SDN to provide new research methods and ideas and provide technical support for improvement of SDN.

2 Related Works

2.1 SDN Introduction

SDN (Software Defined Network) is a new kind of network innovation framework. Its core technology is OpenFlow that controls network device by separating the data, to realize flexible control of network traffic, and provides a good platform for innovation of the core network and application.

From the design point of view, router is controlled by software and hardware data channels. Software control includes management (CLI, SNMP) and Routing Protocol (OSPF, ISIS and BGP) and so on. As SDN network design planning and management provide great flexibility, we can choose a centralized or distributed control, micro-flow (such as a campus network flow) or aggregated flow (such as backbone flow) forwards matching the flow table; you can select a virtual or physical implementation.

2.2 Existing Research

This routing plan is divided into the following three aspects: streams SDN environment feature extraction, demand forecasting, and path planning, including existing research technology and development. Under the environment of SDN, flow feature extraction is a prerequisite for route planning. Current research focuses on the particular flow detection, common practices as follows [1]: through interception method (for example, sFlow random sampling method) detect packet communication, initially screening suspect special flow (carrying large amount of privacy, user or user key or have a harmful effect on network data stream). On this basis, we will filter out suspect special streams SDN controller traffic statistics that is available for the further verification and identification. However, in the context of SDN, stream size definition is ambiguous, both in the traditional five-tuple point of and in a business perspective can never fully express a stream. It is difficult to analyze and identify its convection elements, borders as well as properties accurately, so the current method is not efficient. In terms of data analysis, the traditional methods is based on matching port and the load of analysis [14]. The former through comparison the layer port number of packet transport and IANA assigned port numbers forecast application layer services. However, due to a number of emerging applications in violation of port allocation rules, the method is not exact. The latter detect packets deeply, can achieved session reconstruction, test and application analysis accurately. But the complexity is higher, and it requires a lot of manual operations and advance understanding of application protocols. Another feasible method is clustering analysis, representative studies are as follows: [15] propose a hierarchical clustering analysis model using 3 statistical characteristics (request data, response data size, stop time) to implement the efficient classification of mixed data on the Internet. [16] proposed a clustering fusion probability models with the largest expected data automatic classification methods, and the experimental is 86.5 % accuracy. On the basis of the method, [17] further proposed that the nearest neighbor (NN) classification model is 91 % of the classification accuracy. [18] The design based on data packet (load size, packet flows) collates mechanism for TCP streaming, and use a combination of K-Means, European space EM Clustering cluster 3, hidden Markov spectrum method in testing. The application obtained with the unknown of 90 % and the known of 60 % classification accuracy.

SDN traffic demand forecasting is the routing of pre-planning strategy developed by technical support. In this regard, the current study did not give the feasibility, while supervised classification research and applications provide a theoretical and practical basis for reference. Traditional classification techniques including Bayesian [19], support vector machine [20], rough set [21] and decision tree algorithm [22], BP neural network [23] methods have been widely used in the field of data analysis. In order to meet the higher demand for data processing, the continuous improvement projects which are based on traditional machine learning algorithm are emerging, typical examples include recent SVM (PSVM) [24], least square SVM (LS-SVM) [25] and finite Newton SVM (NLSVM) model [26], and to some extent reduce the difficulty of solving and improve speed. But widespread training is too long, frequent manual intervention and poor generalization ability, which does not apply to large scale data analysis under the environment of network applications.

In recent years, Extreme learning machine (ELM) theory [27] is exploded. Random assignment by single hidden-layer feedforward neural network input value and deviation, calculated output value, the output function $f_L(x) = \sum_{i=1}^{L} \beta_i G(a_i, b_i, x)$ (Where ai and bi are the hidden layer nodes, randomly generated parameters, i is the connection weights of m-dimensional vector, G is the output of the hidden layer neurons). On this basis, get output using least squares weights: $\beta = \arg \min_\beta \|H\beta - H\| = H^+T$ (H output matrices for the hidden layer (h full rank)). After nearly a decade of development, the ELM theory evolves. For example, Fusion B-ELM model of Bayesian inference [28], Fusion theory of fuzzy TSK-ELM model [28], OS-ELM with online training model [29]. ELM series theory has the advantage of speed, without human intervention, access to measuring accuracy and generalization ability higher than (or similar) the traditional classification algorithms. Thus it more suitable for large-scale data analysis under the environment of network applications.

Path planning of route pre-planning is the important guarantee of reliability. Traditional path planning method focus on a particular constraint (for example, latency), and the single-path routing problem (for example, the shortest path first OSPF [30]). The data for specific business requirements (for example, sensitive data) can choose a path as the active path, and a number of other paths as standby paths [31]. Communication on the active path, once its edge or vertex fails, switches to another backup path. For the solution-phase may cause interference problems, k disjoint path selection method can be used, including node-disjoint multipath routing (The path except the source node and the target node, not shared by any other node) and link-disjoint path routing (no common link in the path, but may contain common node) to ensure that it does not intersect the path of. But only a single constraint factor does not consider the practical application of bandwidth, cost, energy consumption and other factors such as security, unable to get the optimal results of the actual application. In the existing studies, multiple constrained routing is an hard NP problem that can be solved using multi objective optimization problems. For example, [32] in 2004 INFOCOM 2 constraints, for the first time proposed the 2 disjoint path calculation method, and proved the effectiveness of methods in theory, approximate ratio is $(1 + 1/k, 1 + k)$. [33] based on the traditional kRSP problem through improved approximation ratios of (2,2) algorithm in the rounding methods, design the approximate ratio of $(1 + 1/k, 1 + lnk)$, approximation algorithms, and calculate at the same time meeting the cost and delay of two constraint factor k disjoint paths. The path planning researches for this project have provided references.

3   Technology Roadmap

Point to the challenges for SDN route planning technology faced, by the methods of machine learning theory, study from feature extraction from a stream, demand forecasting, path planning study, give routing specific programmed pre-planning (Fig. 2).

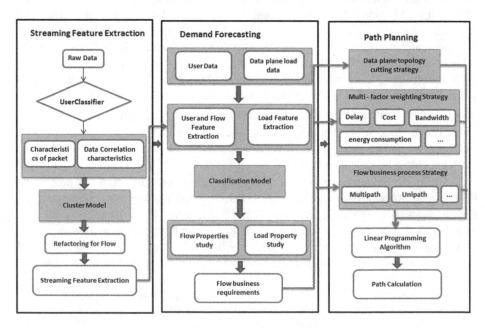

Fig. 2.   Technology roadmap

3.1 SDN Flow Feature Extraction Based Semi-supervised Clustering

According to the collected data, this project aims to develop "user-data" rough taxonomy, for example, IP five-tuple can be used as traditional methods, such as mapping, IP allocation mechanisms, access to historical data for each user. Secondly, the Discover packet characteristics and associated information, packet features are: data length, packet number, total amount of data, packet size, packet intervals and other information associated with the linkages between characteristics of packet properties, for example, by the same IP through a different port, access to the same destination IP and port data can be attributed to the same application. Then, using different clustering models, parameters setting (for example, set cluster number) and artificial determination, classification of data on the network, through accuracy compared to find the most appropriate clustering model and parameter settings. You can use the following two methods:

Gaussian mixture model: By solving the equation $p(\mathrm{x}) = \sum\limits^{k} \pi_k p(\mathrm{x}|k)$ (Where k is the model number, k is weights of K-th Gaussian model, p is K-th Gaussian probability density function) to estimate the probability density distribution of samples. Analysis of sample data in several projection on the Gaussian model is obtained probability belonging to each class in, and select the maximum probability judgments for the class results. Specifically, using maximum likelihood estimation, to turn formulas into:

$$\max \sum_{i=1}^{N} \log(\sum_{k=1}^{k} \pi_k \mathrm{N}\left(\mathrm{x}_i|\mu_k, \sigma_k\right))$$ (Where μ_k is the mean value, σ_k is the variance). Thereafter, by maximum value of algorithm (EM) get iterative solution of Gaussian model parameters.

K-Means clustering: Choose k samples as the initial cluster centers, samples to each other by calculating the distance from the Center (for example, the Cosine distance). The samples are classified, after repeated iterations of each new class center, until Center does not change or changes very little. Ideally, each cluster represents a collection of application data. Therefore, application data, user, time, and other information are used to analyze the constituents of SDN flow structure, and feature extraction.

3.2 SDN-Flow Demand Forecasts Based on Supervised Classification

SDN flow demand forecasting is the routing of pre-planning strategy development support. Therefore, it needs on the basis of 3.1 to introduce supervised classification, and combines with user (For example, the VIP users or not, user behavior, and so on.) and data plane load characteristics (for example, switch device characteristics and flow distribution, characteristics of bearing capacity characteristics, errors). Study SDN flow characteristics of the data plane load and predicted flow SDN business needs. The method combines of batch and iterative, and constantly updates the training data and network structure by optimization of ELM theoretical output weight calculation, on the acquisition of periodic piecewise training data samples (as shown in Fig. 3) and discards old data after the training is complete to avoid repetition of the old and new data.

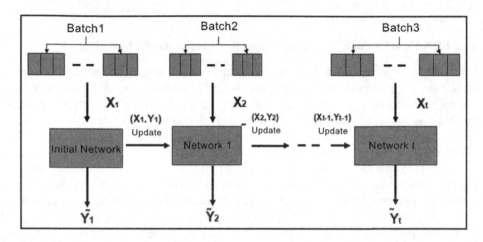

Fig. 3. Block training mode

Finally, take into account the environment of SDN larger amount of sample data, classification algorithm will involve large-scale matrix computation. In order to improve efficiency, this project aims to introduce the matrix decomposition techniques and MapReduce to distribute computing theory. For example, non-negative matrix factorization technique can be used to dimension the hidden layer of the output matrix H_k, $H_k = W^T V$; on this basis, hj is the j-th column of H_k:

$$h_j = \left(w_1^T, \ldots, w_i^T, \ldots, w_m^T\right) \begin{pmatrix} v_{1j} \\ \vdots \\ v_{ij} \\ \vdots \\ v_{mj} \end{pmatrix} = \sum_1^m v_{ij} w_i^T$$

On this basis, calculated using MapReduce patterns, defines the input and output of Map and Reduce functions to further reduce the complexity of hidden layer processing of the output matrix h k. Model output for SDN stream properties to the data plane load forecasting results. Considering it is used to SDN flow analysis and prediction of business needs.

3.3   Personalized SDN Flow Path Planning Based on Policy

Based on Sects. 3.1 and 3.2, according to the SDN business needs, this paper bases on actual network deployment strategies, and develops appropriate data plane topology cropping strategies and current business strategy, as well as many constraint factor weight distribution strategy. For example, properly deleted in the data plane topology weaker, error rate switch nodes higher. Use multipath routing approach to elephant flows, adjust according to the needs of different business related constraint factor weights and so on. Among them, setting up and testing in the constraint factor, use the analytic hierarchy process (AHP) to establish the judgment matrix of factor comparison and consistency checking.

$$A = \begin{pmatrix} a_{11} & \cdots & a_{1N} \\ \vdots & \ddots & \vdots \\ a_{N1} & \cdots & a_{NN} \end{pmatrix}, \ a_{ii} = 1, \ a_{ij} = 1/a_{ji}$$

(a_{ij} is ratio of importance between j and i) Passing this test, the matrix generates a reasonable constraint factor weight.

Based on policy making, the study of path computation is to be carried out and will introduce a linear programming (LP), set a new cost function. A case study of 4 factors:

$$b(e) = \frac{\alpha * c(c)}{C} + \frac{\beta * d(c)}{D} + \frac{\chi * b(c)}{B} + \frac{\delta * c(c)}{E}$$

(Where c, d, b, e representing the network application of total costs, total delay and total bandwidth, total energy consumption maximum constraint value, c (e), d (e), b (e), e (e) representing the constraint factor value corresponding to each side of the path) Subsequently, b (e) as a parameter, combined with the classic collection of shortest paths (SPP) algorithm is iterative routing path (single or k disjoint path), path cost, latency, bandwidth, energy consumption satisfy the corresponding constraint.

4 Conclusion

This paper investigates and proposes efficient SDN routing pre-design solution in three aspects: flow feature extraction, requirement prediction and route selection. First, we analyze and extract data packet and association features from user history data, apply these features into semi-supervised clustering algorithm for efficient data classification, analysis and feature extraction. After that, flow service requirement could be predicted through extraction of user. Flow and data plane load features and implementation of supervised classification algorithm. Furthermore, related to data plane topology, we propose corresponding handling strategies, flow forwarding and multi-constraint weight assignment, and proposes personalized routing selection mechanism.

References

1. Diego, K., Ramos, F.M.V., Esteves, V.P., Esteve, R.C., Siamak, A., Steve, U.: Software-defined networking: a comprehensive survey. Proc. IEEE **103**, 14–76 (2015)
2. 闵应骅, 我所理解的 "软件定义的网络", 视点, 10 (2014)
3. 左青云, 陈鸣, 赵广松, 邢长友, 张国敏, 蒋培成, 基于 Open Flow 的 Sdn 技术研究, **软件学报**, 24 (2013), 1078–109
4. Akyildiz, I.F., Lee, A., Wang, P., Luo, M., Chou, W.: A roadmap for traffic engineering in SDN-openflow networks. Comput. Netw. **71**, 1–30 (2014)
5. Wang, N., Ho, K., Geroge, P., Mark, H.: An overview of routing optimization for internet traffic engineering. IEEE Commun. Surv. Tutorials **10**, 36–56 (2008)
6. Tourrilhes, J., Sharma, P., Banerjee, S., Pettit, J.: SDN and openflow evolution: a standards perspective. Computer **47**, 22–29 (2014)

7. Tootoonchian, A., Ganjali, Y.: Hyperflow: a distributed control plane for openflow. In: Proceedings of the 2010 Internet Network Management Conference on Research on Enterprise Networking, p. 3. USENIX Association, San Jose (2010)

8. Minlan, Y., Rexford, J., Freedman, M.J., Wang, J.: Scalable flow-based networking with difane. SIGCOMM Comput. Commun. Rev. **40**, 351–362 (2010)

9. Koponen, T., Casado, M., Gude, N., Stribling, J., Poutievski, L., Zhu, M., Ramanathan, R., Iwata, Y., Inoue, H., Hama, T., Onix, S.: A distributed control platform for large-scale production networks. In: Proceedings of the 9th USENIX Conference on Operating Systems Design and Implementation, pp. 1–6. USENIX Association, Vancouver (2010)

10. Luo, M., Tian, Y., Li, Q., Wang, J., Chou, W.: Sox – a generalized and extensible smart network openflow controller (X). In: Proceedings of the First SDN World Congress. Damsdadt, Germany (2012)

11. Sun, S.: A survey of multi-view machine learning. Neural Comput. Appl. **23**, 2031–2038 (2013)

12. Vink, J., de Haan, G.: Comparison of machine learning techniques for target detection. Artif. Intell. Rev. **43**, 125–139 (2015)

13. 何清, 李宁, 罗文娟, 史忠植, 大数据下的机器学习算法综述, 模式识别与人工智能 4, 9 (2014)

14. Wang, Y., Xiang, Y., Zhang, J., Zhou, W., Xie, B.: Internet traffic clustering with side information. J. Comput. Syst. Sci. **80**, 1021–1036 (2014)

15. Hernández-Campos, F., Jeffay, K., Smith, F.D.: Statistical clustering of internet communication patterns. In: Proceedings of Symposium on the Interface of Computing Science and Statistics (2003)

16. Zander, S., Nguyen, T., Armitage, G.: Automated traffic classification and application identification using machine learning. In: Proceedings of the IEEE Conference on Local Computer Networks 30th Anniversary, pp. 250–257. IEEE Computer Society (2005)

17. Erman, J., Mahanti, A., Arlitt, M.: Internet Traffic Identification Using

18. Bernaille, L., Teixeira, R.: Early recognition of encrypted applications. In: Uhlig, S., Papagiannaki, K., Bonaventure, O. (eds.) PAM 2007. LNCS, vol. 4427, pp. 165–175. Springer, Heidelberg (2007)

19. 李笛, 胡学钢, 胡春玲, 主动贝叶斯分类方法研究, 计算机研究与发展, 47–51 (2007)

20. Yue, S., Li, P., Hao, P.: SVM classification: its contents and challenges. Appl. Math. A J. Chinese Univ. **18**, 332–342 (2003)

21. 黄金杰, 挛士勇, 广义粗糙集模型及应用, 模式识别与人工智能 17, 184–189 (2004)

22. Chen, Y.-L., Hsu, C.-L., Chou, S.-C.: Constructing a multi-valued and multi-labeled decision tree. Expert Syst. Appl. **25**, 199–209 (2003)

23. Chen, D.S., Jain, R.C.: A robust backpropagation learning algorithm for function approximation. IEEE Trans. Neural Netw. **5**, 467–479 (1994)

24. Vijayan, S., Ramachandran, K.I.: Effect of number of features on classification of roller bearing faults using SVM and PSVM. Expert Syst. Appl. **38**, 4088–4096 (2011)

25. Suykens, J.A.K., Vandewalle, J., De Moor, B.: Optimal control by least squares support vector machines. Neural Netw. **14**, 23–35 (2001)

26. Fung, G., Mangasarian, O.L.: Finite newton method for lagrangian support vector machine classification. Neurocomputing **55**, 39–55 (2003)

27. Zhu, Q.-Y., Qin, A.K., Suganthan, P.N., Huang, G.-B.: Evolutionary extreme learning machine. Pattern Recogn. **38**, 1759–1763 (2005)

28. Emilio, S.-O., Juan, G.-S., Marcelino, M., Rafael, M., Antonio, S., et al.: Belm: Bayesian extreme learning machine. IEEE Trans. Neural Netw. **22**, 505–509 (2011)

29. Liang, N.-Y., Huang, G.-B., Paramasivan, S., Narasimhan, S.: A fast and accurate online sequential learning algorithm for feedforward networks. IEEE Trans. Neural Netw. **17**, 1411–1423 (2006)
30. Narsingh, D., Chi-Yin, P.: Shortest-path algorithms: taxonomy and annotation. Networks **14**, 275–323 (1984)
31. Surballe, J.W., Tarjan, R.E.: A quick method for finding shortest pairs of disjoint paths. Networks **14**, 325–336 (1984)
32. Orda, A., Sprintson, A.: Efficient algorithms for computing disjoint Qos paths. In: Twenty-third Annual Joint Conference of the IEEE Computer and Communications Societies, INFOCOM 2004, p. 738 (2004)
33. Guo, L., Shen, H., Liao, K.: Improved approximation algorithms for computing k disjoint paths subject to two constraints. J. Comb. Optim. **29**, 153–164 (2015)
34. Chen, J., Zheng, X., Guo, W.: A survey on software defined networking. In: 2015 IEEE International Conference on Cloud Computing and Big Data, CloudCom-Asia 2015, Huangshan
35. Zheng, X., Ye, H., Tang, C., Rong, C., Chen, G.: A survey on cloud accountability. In: IEEE International Conference on Cloud Computing and Big Data, CloudCom-Asia 2013, Fuzhou, China, December 2013
36. Zheng, X., An, D., Guo, W.: Interest prediction in social networks based on Markov chain modeling on clustered users. In: 2015 IEEE International Conference on Cloud Computing and Big Data, CloudCom-Asia 2015, HuangShan, China (2015)
37. Zheng, X., Zeng, Z., Chen, Z., Yu, Y., Rong, C.: Detecting spammers on social networks. Neurocomputing (2015). http://dx.doi.org/10.1016/j.neucom.2015.02.047
38. Zheng, X., Zhang, X., Yu, Y., Kechadi, T.: Extreme learning machine based spammer detection in social networks. In: Proceedings of 2015 IEEE International Conference on Cloud Computing and Big Data, China (2015)
39. Zheng, X., Chen, N., Chen, Z., Rong, C., Guo, W.: Mobile cloud based framework for remote-resident multimedia discovery and access. J. Internet Technol. **15**, 1043–1050 (2014)

Imbalanced Extreme Learning Machine Based on Probability Density Estimation

Jü Yang, Hualong Yu$^{(\boxtimes)}$, Xibei Yang, and Xin Zuo

School of Computer Science and Engineering,
Jiangsu University of Science and Technology,
Zhenjiang 212003, Jiangsu, People's Republic of China
yangju_justcs@126.com, yuhualong@just.edu.cn,
yangxibei1980@sina.com, 632343650@qq.com

Abstract. Extreme learning machine (ELM) is a fast algorithm to train single-hidden layer feedforward neural networks (SLFNs). Like the traditional classification algorithms, such as decision tree, Naïve Bayes classifier and support vector machine, ELM also tends to provide biased classification results when the classification tasks are imbalanced. In this article, we first analyze the relationship between ELM and Naïve Bayes classifier, and then take the decision outputs of all training instances in ELM as probability density representation by kernel probability density estimation method. Finally, the optimal classification hyperplane can be determined by finding the intersection point of two probability density distribution curves. Experimental results on thirty-two imbalanced data sets indicate that the proposed algorithm can address class imbalance problem effectively, as well outperform some existing class imbalance learning algorithms in the context of ELM.

Keywords: Extreme learning machine · Class imbalance learning · Probability density estimation · Naïve Bayes classifier

1 Introduction

Extreme learning machine proposed by Huang *et al.*, [1] has become a popular research topic in machine learning in recent years [2]. It is proved that single-hidden layer feedforward neural networks (SLFNs) with arbitrary hidden parameters and continuous activation function can universally approximate to any continuous functions [1]. Some recent research [3–6], however, indicated that the performance of ELM could be destroyed by class imbalance distribution, which is similar with some traditional classifiers, such as support vector machine, Naïve Bayes classifier and decision tree. In class imbalance scenario, the accuracy of the minority class always tends to be underestimated, causing meaningless classification results [7]. Therefore, it is necessary to adopt some strategies to make the classification model provide impartial classification results.

In the context of ELM, some researchers have presented several class imbalance learning algorithms. Weighted extreme learning machine (WELM) appoints different penalty parameters for the training errors belonging to the instances in different categories, decreasing the possibility of misclassifying the minority class samples [3]. The

A. Bikakis and X. Zheng (Eds.): MIWAI 2015, LNAI 9426, pp. 160–167, 2015.
DOI: 10.1007/978-3-319-26181-2_15

penalty parameters, however, can be only allocated empirically. A similar algorithm called Fuzzy ELM (FELM) was proposed in [4], which changes the distributions of penalty parameters by inserting a fuzzy matrix. As two well-known data-layer class imbalance learning algorithms, random oversampling (ROS) and synthetic minority oversampling technology (SMOTE) have also be integrated into ELM to deal with practical class imbalance applications [5, 6].

In this article, we try to present a novel algorithm to deal with class imbalance problem in the context of ELM. First, we analyze the relationship between ELM and Naïve Bayes classifier, and indicate that the decision output in ELM approximately equals to the posterior probability in Naïve Bayes classifier. Then, on the decision output space, we estimate the probability density distributions for two different classes, respectively. Finally, the optimal position of the classification hyperplane can be determined by finding the intersection point of two probability density distribution curves. We compare the proposed algorithm with several popular class imbalance learning algorithms, and the experimental results indicate its superiority.

2 Theories and Methods

2.1 Extreme Learning Machine

Considering a supervised learning problem where we have a training set with N training instances and m classes, $(x_i, t_i) \in R^n \times R^m$. Here, x_i is an $n \times 1$ input vector and t_i is the corresponding $m \times 1$ target vector. ELM aims to learn a decision rule or an approximation function based on the training data. In other words, ELM is used to create an approximately accurate mapping relationship between x_i and t_i.

Unlike the traditional back propagation (BP) algorithm [8], ELM provides the hidden parameters randomly to training SLFNs. Suppose there are L hidden layer nodes, then for an instance x, the corresponding hidden layer output can be presented by a row vector $h(x) = [h_1(x), \ldots, h_L(x)]$, thus the mathematical model of ELM is:

$$H\beta = T \tag{1}$$

where $H = [h(x_1), \ldots, h(x_N)]^T$ is the hidden layer output matrix for the whole training set, β is the output weight matrix and T is the target vector. Here, only the output weight matrix β is unknown. Then we can adopt least square method to acquire the solution of β that can be described as follows:

$$\begin{cases} \beta = \hat{H}^\dagger T = H^T(\frac{I}{C} + HH^T)^{-1}T, \text{when } N \leq L \\ \beta = \hat{H}^\dagger T = (\frac{I}{C} + H^TH)^{-1}H^TT, \text{when } N > L \end{cases} \tag{2}$$

Here, $\hat{H}^\dagger$ is the Moore-Penrose "generalized" inverse of the hidden layer output matrix H, which can guarantee the solution is least norm least square solution of Eq. (1). C is the penalty parameter to mediate the balance relationship between the training errors and the generalization ability.

2.2 Relationship Between ELM and Naïve Bayes Classifier

According to some previous work, the decision outputs of SLFNs trained by BP algorithm [8] can be regarded as an approximation of posteriori probability functions in Naïve Bayes classifier [9, 10]. Suppose there are lots of training instances, and each of them belongs to one of m classes. We can train an SLFNs to obtain the output weight matrix w. Let $f_k(x, w)$ be the output of the kth output node of the SLFNs, i.e., the discriminant function corresponding to the kth class w_k, then we can recall Bayes formula,

$$P(w_k|x) = \frac{P(x|w_k)P(w_k)}{\sum_{i=1}^{m} P(x|w_i)P(w_i)} = \frac{p(x, w_k)}{P(x)}$$

and the Bayes decision for any instance x: choosing the class w_k which has the largest discriminant function $f_k(x) = P(w_k|x)$. Without loss of generality, suppose the training outputs are restricted as $\{0, 1\}$, where 1 denotes the output of the corresponding class and 0 denotes the outputs of the other classes. The contribution to the criterion function based on a single output unit k for finite number of training samples x is:

$$J(w) = \sum_x (f_k(x, w) - t_k)^2$$

$$= \sum_{x \in w_k} (f_k(x, w) - 1)^2 + \sum_{x \notin w_k} (f_k(x, w) - 0)^2$$

$$= n\left\{\frac{n_k}{n}\frac{1}{n_k}\sum_{x \in w_k} (f_k(x, w) - 1)^2 + \frac{n - n_k}{n}\frac{1}{n - n_k}\sum_{x \notin w_k} (f_k(x, w) - 0)^2\right\} \quad (3)$$

where n denotes the number of training instances, while n_k stands for the number of instances belonging to the class w_k. In the limit of infinite data, we can use Bayes formula to express Eq. (3) as:

$$\lim_{n \to \infty} \frac{1}{n} J(w) = \tilde{J}(w)$$

$$= p(w_k)\int (f_k(x, w) - 1)^2 P(x|w_k)dx + p(w_{i \neq k})\int f_k(x, w)^2 P(x|w_{i \neq k})dx$$

$$= \int f_k^2(x, w)p(x)dx - 2\int f_k(x, w)p(x, w_k)dx + \int p(x, w_k)dx \quad (4)$$

$$= \int (f_k(x, w) - p(w_k|x))^2 p(x)dx + \int p(w_k|x)p(w_{i \neq k}|x)p(x)dx$$

Obviously, the right-hand side in Eq. (4) is irrelevant with the weight w, thus SLFNs is to minimize:

$$\int (f_k(x, w) - p(w_k|x))^2 p(x)dx \quad (5)$$

Because this is true for each class, SLFNs minimizes the sum:

$$\sum_{k=1}^{m} \int (f_k(\mathbf{x}, \mathbf{w}) - p(w_k|\mathbf{x}))^2 p(\mathbf{x}) dx \qquad (6)$$

Therefore, in the limit of infinite data, the outputs of the trained SLFNs will approximate the true posterior probabilities in a least-squares sense, i.e., $f_k(\mathbf{x}, \mathbf{w}) = p(w_k|\mathbf{x})$.

As we know, like BP algorithm, ELM also provides approximate least squares solution of Eq. (1) though it simultaneously minimizes the norm of the weight matrix. Therefore, the decision outputs in ELM reflect the posteriori probabilities of different classes in Naïve Bayes classifier to some extent.

2.3   Probability Density Estimation

As described above, the decision outputs in ELM can reflect posteriori probabilities of different classes, thus for binary-class problem, we can map all instances from original feature space to an one-dimensional space. Here, ELM is used as a feature extraction tool. Then, we regard to estimate the probability density distributions of two different classes on the compressed one-dimensional feature space. Specifically, kernel probability density estimation [11], which is a nonparametric way to estimate the probability function of a random variable, is adopted.

Figure 1 shows a schematic diagram about probability density distributions of a binary-class problem on the one-dimensional decision output space acquired from ELM. From Fig. 1, we observe that after estimating probability density distributions, the prior probabilities for the two classes $p(w_+)$ and $p(w_-)$, and the corresponding conditional distribution probabilities $p(\mathbf{x}|w_+)$ and $p(\mathbf{x}|w_-)$ can be both obtained, then recall Bayes function, the posterior probabilities of two classes can be calculated as:

$$p(w_+|\mathbf{x}) = \frac{p(\mathbf{x}|w_+)p(w_+)}{p(\mathbf{x})}, p(w_-|\mathbf{x}) = \frac{p(\mathbf{x}|w_-)p(w_-)}{p(\mathbf{x})} \qquad (7)$$

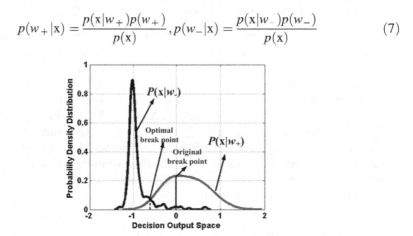

Fig. 1. Probability density distributions, original and optimal break points for a binary-class imbalanced problem, where w_+ denotes the positive class (minority class), and where w_- denotes the negative class (majority class).

It is clear that when $p(w_+|x) = p(w_-|x)$, i.e., when $p(x|w_+)p(w_+) = p(x|w_-)p(w_-)$, the corresponding x value is selected as break point in Bayes decision. For class imbalance data, however, because $p(w_+) \ll p(w_-)$, to guarantee $p(x|w_+)p(w_+) = p(x|w_-)p(w_-)$, the actual break point will be pushed towards the minority class. Therefore, Fig. 1 shows that using the original break point acquired from ELM will seriously destroy the performance of the minority class, but when we find the horizon axis corresponding to the intersection point of two density distribution curves, it can be seen as the optimal break point for the classification task.

2.4   Description of the Proposed Algorithm

The detailed computational procedure of the proposed algorithm is described as follows:

Input: A binary-class training set $S = (x_i, t_i) \in R^n \times R^1$, where i=1,..., N, $t_i \in \{-1, 1\}$; the number of hidden layer nodes L; the penalty parameter C existing in Eq.(2).

Output: The shift s between the original and optimal break points, and the modified ELM classifier F'.

Process:

a. Adopt the training set S and the penalty parameter C to train an ELM classifier F which has L hidden layer nodes;
b. Acquire the decision output of each training instance, and then map them into an one-dimensional feature space;
c. Obtain probability density distributions of two classes by using kernel probability density estimation approach, respectively;
d. Find the horizon axis x which corresponds to the intersection point of two probability density distribution curves, and then calculate the shift s by using the formula s=0-x;
e. Update ELM classifier from F to F' by adding s on each original decision output.

3   Experiments

3.1   Datasets and Parameters Settings

The experiments are carried out on thirty-two binary-class imbalanced data sets acquired from Keel data repository [12]. The detailed information of these data sets are summarized in Table 1, where IR denotes the class imbalance ratio.

To present the superiority of the proposed algorithm, we compared it with some other class imbalance learning algorithms in the context of ELM, including ELM [1], WELM1 [3], WELM2 [3], ELM-RUS, ELM-ROS [5] and ELM-SMOTE [6]. In addition, to guarantee the impartiality of the comparative results, grid search was adopted to search the optimal parameters, where sigmoid function was used as activation function at the hidden level, and two other parameters L and C were selected from $\{10, 20, ..., 200\}$ and $\{2^{-20}, 2^{-18}, ..., 2^{20}\}$, respectively. As for the performance evaluation, G-mean metric was used.

3.2   Results and Discussions

Considering the randomness of ELM, five fold cross-validation was adopted, and each experiment was randomly executed 10 times, finally the average classification results

Table 1. Data sets used in the experiments.

Data set	Number of features	Number of instances	IR
ecoli3	7	336	8.6
glass1	9	214	1.82
haberman	3	306	2.78
new_thyroid1	5	215	5.14
pima	8	768	1.87
vehicle1	18	846	2.9
wisconsin	9	683	1.86
yeast3	8	1484	8.1
abalone9_18	8	731	16.4
ecoli4	7	336	15.8
shuttle_c0_vs_c4	9	1829	13.87
vowel0	13	988	9.98
yeast4	8	1484	28.1
yeast5	8	1484	32.73
page_blocks0	10	5472	8.79
ecoli_0_1_vs_2_3_5	7	244	9.17
ecoli_0_1_vs_5	6	240	11
ecoli_0_3_4_vs_5	7	200	9
ecoli_0_6_7_vs_3_5	7	222	9.09
ecoli_0_6_7_vs_5	6	220	10
led7digit_0_2_4_5_6_7_8_9_vs_1	7	443	10.97
yeast_0_2_5_7_9_vs_3_6_8	8	1004	9.14
yeast_0_3_5_9_vs_7_8	8	506	9.12
cleveland_0_vs_4	13	177	12.62
shuttle_2_vs_5	9	3316	66.67
shuttle_6_vs_2-3	9	230	22
winequality-red_4	11	1599	29.17
winequality-red_3_vs_5	11	691	68.1
iris0	4	150	2
page-blocks_1_3_vs_4	10	472	15.86
vehicle0	18	846	3.25
glass_0_1_5_vs_2	9	172	9.12

were given. Table 2 provides the G-mean values of various algorithms, where in each row, the bold denotes the best result, the underline labels the second best and the italic stands for the worst one.

From Table 2, we observe that nearly all other algorithms outperform the original ELM algorithm, demonstrating each bias correction strategy can effectively alleviate the class imbalance problem. We also note that sampling and weighting technologies have quitely similar classification performance. As we know, WELM1 and WELM2 adopt different weights to punish the training errors, but there seems no a clear winner between them, as they have acquired the highest G-mean values on five data sets,

Table 2. G-mean values of various algorithms on 32 imbalanced data sets.

Data set	ELM	WELM1	WELM2	ELM-RUS	ELM-ROS	ELM-SMOTE	Proposed algorithm
ecoli3	*0.7182*	0.8792	**0.8883**	0.8733	0.8780	0.8602	0.8800
glass1	0.6712	0.6923	*0.6457*	0.6976	**0.7030**	0.6932	0.7020
haberman	*0.4703*	0.6262	0.5300	0.6305	0.6310	0.6103	**0.6419**
new_thyroid1	*0.8540*	0.9683	0.9327	0.9787	**0.9944**	0.9743	0.9831
pima	*0.6936*	0.7442	0.7127	0.7296	0.7314	**0.7483**	0.7474
vehicle1	*0.7709*	0.8248	0.7751	0.8049	0.8285	**0.8327**	0.8293
wisconsin	*0.9570*	**0.9711**	0.9667	0.9615	0.9597	0.9667	0.9707
yeast3	*0.7612*	0.9072	0.8968	0.9199	0.9197	0.9207	**0.9221**
abalone9_18	*0.3536*	0.8498	0.8267	0.7954	0.7851	0.8001	**0.8729**
ecoli4	*0.7966*	**0.9719**	0.8532	0.8891	0.9509	0.8292	0.9661
shuttle_c0_vs_c4	*0.9930*	0.9932	**0.9962**	*0.9930*	0.9932	0.9932	*0.9930*
vowel0	0.9858	0.9860	0.9888	*0.9672*	**0.9981**	0.9978	0.9909
yeast4	*0.2969*	0.8041	0.8088	0.8007	0.8090	0.8133	**0.8167**
yeast5	*0.5668*	0.9623	**0.9684**	0.9414	0.9615	0.9425	0.9680
page_blocks0	*0.7914*	0.8640	0.8510	0.9123	0.9190	0.9199	**0.9265**
ecoli_0_1_vs_2_3_5	0.8624	0.8844	0.8849	*0.8140*	0.8406	**0.8973**	0.8891
ecoli_0_1_vs_5	*0.8737*	0.9072	0.9129	**0.9213**	0.8942	0.8842	0.9181
ecoli_0_3_4_vs_5	*0.8398*	0.8761	0.8765	0.8986	**0.9394**	0.9283	0.9352
ecoli_0_6_7_vs_3_5	*0.7809*	0.8649	**0.8900**	0.7970	0.8112	0.8697	0.8817
ecoli_0_6_7_vs_5	*0.8481*	0.8878	0.8962	0.8629	0.8853	0.8966	**0.8996**
led7digit_0_2_4_5_6_7_8_9_vs_1	**0.8809**	0.8674	0.8653	*0.8223*	0.8331	0.8667	0.8514
yeast_0_2_5_7_9_vs_3_6_8	*0.8497*	0.9072	0.9061	0.8913	0.8987	**0.9082**	0.9063
yeast_0_3_5_9_vs_7-8	*0.4346*	**0.6934**	0.6690	0.6486	0.6526	0.6630	0.6847
cleveland_0_vs_4	0.6549	0.7934	0.7073	**0.8270**	*0.6348*	0.7079	0.8189
shuttle_2_vs_5	*0.9140*	0.9609	0.9389	0.9973	**0.9990**	0.9988	0.9986
shuttle_6_vs_2-3	0.9573	0.9573	0.9573	*0.9357*	0.9573	0.9707	**1.0000**
winequality-red_4	*0.0000*	**0.6936**	0.6405	0.6408	0.6630	0.6329	0.6733
winequality-red_3_vs_5	*0.0000*	0.4057	0.4823	0.6388	0.3562	0.2688	**0.6470**
iris0	**1.0000**	**1.0000**	**1.0000**	**1.0000**	**1.0000**	**1.0000**	**1.0000**
page-blocks_1_3_vs_4	*0.8186*	0.9770	0.9608	0.9338	0.9549	**0.9841**	0.9782
vehicle0	0.9756	0.9744	*0.9662*	0.9703	0.9716	**0.9757**	0.9719
glass_0_1_5_vs_2	*0.1146*	0.7645	0.7308	0.6829	0.7606	**0.7730**	0.7555

respectively. We consider that the optimal weight settings should be closely related with the practical instance distributions. In sampling series algorithms, oversampling performs better than undersampling on majority data sets, especially on those highly skewed ones. On these data sets, the instances in the minority class are quitely sparse, causing much useful information loss by using RUS algorithm. Moreover, ELM-SMOTE obviously outperforms ELM-ROS as it have acquired two more best results and seven more second best results.

In contrast with six other algorithms, the proposed algorithm performs best, because it has acquired the highest G-mean value on nine data sets and the second highest G-mean value on fifteen ones. The results demonstrate that exploring the prior information about data distribution is helpful for improving classification performance in class imbalance tasks more or less.

4 Conclusions

In this article, a probability density estimation-based ELM classification algorithm is proposed to classify imbalanced data. Unlike other class imbalance learning algorithms, the proposed algorithm does not need to change the original data or weight distributions, but only to estimate the probability density distribution of the decision outputs acquired from ELM and then to find the optimal position to place the classification hyperplane. Experimental results on thirty-two benchmark data sets verified the effectiveness and superiority of the proposed algorithm.

Acknowledgements. This work was supported in part by National Natural Science Foundation of China under grant No. 61305058, Natural Science Foundation of Jiangsu Province of China under grant No. BK20130471, and China Postdoctoral Science Foundation under grant No. 2013M540404 and No. 2015T80481.

References

1. Huang, G.B., Zhou, H., Ding, X., Zhang, R.: Extreme learning machine for regression and multiclass classification. IEEE Trans. Syst. Man Cybern. Part B Cybern. **42**, 513–529 (2012)
2. Huang, G., Huang, G.B., Song, S., You, K.: Trends in extreme learning machine: a review. Neural Netw. **61**, 32–48 (2015)
3. Zong, W., Huang, G.B., Chen, Y.: Weighted extreme learning machine for imbalance learning. Neurocomputing **101**, 229–242 (2013)
4. Zhang, W.B., Ji, H.B.: Fuzzy extreme learning machine for classification. IET Electron. Lett. **49**, 448–450 (2013)
5. Vong, C.M., Ip, W.F., Wong, P.K., Chiu, C.C.: Predicting minority class for suspended particulate matters level by extreme learning machine. Neurocomputing **128**, 136–144 (2014)
6. Sun, S.J., Chang, C., Hsu, M.F.: Multiple extreme learning machines for a two-class imbalance corporate life cycle prediction. Knowl. Based Syst. **39**, 214–223 (2013)
7. He, H., Garcia, E.A.: Learning from imbalanced data. IEEE Trans. Knowl. Data Eng. **21**, 1263–1284 (2009)
8. Rumelhart, D.E., Hinton, G.E., Williams, R.J.: Learning representations by back-propagation errors. Nature **323**, 533–536 (1986)
9. Ruck, D.W., Rogers, S.K., Kabrisky, M., Oxley, M.E., Suter, B.W.: The multilayer perceptron as an approximation to a Bayes optimal discriminant function. IEEE Trans. Neural Netw. **1**, 296–298 (1990)
10. Wan, E.A.: Neural network classification: a Bayesian interpretation. IEEE Trans. Neural Netw. **1**, 303–305 (1990)
11. Parzen, E.: On estimation of a probability density function and mode. Ann. Math. Stat. **33**, 1065–1076 (1962)
12. Alcalá-Fdez, J., Fernandez, A., Luengo, J., Derrac, J., García, S., Sánchez, L., Herrera, F.: KEEL data-mining software tool: data set repository, integration of algorithms and experimental analysis framework. J. Multiple-Valued Logic Soft Comput. **17**, 255–287 (2011)

Computer Vision

Integrating Simplified Inverse Representation and CRC for Face Recognition

Yingnan Zhao[1,2,3(✉)], Xiangjian He[3], Beijing Chen[1,2],
and Xiaoping Zhao[1,2]

[1] Jiangsu Engineering Center of Network Monitoring,
Nanjing University of Information Science & Technology, Nanjing 210024, China
yingnan.zhao@uts.edu.au, nbutimage@126.com,
xp_zhao@163.com
[2] School of Computer & Software,
Nanjing University of Information Science & Technology, Nanjing 210044, China
[3] School of Computing and Communications,
University of Technology Sydney, Sydney, Australia
xiangjian.he@uts.edu.au

Abstract. The representation based classification method (RBCM) has attracted much attention in the last decade. RBCM exploits the linear combination of training samples to represent the test sample, which is then classified according to the minimum reconstruction residual. Recently, an interesting concept, Inverse Representation (IR), is proposed. It is the inverse process of conventional RBCMs. IR applies test samples' information to represent each training sample, and then classifies the training sample as a useful supplement for the final classification. The relative algorithm CIRLRC, integrating IR and linear regression classification (LRC) by score fusing, shows superior classification performance. However, there are two main drawbacks in CIRLRC. First, the IR in CIRLRC is not pure, because the test vector contains some training sample information. The other is the computation inefficiency because CIRLRC should solve C linear equations for classifying the test sample respectively, where C is the number of the classes. Therefore, we present a novel method integrating simplified IR (SIR) and collaborative representation classification (CRC), named SIRCRC, for face recognition. In SIRCRC, only test sample information is fully used in SIR, and CRC is more efficient than LRC in terms of speed, thus, one linear equation system is needed. Extensive experimental results on face databases show that it is very competitive with both CIRLRC and the state-of-the-art RBCM.

Keywords: Face recognition · Inverse representation · Collaborate recognition classification

1 Introduction

The representation based classification method (RBCM) has emerged as a powerful tool in a wide range of application fields, especially in signal processing [1, 2], image processing [3–5] and visual tracking [6, 7]. It plays an important role in biometrics recognition, such as recognitions of faces [8–12], palmprints [13], ears [14],

© Springer International Publishing Switzerland 2015
A. Bikakis and X. Zheng (Eds.): MIWAI 2015, LNAI 9426, pp. 171–183, 2015.
DOI: 10.1007/978-3-319-26181-2_16

fingerprints [15] and irises [16]. RBCM requires a test sample to be sparsely represented by a weighted sum of all the training samples. According to the types of norm regularization, RBCM can be classified into three categories, which are, l_1 norm, l_2 norm and l_p norm $(0 < p < 1)$.

It has been studied that sparsity can be measured by l_0 norm. Nevertheless, it is not tractable in real applications due to its NP-hard property. l_1 minimization, the closest convex function to l_0 norm, is thereby widely used in RBCM. Sparse representation based Classification (SRC) [8] is regarded as a milestone of RBCM in face recognition, which has boosted the corresponding research, and extensive RBCMs with l_1 norm minimization have been proposed [17, 18]. SRC codes the query face image as a sparse linear combination of all the training samples via l_1 norm minimization. It not only yields high recognition accuracy in face databases, but also is robust to face occlusion and corruption. However, Zhang et al. pointed out that it was collaborative representation (CR) rather than l_1 norm sparsity that contributes to the final classification accuracy [9]. Based on the non-sparse l_2 norm, collaborate representation classification (CRC) [9] could lead to similar recognition but a significantly highly computational speed. It is noted that all the feature elements both in SRC and CRC share the same coding vector over their associated sub-dictionaries. This requirement ignores the fact that the feature elements in a pattern not only share similarities but also have differences. Therefore, Yang et al. presented a relaxed collaborative representation (RCR) [11] model to effectively exploit the similarity and distinctiveness of features. Liner Regression Classification (LRC) [12] can be referred as a l_2 norm based on the linear regression model. In addition to the l_1 norm and l_2 norm minimizations, some researchers have been trying to solve the sparse representation problem with the l_p norm for $0 < p < 1$, especially $p = 0.1$, 1/2, 1/3, or 0.9 [19–21]. More information about RBCM can be found in a review [22].

More recently, an interesting concept, Inverse Representation (IR), was proposed. IR tries to classify a training sample using test samples' information. It is the inverse process of RBCM by essence. Integrating IR and LRC by score fusing, CIRLRC [23] has superior classification performance. However, there are two main drawbacks in CIRLRC. First, the IR in CIRLRC is not a pure one, because the IR test vector contains some training sample information. It accordingly increases the relativity between IR and LRC, which is not good for core fusing. Second, CIRLRC is computational inefficiency that it needs to solve C linear equations, where C is the number of the classes. Therefore, we present a novel method integrating simplified IR (SIR) and collaborative representation classification (CRC), named SIRCRC, for face recognition.

The rest of this paper is organized as follows. Section 2 gives a brief review of some related works. Section 3 presents the proposed SIRCRC method and some analysis. Section 4 performs experiments and Sect. 5 concludes the paper.

2   Related Works

2.1   CIRLRC

The CIRLRC [23] exploits the conventional LRC and IR to generate two kinds of scores, and then combines them by score fusing to recognize a face. The comparison of

LRC and CRC is presented in Sect. 2.2. In CIRLRC, the IR is the inverse process of LRC and mirror samples are used to form the visual samples. Assume that there are C classes and each class has n training samples. We denote $X = [x_1, \cdots, x_N]$ as all N training samples, and $y \in Y$ as the test sample. The main steps in CIRLRC are as follows.

Step 1. If screenshots are necessary, please make sure that you are happy with the print quality before you send the files. Mirror sample creation. Calculate the mirror samples [24] of training samples x and test sample y, respectively, denoted by x^v and y^v.

Step 2. LRC process. Let the training sample vector contain original samples and their mirror images, denoted by $X' = [x_1 \cdots x_n x_1^v \cdots x_n^v \cdots x_{(C-1)n+1} \cdots x_N x_{(C-1)n+1}^v \cdots x_N^v]$ ($N = Cn$). The linear system is $y = X_i' \alpha_i$, $i = 1, \cdots, C$. Using the LRC method to classify each test sample and yield one score named S_{LRC}.

Step 3. IR process.

(a) Test vector definition. For the j-th class, combine all the naive training samples from the other classes, the naive test sample and virtual test sample to form the test vector Z, i.e., $Z = [X_1' \cdots X_{j-1}' \ X_{j+1}' \cdots X_C' \quad y \quad y^v]$.

(b) IR classification. Z acts as the training vector and each naive training sample from the j-th class acts as the test sample. Then, apply inverse LRC to classify each naive training sample from the j-th class. Here, the residual is calculated using $d_k^0 = \left\| X - \hat{\beta}_k^{(C-1)n+1} y - \hat{\beta}_k^{(C-1)n+2} y^v \right\|$. For the j-th class, the mean of d_k^0 is used as the distance between the test sample and the j-th class and is denoted by d_j. Based on the inverse LRC method, another score, named S_{IR}, is yield.

Step 4. Score fusing. The final score is calculated by $S = \omega_1 S_{LRC} + \omega_2 S_{IR}$, where ω_1 and ω_2 are fusing coefficients, and $\omega_1 + \omega_2 = 1$. According to the ultimate score, classify the test sample finally.

It can be observed that the test vector contains some training sample information in Step 3. Therefore, it is not a pure IR. In the proposed SIRCRC, we simplify the IR definition, which contains only test sample information for better classification accuracy.

2.2 Comparison of LRC and CRC

In this subsection, we present a comparison of LRC and CRC. Tables 1 and 2 show the LRC and CRC algorithms respectively.

During the residual calculation process, both LRC and CRC fall into the l_2 minimization category according to Eq. (2) and Eq. (4). It can be seen from the tables that the main difference is in Eq. (1) and Eq. (3). Indeed, LRC and CRC are based on different ideas. The former is based on the linear regression model, and the latter is the collaborative representation. While, they can both be attributed to the minimum squired error (MSE) problem. For example, Eq. (1) can be rewritten as the following, aiming to

Table 1. LRC Algorithm.

1. $\hat{\beta}_i \in R^{P_i \times 1}$ is evaluated against each class model,

$$\hat{\beta}_i = (X_i^T X_i)^{-1} X_i^T y, i = 1, 2, \cdots, C. \tag{1}$$

2. Calculate the distance between original and predicted response variable

$$d_i(y) = \left\| y - X_i \hat{\beta}_i \right\|_2, i = 1, 2, \cdots, C. \tag{2}$$

3. If $k = \arg\min_i d_i(y)$, then the test sample is assigned to the

k-th class.

Table 2. CRC Algorithm.

1. Normalize the columns of X to have unit l_2 -norm.
2. Code y by

$$\hat{\rho} = (X^T X + \lambda I)^{-1} X^T y. \tag{3}$$

3. Compute the regularized residuals.

$$r_i = \left\| y - X_i \hat{\rho}_i \right\|_2, i = 1, 2, \cdots, C. \tag{4}$$

4. Output the identity of y as Identity(y) $= \arg\min_i \{r_i\}$.

make the least square solution stable and to impose a weaker sparsity constraint on the solution in real application:

$$\hat{\beta}_i = (X_i^T X_i + \lambda I)^{-1} X_i^T y, \quad i = 1, 2, \cdots, N \tag{5}$$

where λ is a positive constant, which is the same as in Eq. (3). However, in Eq. (1), LRC needs to solve C linear equations for classifying the test sample, where C is the number of the classes. Equation (3) in CRC is more efficient than LRC in terms of speed, that is, only one linear equation system is needed. Therefore, we prefer CRC in the proposed method.

3   Description of Proposed Method SIRCRC

3.1   SIRCRC Framework

In this subsection, we illustrate the proposed SIRCRC in detail. Comparing to the CIRLRC, we make two improvements. One is the simplification of the IR definition, which is more reasonable and leads to higher classification accuracy. The other is in integrating the presented SIR to CRC instead of LRC, which produces higher computation speed. The number of training samples is always bigger than that of test samples in real applications. When we try to consider the test vector as the training samples according to the IR definition, we need to enlarge its number. In other words, we need to construct visual test samples, and constructing mirror images is a simple but an effective method [24]. Let $X = [x_1, \cdots, x_N]$ stand for the training sample vector, and $Y = [y_1, \cdots, y_M]$ be the test sample vector. Suppose that there are C classes and each class has n training samples and m testing samples. Obviously, $N = Cn$ and $M = Cm$. $x_{(i-1)n+k}$ stands for the k-th training sample of the i-th class. Similarly, $y_{(j-1)m+p}$ stands for the p-th training sample of the j-th class.

The proposed method is described as follows. First, it forms the test vector by the test sample and its mirror image. Note that all the test sample information is in the test vector, which contains no training information at all unlike CIRLRC. We believe that it is more natural according to the IR essence. Furthermore, for score fusing, the more uncorrelated each item is, the better recognition rate it will yield. The SIR definition is more independent than IR. Second, we obtain the optimal linear training samples from every class to represent the test sample, and calculate the score of each class. Here, the classical CRC method is applied. Third, we conduct SIR on the base of CRC. Finally, it fuses the scores produced from the second and third steps for the ultimate classification. These steps are presented in detail below.

Step 1. Produce the test vector. It contains two parts, the original test samples and their mirror images. For a test sample y, the virtual test sample is defined as

$$y^v(t, s) = y(t, S - s + 1), \quad t = 1, \cdots, T, \quad s = 1, \cdots, S \tag{6}$$

where T is the number of rows and S is the number of columns of the face image matrix.

Step 2. CRC procedure. We first establish the linear system

$$y = XA, \tag{7}$$

and then solve A using

$$\hat{A} = (X^T X + \lambda I)^{-1} X^T Y, \tag{8}$$

where λ and I stand for a small positive constant and the identity matrix, respectively. We calculate the CRC-residual between the test sample y and the i-th class, i.e. $r_i = \left\| y - X_i \hat{A} \right\|$.

Step 3. SIR procedure. First define the test vector Z as following

$$Z = [Y \quad Y^v]. \tag{9}$$

Then, we establish a linear system for each original training sample as

$$X = Z\beta. \tag{10}$$

β is solved by

$$\hat{\beta} = (Z^T Z + \lambda I) Z^T X, \tag{11}$$

where λ and I are the same as Eq. (8). We use the following

$$d_i = \left\| X_i - \hat{\beta} Z \right\| \tag{12}$$

as the SIR-distance of between the test sample X and the training sample Z.

Step 4. Score fusing. For a test sample y, we first normalize its CRC-residual and SIR-distance, and then use $t_j = \omega_1 r_j' + \omega_2 d_j'$ to calculate the ultimate score with respect to the j-th class. ω_1 and ω_2 are the weights and $\omega_1 + \omega_2 = 1$. If $k = \arg\min_j t_j$, then test sample y is assigned to the k-th class.

3.2   Analysis of SIRCRC

The proposed SIRCRC method has two main attributes. One is the simplification of IR. The other is that we apply CRC instead of LRC.

It can be seen from Step 3 in the referenced CIRLRC that the test vector contains three parts: the original training sample, the test sample and its mirror image. In the proposed SIRCRC, we simplify it by getting rid of the naive training sample. The essence of IR is the inverse procedure of conventional RBCM. Hereafter, in SIRCRC we make the best of the test sample information. Another drawback of the IR definition in CIRLRC is in the distance calculation phase. Owing to the training sample part in the test vector of IR, we could not calculate them to the naive training sample. Therefore, CIRLRC calculates the other two parts respectively and then averages them. On the other hand, SIR in the proposed SIRCRC, all the information in the test vector is about the test sample, and we can directly compute the residual difference. Therefore, it is natural, simple, and efficient.

In CIRLRC [23], as in real-world applications, an error exists in both the test sample and the training sample. The conventional RBCM is based on the least-squares algorithm. Actually, Eq. (7) can be rewritten as

$$XA = y = y_0 + \Delta y, \tag{13}$$

where y_0 and Δy stand for the true test sample and error, respectively. Equation (14) generates the following objective function:

$$\{\hat{A}, \Delta y\} = \arg\min(\|\Delta y\| + \lambda \|A\|) \qquad s.t. \quad XA = y_0 + \Delta y. \tag{14}$$

It is obvious in Eq. (15) that conventional RBCM only takes into account the error in the test sample. As for IR in CIRLRC, the inverse presentation of Eq. (7), $Z\beta = x$, can be rewritten as

$$Z\beta = x = x_0 + \Delta x, \tag{15}$$

where x_0 and Δx stand for the true train sample and error, respectively. It can be easily seen that the relative objective function is as follows:

$$\{\hat{\beta}, \Delta x\} = \arg\min(\|\Delta x\| + \lambda\|\beta\|) \quad s.t. \quad Z\beta = x_0 + \Delta x. \tag{16}$$

Hence, IR takes into account the error in the training sample. While using the fusing strategy, CIRLRC allows the errors both in the test sample and the training sample to be simultaneously considered and processed. This will be beneficial to achieve good face recognition performance. However, Eq. (14) takes account of the error in the training sample on the right side, but the error of the training sample in the test vector Z is neglected. When we use the proposed SIR, this drawback will be tackled.

Both CIRLRC and the presented SIRCRC finally adopt the integrating technique and score fusing. In score level fusion, if the correlation coefficient between the two kinds of scores is low, the fusion result is usually good. That is to say, a smaller correlation coefficient allows the fusion to have better accuracy. Here, we adopt the general correlation coefficient calculation as in [23]. Table 3 shows the mean of all the correlation coefficients of the CRC-residual and SIR-distance of the test samples from the ORL database. The highest mean of correlation coefficients of CIRLRC is 0.5469, while that in SIRCRC is 0.4939. It implies that SIRCRC has lower difference than CIRLRC, and hence the better fusion performance.

Table 3. Means of all the correlation coefficients of the residuals from Step 2 and Step 3 of CIRLRC and the proposed SIRCRC in the ORL face database.

Training sample per class	3	4	5
CIRLRC [23]	0.5469	0.5266	0.5024
SIRCRC	0.4939	0.4863	0.4733

Figure 1 depicts the CRC-residual and the SIR-distance of the last test sample, using Steps 2 and 3 in the case where the first 5 face images of each subject in the ORL database are used as training samples and the others are taken as test samples. Figure 1(a) is SIRCRC and Fig. 1(b) is CIRLRC. We can observe that, in the former, there are 3 points with similar data, while the latter has 9 such points. Hence, for SIRCRC, the difference of the correlation coefficient of scores in CRC and distance in SIR is greater than that in CIRLRC. Figure 2 depicts the CRC-residual and the SIR-distance, of the last test sample from the subset of the FERET database, using Steps 2 and 3 in the case where the first 4 images of each subject are used as training samples and the rest as the test samples. We can easily draw a similar conclusion to Fig. 1.

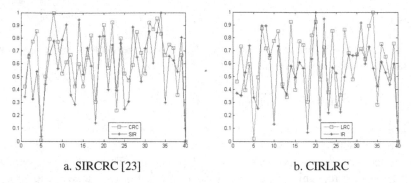

a. SIRCRC [23]   b. CIRLRC

Fig. 1. Residuals of the last test sample obtained using steps 2 and 3 of the methods. Figure 1(a) is SIRCRC and Fig. 1(b) is CIRLRC. The first 5 images of each subject in the ORL database are used as training samples and the others are taken as test samples. The vertical axis shows the values of residuals and the horizontal axis shows the no. of the component of the normalized score vector and distance vector.

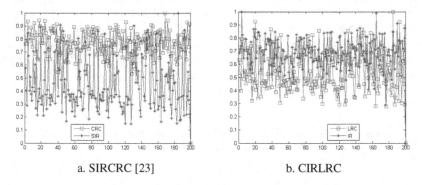

a. SIRCRC [23]   b. CIRLRC

Fig. 2. Residuals of the last test sample obtained using steps 2 and 3 of the methods. Figure 2(a) is SIRCRC and Fig. 2(b) is CIRLRC. The first 4 images of each subject in the FERET database are used as training samples and the others are taken as test samples. The vertical axis shows the values of residuals and the horizontal axis shows the no. of the components of the normalized score vector and distance vector.

4   Experiments

We use the ORL, FERET and Georgia Tech (GT) databases to test the presented SIRCRC. We compare it with CIRLRC [23], and some classical state-of-art face recognition methods as well, such as SRC [8], LRC [12], CRC [9] and RCR [11]. In SIRCRC, the weights of score fusing satisfy $\omega_1 + \omega_2 = 1$. Therefore, we just show the value of ω_1. While in CIRLRC, we only use the optimal weights. The parameter λ in Eqs. (8) and (11) is set to 0.001.

4.1   Experiments on the ORL Face Database

We use the ORL face database [24] to evaluate our method. There are 400 gray images from 40 subjects, and each subject provides 10 images. For some subjects, the images were taken at different times, with varying lighting, facial expressions and facial details. Each image was also resized to an image one half of the original size by using the down-sampling algorithm. We take the first 2, 3 and 4 face images of each subject as original training samples and treated the remaining face images as test samples. The experimental results are shown in Table 4. It shows that both CIRLRC and SIRCRC are able to perform better than the classical RBCMs, such as SRC, RCR, and LRC and CRC as well. In the row of CIRLRC, we also give the recognition error rates of LRC and CIR, which are the two parts of score fusing. In the same way, we list them, CRC and SIR, in the SIRCRC. Comparing CRC and LRC, it can been seen that CRC performs well on the ORL face database. But, SIR is not as good as CIR in terms of recognition performance. As we analyzed in Sect. 3.2, the two parts in SIRCRC have greater irrelevance than in CIRLRC. Hereafter, we obtain lower final error rates than CIRLRC on the ORL face database.

Table 4.   Recognition error rates (%) of different methods on the ORL face database.

Number of the original training samples per class	2	3	4
SRC [8]	22.25	11.93	09.92
RCR [11]	21.77	18.86	17.52
LRC [12]	21.56	18.93	15.42
CIR	20.26	18.57	17.92
CIRLRC ($\omega_1 = 0.6$) [23]	09.64	08.33	11.00
CRC [9]	16.56	15.00	11.25
SIR	21.38	20.43	18.33
SIRCRC ($\omega_1 = 0.8$)	09.93	08.18	07.33
SIRCRC ($\omega_1 = 0.7$)	09.56	07.94	08.15
SIRCRC ($\omega_1 = 0.6$)	08.60	07.49	09.50

4.2   Experiments on the FERET Face Database

The FERET database is one of the standard facial image databases specially used for face recognition algorithms [25]. We use a subset of it. It is composed of 1400 images from 200 individuals with each subject providing 7 images. This subset includes face images whose names contain two-character string: "ba", "bj", "bk", "be", "bf", "bd" and "bg". The images in this subset have pose variations of ± 15°, ± 25°, and also variations of illumination and expression. Before the experiments, we use the down-sampling algorithm to resize each image into 40 × 40 pixels. Table 5 shows that our proposed method usually classifies more accurately than CIRCLC and the classical RBCM.

Table 5. Recognition error rates (%) of different methods on the FERET face database.

Number of the original training samples per class	1	2	3
SRC [8]	64.90	52.77	56.00
RCR [11]	79.99	80.18	88.52
LRC [12]	80.18	59.97	60.87
CIR	81.47	61.23	63.83
CIRLRC ($\omega_1 = 0.8$) [23]	64.39	53.97	54.77
CRC [9]	55.67	41.60	55.63
SIR	57.75	49.60	53.25
SIRCRC ($\omega_1 = 0.8$)	52.83	37.20	43.13
SIRCRC ($\omega_1 = 0.7$)	52.58	38.30	45.62
SIRCRC ($\omega_1 = 0.6$)	50.83	36.50	43.13

4.3   Experiments on the GT Face Database

We use the Georgia Tech (GT) face database [26] to test our method. It was built at the Georgia Institute of Technology, which contains images of 50 people taken in two or three sessions. All people in the database are represented by 15 color JPEG images with cluttered background taken at the resolution of 640 × 480 pixels. The pictures show frontal or tilted faces with different facial expressions, lighting conditions, and scale. Each image was manually labeled to determine the position of the face in the image. We use the face images with removed background and each of these face images has a resolution of 40 × 30 pixels. They are all converted into gray images in advance. The first 2, 3 and 4 face images of each subject are used as training samples and the remaining images are taken as test samples. Table 6 shows again that CIRLRC and SIRCRC are better than SRC, RCR, LRC and CRC. Meanwhile, CIRLRC and SIRCRC have similar recognition error rates.

Table 6. Recognition error rates (%) of different methods on the GT face database.

Number of the original training samples per class	2	3	4
SRC [8]	45.26	42.17	40.24
RCR [11]	63.71	61.57	58.21
LRC [12]	54.15	49.17	44.36
CIR	64.15	64.33	62.55
CIRLRC ($\omega_1 = 0.9$) [23]	46.92	42.83	39.45
CRC [9]	57.54	54.50	52.55
SIR	58.31	57.50	55.09
SIRCRC ($\omega_1 = 0.8$)	45.69	43.50	40.00
SIRCRC ($\omega_1 = 0.7$)	45.23	42.33	40.18
SIRCRC ($\omega_1 = 0.6$)	45.26	42.17	40.24

4.4   Running Time

We compare the computation efficiency between CIRLRC and SIRCRC on the ORL, FERET and GT face databases. The first 5, 3 and 4 face images of each subject are used as training samples and the remaining images are taken as test samples in the ORL, FERET and GT. Table 7 presents the speed of CIRLRC and SIRCRC on face databases. It is obvious that the presented SIRCRC is more computationally efficient than CIRLRC by 11.81 averaging speed-up times.

Table 7. Speed on the face databases (Time s).

Heading level	Number of training samples per class	CIRLRC [23]	SIRCRC	Speed-up (times)
ORL	5	1452.51	139.93	10.38
FERET	3	16495.38	1138.77	14.49
GT	4	602.26	57.11	10.55

5 Conclusions

The representation based classification method (RBCM) has attracted much attention in the last decade. It exploits the linear combination of training samples to represent the test sample and then classifies the test sample according to the minimum reconstruction residual. Among the RBCMs, a novel concept, Inverse Representation (IR), was recently proposed. It uses the test sample to represent each training sample subject. The relative CIRLRC algorithm integrates IR and LRC by score fusing and shows superior classification performance. However, it suffers from two aspects. One is that the test vector contains some training sample information. The other is the computation inefficiency that CIRLRC should solve C linear equations for classifying the test sample, where C is the number of the classes. Therefore, we have presented a novel method integrating simplified IR (SIR) and collaborative representation classification (CRC) for face recognition, named SIRCRC. In SIRCRC, only test sample information has been fully used in SIR, and CRC is more efficient than LRC in terms of speed, because only one linear equation system is needed. We have done extensive experiments on ORL, FERET and GT databases. As for ORL and FERET face databases, SIRCRC is superior than CIRLRC in terms of recognition, and they have similar error rates on GT database. According to the experimental results, the presented SIRCRC is faster than CIRLRC by more than 10 times. In one word, the proposed SIRCRC is feasible and efficient for face recognition.

Acknowledgments. This work is supported in part by the Priority Academic Program Development of Jiangsu Higher Education Institutions, Natural Science Foundation of China (No. 61572258, No. 61103141 and No. 51405241) and Student Innovation Training Program of NUIST(No. 201410300190 and No. 201410300178).

References

1. Li, Y., Zhu, L., Bi, N., Xu, Y.: Sparse representation for brain signal processing: a tutorial on methods and applications. Sig. Process. **31**(3), 96–106 (2014)

2. Boufounos, P., Kutyniok, G., Rauhut, H.: Sparse recovery from combined fusion frame measurements. IEEE Trans. Inf. Theor. **57**(6), 3864–3876 (2011)
3. Arbelaez, P., Maire, M., Fowlkes, C., Malik, J.: Contour detection and hierarchical image segmentation. IEEE Trans. Pattern Anal. Mach. Intell. **33**(5), 898–916 (2011)
4. Yang, J., Wright, J., Huang, T.S., Ma, Y.: Image super-resolution via sparse representation. IEEE Trans. Image Proc. **19**(11), 2861–2873 (2010)
5. Mairal, J., Elad, M., Sapiro, G.: Sparse representation for color image restoration. IEEE Trans. Image Proc. **17**(1), 53–69 (2008)
6. Mei, X., Ling, H.: Robust visual tracking and vehicle classification via sparse representation. IEEE Trans. Pattern Anal. Mach. Intell. **33**(11), 2259–2272 (2011)
7. Jia, X., Lu, H., Yang, M.-H.: Visual tracking via adaptive structural local sparse appearance model. In: IEEE Conference on Computer vision and pattern recognition (CVPR), pp. 1822–1829 (2012)
8. Wright, J., Yang, A.Y., Ganesh, A., Sastry, S.S., Ma, Y.: Robust face recognition via sparse representation. IEEE Trans. Pattern Anal. Mach. Intell. **31**(2), 210–227 (2009)
9. Zhang, D., Yang, M., Feng, X.: Sparse representation or collaborative representation: which helps face recognition?. In: IEEE International Conference on Computer Vision (ICCV), pp. 471–478 (2011)
10. Xu, Y., Zhang, D., Yang, J., Yang, J.-Y.: A two-phase test sample sparse representation method for use with face recognition. IEEE Trans. Circuits Syst. Video Technol. **21**(9), 1255–1262 (2011)
11. Yang, M., Zhang, D., Shenlong, W.: Relaxed collaborative representation for pattern classification. In: IEEE Conference on Computer Vision and Pattern Recognition (CVPR), pp. 2224–2231 (2012)
12. Naseem, I., Togneri, R., Bennamoun, M.: Linear regression for face recognition. IEEE Trans. Pattern Anal. Mach. Intell. **32**(11), 2106–2112 (2010)
13. Xu, Y., Fan, Z., Qiu, M., Zhang, D., Yang, J.-Y.: A sparse representation method of bimodal biometrics and palmprint recognition experiments. Neurocomputing **103**, 164–171 (2013)
14. Naseem, I., Togneri, R., Bennamoun, M.: Sparse representation for ear biometrics. In: Bebis, G., et al. (eds.) ISVC 2008, Part II. LNCS, vol. 5359, pp. 336–345. Springer, Heidelberg (2008)
15. Shekhar, S., Patel, V.M., Nasrabadi, N.M., Chellappa, R.: Joint sparse representation for robust multimodal biometrics recognition. IEEE Trans. Pattern Anal. Mach. Intell. **36**(1), 113–126 (2014)
16. Pillai, J.K., Patel, V.M., Chellappa, R., Ratha, N.K.: Secure and robust iris recognition using random projections and sparse representations. IEEE Trans. Pattern Anal. Mach. Intell. **33**(9), 1877–1893 (2011)
17. Yang, J., Yu, K., Gong, Y., Huang, T.: Linear spatial pyramid matching using sparse coding for image classification. In: IEEE Conference on Computer Vision and Pattern Recognition, CVPR, pp. 1794–1801 (2009)
18. Yang, M., Zhang, D., Yang, J.: Robust sparse coding for face recognition. In: Computer Vision and Pattern Recognition (CVPR), pp. 625–632 (2011)
19. Lyu, Q., Lin, Z., She, Y., Zhang, C.: A comparison of typical l_p minimization algorithms. Neurocomputing **119**, 413–424 (2013)
20. Xu, Z., Chang, X., Fengmin, X., Zhang, H.: $l_{1/2}$ regularization: a thresholding representation theory and a fast solver. IEEE Trans. Neural Networks Learning Syst. **23**(7), 1013–1027 (2012)
21. Guo, S., Wang, Z., Ruan, Q.: Enhancing sparsity via l_p minimization for robust face recognition. Neurocomputing **99**, 592–602 (2013)

22. Zhang, Z., Xu, Y., Yang, J., Li, X., Zhang, D.: A survey of sparse representation: algorithms and applications. doi:10.1109/ACCESS.2015.2430359
23. Xu, Y., Li, X., Yang, J., Lai, Z., Zhang, D.: Integrating conventional and inverse representation for face recognition. IEEE Trans. Cybernetics **44**(10), 1738–1746 (2014)
24. Available: http://www.cl.cam.ac.uk/research/dtg/attarchive/facedatabase.html
25. Available: http://www.itl.nist.gov/iad/humanid/feret
26. Available: http://www.anefian.com/research/face_reco.html

Segmentation of Motion Capture Data Based on Measured MDS and Improved Oblique Space Distance

Dandan Song, Jing Dong, and Qiang Zhang$^{(\boxtimes)}$

Key Laboratory of Advanced Design and Intelligent Computing,
Dalian University, Ministry of Education, Dalian 116622, China
zhangq26@126.com

Abstract. The segmentation of motion capture data is essential for the synthesis of motion data, its purpose is to split long movement sequence data into many different independent semantic motion clips, and it requires that the segmentation of motion capture data is effective and accurate. This paper proposed a segmentation algorithm of motion capture data based on measured MDS and improved oblique space distance. The proposed approach used the multidimensional scaling (MDS) to achieve the space mapping from original high-dimensional data to low-dimensional, and then calculated the improved oblique space distance between frames in the specified windows and the preceding section in the low-dimensional space, and obtained the final segmentation points by similarity detection. Finally we obtained the independent semantic motion clips, and we verified the feasibility of the algorithm through experiments, and the accuracy rate of our method is improved compared with the traditional algorithm.

Keywords: Motion capture · Segmentation · MDS

1 Introduction

In recent years, human motion data is applied to various fields more and more widely, but the human body movements are usually composed of multiple movements, and they are long and complicated sequences, so dealing with long sequences of motion capture data directly is complicated and time-consuming. Therefore, we need to split long motion sequences into independent semantic fragments in order to reuse them later. So far, many domestic and foreign scholars have carried on the research to the long serial human motion segmentation techniques, and have obtained a series of research results in theory and practice.

(1) Segmentation methods based on clustering algorithms
Kahol et al.[1, 2] proposed a hierarchical behavioral segmentation algorithm, in this paper they used a dynamic hierarchical data structure to represent the human motion data, and then used the simple Bayesian classifier to complete motion segmentation, but this method has great dependence on training samples and the empirical data from experts. Zhou et al. [3] proposed a combined clustering

© Springer International Publishing Switzerland 2015
A. Bikakis and X. Zheng (Eds.): MIWAI 2015, LNAI 9426, pp. 184–196, 2015.
DOI: 10.1007/978-3-319-26181-2_17

algorithm (ACA), which is extended to the k-means by changing the characteristic number of clustering means and using the DTW technique. Balci et al. [4] proposed that the mass points of the limps were used as the characteristic vectors to cluster the motion postures. The mass points have a smaller dimension, so the method can reduce the complexity of clustering.

(2) Segmentation methods based on special dimension reduction techniques

J. Barbic et al. [5] reduced the dimension of the long motion sequences by principal component analysis (PCA) and probabilistic principal component analysis (PPCA). But PCA is a linear analysis method, so it cannot split the non-linear human motion data accurately. Qu et al. [6] used the Gaussian model to reduce the dimension of motion capture data, mapped them to a low dimensional hidden space, and detected segmentation points in the low dimensional space. Jing Shuxu et al. [7] considered that different movements have different internal structures, so they used PCA algorithm to find the internal structure of the motion data, then constructed the similarity matrix, searched diagonal line of the similarity matrix to find the movement group, and achieved the segmentation of the motion sequences. However, this method has some limitations, and is not suitable for the homologous motions, needs for further processing.

(3) Segmentation methods based on geometrical features

Peng Shujuan et al. [8] took the center distance of human root node and other members of the joint points for the distance features, used PCA method to obtain the main features of human motion and then proceeded low-pass filtering and noise reduction, thus realize the segmentation of human motion capture data. Xiaojun et al. [9, 10] extracted the angle of the major bones between the limb bones and the skeletal center, and used it as geometric features of motion data. Then they detected the segmentation points in heuristic way, realized the interactive accurate segmentation of motion data. The essence of the segmentation methods based on geometrical features is that they reduce the dimension of the motion data first, and then realize the segmentation of motion sequences.

(4) The other segmentation methods

Chattopadhyay and Bhandarkar [11] proposed a Body Animation Parameter (BAP) indexing algorithm for decomposing the motion data sequence into smaller motion data sequences. Samer Salamah et al. [12] presented a novel simple and efficient method for segmentation by classification of motion capture data automatically and efficiently. Lin et al. [13] segmented motion sequences at velocity peaks and zero velocity crossings, and used HMM to refine and classify the resulted segments. Carlos Orrite and Mario Rodriguez et al. [14] proposed an approach based on Observable Markov Models (OMMs) to construct a network models for action states and middle states. In this model, the recognition of a complex unknown sequence is accomplished by decoding the state at each particular time step.

To sum up, it is the key to look for the segmentation features of different human motion sequences, and we can see that the reduction of the dimension technology is widely used in the segmentation algorithms. Human motion capture data is a kind of multi-dimension data with high complexity and most of them are nonlinear, so dealing with long sequences of motion capture data directly is laborious and time-consuming. In order to reduce the difficulty of segmentation, under the premise of keeping the original topological structure and correlation of original data, we present a measured MDS [15] and improved oblique space distance of motion capture data segmentation method. The flowchart of our method is shown in Fig. 1.

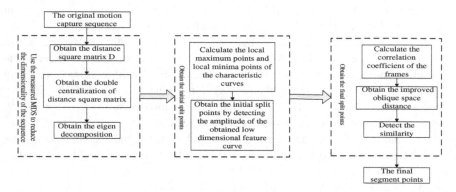

Fig. 1. The flowchart of our method

2 Proposed Approach

2.1 The Data Dimension Reduction

We considered the internal particular topological structure of the motion capture data, and the correlation between frames of each sequence, so the data dimension reduction cannot damage the original topology structure and correlation, which is the prerequisite of the following sequence segmentation.

The main idea of the metric MDS in this paper is to map the original multi-dimensional data points to Euclidean space, and use the appropriate dot pitch to approximate the distance between the original data points in the Euclidean space. Let $X = (X_1, X_2, \ldots, X_N)^T$ be a vector points set including N input data in the D-dimensional space, $d(X_i, X_j)$ represents the Euclidean distance between the data points, namely:

$$d(X_i, X_j) = \|X_i - X_j\|_2 = \left(\sum_{k=1}^{D} (x_{ki} - x_{kj})^2 \right)^{1/2}$$

The measured MDS reconstructs the coordinates of X from the distance set, in order to make the center of the low dimensional embedding coordinates be in the origin and align with the axis better, the original data can be assumed to have got central-ization, that is $\sum_{i=1}^{N} X_i = 0$. The algorithm is mainly divided into three steps:

Obtain the distance square matrix D:

$$D = \left(d^2 \left(X_i, X_j \right) \right)_{N \times N} = \left[\left(X_i - X_j \right)^2 \right]$$

$$= \left[\|X_i\|^2 - 2X_i X_j^T + \|X_j\|^2 \right] = Be^T - 2XX^T + eB^T$$

Where $i, j = 1, 2, \ldots, N$, e is a N × 1 matrix full of 1, $B = \left(\|X_1\|^2, \ldots, \|X_N\|^2 \right)^T$.

Obtain the double centralization of distance square matrix: assume $J = I - ee^T/N$, I is a N × N unit matrix, then $Je = 0, e^T J = 0, J^T = J$, for any input including N sample points $X = (X_1, X_2, \ldots, X_N)^T$, we can get $JX = [X_i - \mu], \mu - 1/N \sum_{i=1}^{N} X_i$. Use J to obtain the double centralization of distance square matrix:

$$JDJ = JBe^T J - 2JXX^T J + JeB^T J = -2JXX^T J$$

$$= -2JXX^T J^T = -2(JX)(JX)^T = -2[X_i - \mu][X_i - \mu]^T$$

Let $V = JX = [X_i - \mu]$, then $JDJ = -2VV^T$.

(3) Low dimensional coordinate representation of the original data is obtained: let $H = VV^T = -JDJ/2$, obtain the eigen decomposition of H, $H = U\Lambda U^T$. Let $\lambda_1 \geq \lambda_2 \geq \ldots \geq \lambda_N \geq 0$ be the eigenvalues of H, let $U_1, U_2, \ldots, U_N \in R^N$ be the eigenvectors of H, take the first d eigenvalues $\lambda_1, \lambda_2, \ldots, \lambda_d$ and its corresponding d eigenvectors $U_1, U_2, \ldots, U_d$, then we can get the low dimensional coordinate representation of X, as follows:

$$Y = daig \left(\sqrt{\lambda_1}, \sqrt{\lambda_2}, \ldots, \sqrt{\lambda_d} \right) [U_1, U_2, \ldots, U_d]^T$$

Take Y for the result of the reduced dimension output, so as to achieve the purpose of reducing the dimension of the original motion capture data.

2.2   The Segmentation of the Motion Sequences

We can obtain a smooth characteristic curve by using measured MDS algorithm to reduce the dimension of motion sequences. First of all, obtain the initial split points by detecting the amplitude of the obtained low dimensional feature curve, and the process can be divided into three steps:

(1) Calculate the local maximum points and local minima points of the characteristic curves.
(2) Measure the difference of the amplitude between every local maximum (minima) point and the local minima (maximum) point of its both sides, if the small D-value is λ ($0 < \lambda < 1$) times smaller than the large D-value, this extreme point is positioned as an initial segmentation point.

(3) In order to ensure that the initial segmentation points cannot be too dense, set a threshold T = 150 to judge whether the length of the frames between two adjacent initial segmentation points is larger than the threshold T. If the condition is satisfied, two frames are used as the initial segmentation points, otherwise remove the large point. Stop the judgment when reach the final frame.

Secondly, confirm the exact split points further, the detection process can be divided into five steps:

(1) $m = 1$.
(2) $m = m + 1$, take the m-th initial segmentation point as the center, set up a window of length $L = 2\alpha + 1$, they are the frames $[p_m - \alpha, p_m + \alpha]$, then calculate the frame distance between each frame in the window and the previous segment. Because there are different relevance between the variables, using orthogonal space distance is easy to deform, select improved oblique space distance as the frame distance. The improved oblique space distance is defined as follows:

$$D(i,j) = \frac{1}{n}\sqrt{r_{i,j}\sum_{k=1}^{n}\sum_{l=1}^{n}(x_{i,k} - x_{j,k})(x_{i,l} - x_{j,l})}$$

Where $n = 93$, $r_{i,j}$, is the correlation coefficient for the i-th frame and j-th frame of the motion sequence, its formula is as follows:

$$r_{i,j} = \frac{\sum_{k=1}^{n}(x_{ik} - \bar{x}_i)(x_{jk} - \bar{x}_j)}{\sqrt{[\sum_{k=1}^{n}(x_{ik} - \bar{x}_i)^2][\sum_{k=1}^{n}(x_{jk} - \bar{x}_j)^2]}}$$

When the frames are not related, the improved oblique space distance is degenerated to the Euclidean distance.

(3) Judge the similarity of the frames. The inter-frame distance is stored in the matrix D. Define a threshold β here, $\beta = 5\min(D)$, if the inter-frame distance is smaller than β, this frame belongs to the former segment, then mark 1 in the corresponding window, otherwise this frame belongs to the next segment, then mark 0 in the corresponding window.
(4) Judge whether the current frame is the last frame, if so go to the step (5), if not go to the step (2).
(5) Take M as the number of the data marked 1, redefine the precise split point as $p_m - \alpha + M$.

3   The Analysis of Our Experiments

We verify the feasibility of the algorithm through experiments, and analyze the difference of the dimensionality reduction performance and error rate with other different algorithms.

3.1   The Dimension Reduction Performance of Motion Capture Data

In this paper, the experience uses the motion capture data of BVH format. Select a non-regular motion sequence composed of 2465 frames, it can be divided into six parts: "jumping", "turning around", "kicking", "squatting", "jogging" and "standing". Reduce the dimension of the data by using the traditional MDS, ISOMAP and the measured MDS, the dimension curve is shown in Fig. 2.

Many experiments show that the traditional MDS dimensionality reduction and the Isomap dimensionality reduction methods are not stable, and we cannot distinguish the original motion clips of a motion sequence well in the dimensionality reduction curve. While the curve shows long-term non-stationary and short-term stationary by the measured MDS algorithm, then we can distinguish different semantic segment better. As it can be seen in Fig. 2, in the curve obtained by the traditional MDS algorithm of motion sequences dimensionality reduction, there are many great differences with filter and without filter, but in the curve obtained by the measured MDS algorithm, the result with filter is almost smooth as same as the result without filter. So the filtering operation can be omitted after measured MDS algorithm, but the traditional MDS algorithm dimension reduction requires smoothing and de-noising in order to further analyze the data processing.

The MDS [16] is a nonlinear dimensionality reduction method that keeps the similarity or the distance between the data points. The traditional MDS algorithm [17] uses the iterative method and the sequential relationship of the original motion capture data to find the representation in the low dimensional space, but it reduces the dimensionality of data by a random matrix, resulting in inconsistent dimensionality reduction of high-dimensional data every time, thus it makes the segmentation points are not identified, its practicality is not high. The Isomap algorithm is time-consuming, and the curve after dimensionality reduction is close to a straight line, it is difficult to find the extreme points, so it cannot continue to segment the data after dimension reduction. This paper utilizes the similarity or distance between the data points to obtain the geometric representation of the data in the low dimensional space by referencing the measured MDS. The square matrix of the distance is converted into the inner product matrix, and finally we obtain the eigenvalues and eigenvectors of the inner product matrix to get the low dimensional representation with high stability.

By contrast, we can find out that the proposed algorithm obtained the effective dimensionality reduction and kept the original topology structure and correlation of the original motion sequences at the same time. The results reflect that the measured MDS method is applied to the dimensionality reduction of the motion sequences.

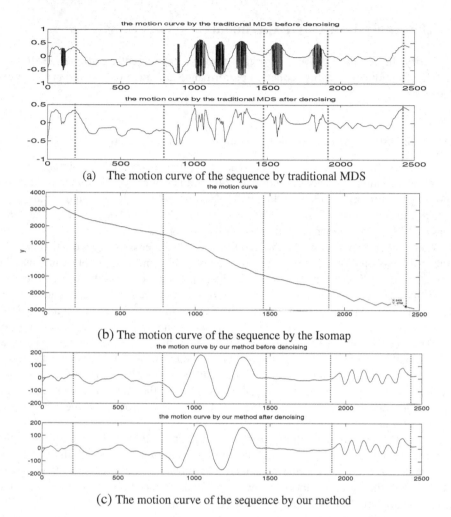

(a) The motion curve of the sequence by traditional MDS

(b) The motion curve of the sequence by the Isomap

(c) The motion curve of the sequence by our method

Fig. 2. The curve of the motion sequence after dimensionality reduction by three different methods

3.2   The Results of the Segmentation of the Motion Capture Data

In this experiment, we choose two human motion capture sequences with different semantic and different length segmentation, segment them and compare this proposed algorithm with the traditional MDS algorithm and the manual segmentation. In this paper, we use the multiple manual segmentations segmented by different people to eliminate the influence of individual subjective factors, and ensure the accuracy of the manual segmentation.

(1) Select a set of simple motion sequence. The length of the sequence is 824 frames. The motion is non-regular and the motion clips can be divided into four parts: "going forward", "picking up the goods", "dropping the goods" and "going back", as shown in Fig. 3.

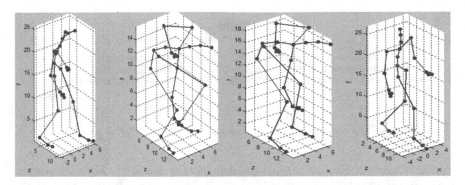

Fig. 3. The motion "going forward", "picking up the goods", "dropping the goods", "going back"

We reduce the dimensionality of this motion sequence by traditional MDS and our algorithm respectively, the characteristic curve of the dimensionality reduction and the marks of the segmentation points are shown in Figs. 4 and 5. Figure 6 shows the segmentation of the motion sequence by different methods. Comparing our algorithm, the traditional MDS segmentation method and the manual segmentation, we can find that the proposed algorithm is more closed to the manual segmentation results than the traditional segmentation algorithm. The first motion semantic segment is moving forward slowly, its motion characteristic is not obvious, so there are some errors in our method compared with the manual segmentation.

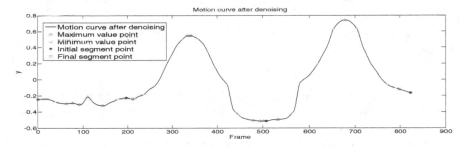

Fig. 4. The motion curve and the labeling of the sequence by traditional MDS

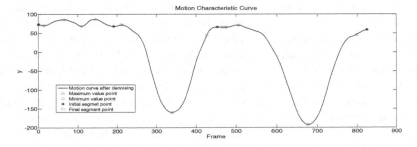

Fig. 5. The motion curve and the labeling of the sequence by our method

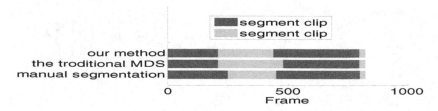

Fig. 6. The segmentation of the motion sequence by different methods

In this paper, we take the manual segmentation results as the reference template, use the error rate ER [18] to measure the effectiveness of the algorithm. ER is defined as follows:

$$ER = \frac{N_{miss} + N_{extra}}{N} \times 100\%$$

Where N_{miss} expresses the number of the lost frames after segmentation, N_{extra} expresses the number of the extra frames after segmentation, N is the length of the semantic sequence after segmentation. The error rate is shown in Table 1 by using different algorithms to segment the motion sequence.

Table 1. The error rate of different algorithms to segment the motion sequence

Method	Going forward	Picking up the goods	Dropping the goods	Going back
Traditional MDS	4.1 %	16.5 %	8.9 %	8.3 %
Our method	4.1 %	9.6 %	4.3 %	8.3 %
Difference	0	6.9 %	4.6 %	0

It can be found that our algorithm has lower error rate than the traditional MDS algorithm and reaches better segmentation effect, so it completes the segmentation of human motion capture sequences better.

(2) Select a set of complex motion sequence, the length of the sequence is 2465 frames. The motion clips can be divided into six parts: "jumping", "turning right and left", "kicking", "squatting", "jogging" and "standing", as shown in Fig. 7.

Figures 8 and 9 show the motion curve and the labeling of the sequence by traditional MDS and our method. Figure 10 shows the segmentation of the motion sequence by different methods. By segmenting the same motion sequences in the same conditions several times, it can be proved that the traditional MDS dimension reduction method is not stable, the low dimensional curve obtained by it is not identified every time, which makes it difficult to judge the best segmentation points. While the measured MDS dimension reduction method is stable, the results of many experiments are consistent, it achieves the satisfying effect of the dimension reduction. As shown in the dotted circle in Fig. 8, it can be seen that there are redundant segmentation points after the traditional MDS, which makes the incorrect segmentation of the sequence.

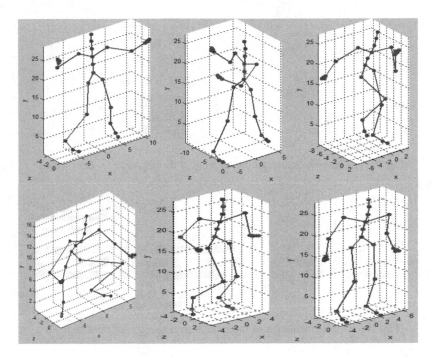

Fig. 7. The motion "jumping", "turning right and left", "kicking", "squatting", "jogging" and "standing"

The error rate is shown in Table 2 by using different algorithms to segment the motion sequence.

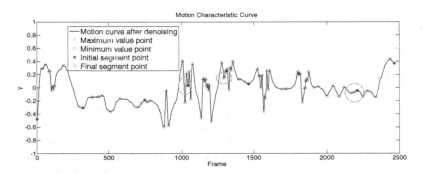

Fig. 8. The motion curve and the labeling of the sequence by traditional MDS

Compared with the traditional MDS algorithm, the proposed algorithm improves the accuracy and stability of the segmentation, and realizes the segmentation of the human motion capture sequence better.

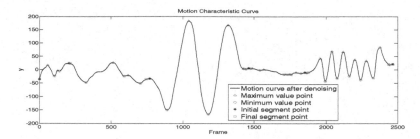

Fig. 9. The motion curve and the labeling of the sequence by our method

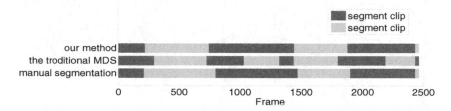

Fig. 10. The segmentation of the motion sequence by different methods

Table 2. The error rate of different algorithms to segment the motion sequence

Methods	Jumping	Turning right and left	Kicking	Squatting	Jogging	Standing
Traditional MDS	27.9 %	22.5 %	15.5 %	30.9 %	19.3 %	0
Our method	4.3 %	9.7 %	12.7 %	9.5 %	3.6 %	0
Difference	23.6 %	12.8 %	2.8 %	21.4 %	15.7 %	0

4   Conclusions

This paper proposed a segmentation algorithm of motion capture data based on measured MDS and improved oblique distance. The proposed approach achieved the space mapping from original high-dimensional data to low-dimensional by the measured MDS, then calculated the improved oblique space distance between the fixed frames in the low-dimensional space, at last obtained the final segmentation points by similarity detection. We obtained the independent semantic motion clips, and verified the validity and feasibility of the algorithm through experiments. Finally we analyzed the results of our experiments to verify our method improved the accuracy of the segmentation of motion sequences.

In future study and work, we will make further research and improvement on the algorithm proposed in this paper, which mainly includes two aspects: (1) We will try to adopt the clustering method for the detection of segmentation points. (2) In this paper, we set an empirical function to determine the similarity between frames after several multiple experiments, in future we will find a better adaptive method to make the split points more accurate.

Acknowledge. This work is supported by the National Natural Science Foundation of China (No.61370141, 61300015), Natural Science Foundation of Liaoning Province (No. 2013020007), the Scientific Research Fund of Liaoning Provincial Education Department (No. L2013459, L2015015), the Program for Science and Technology Research in New Jinzhou District (No. KJCX-ZTPY-2014-0012).

References

1. Kahol, K., Tripathi, P., Panehanathan, S.: Gesture segmentation in complex motion sequences. In: Proceedings of IEEE International Conference on Image Processing, pp. II–105–108. IEEE (2003)
2. Kahol, K., Tripathi, P., Panchanathan, S.: Automated gesture segmentation from dance sequences. In: 6th IEEE International Conference, pp. 883–888. IEEE (2004)
3. Zhou, F., Torre, F.D., Hodgins, J.K.: Aligned cluster analysis for temporal segmentation of human motion. In: 8th IEEE International Conference on Automatic Face and Gesture Recognition, pp. 1–7. IEEE (2008)
4. Balci, K., Akarun, L.: Clustering poses of motion capture data using limb centroids. In: 23rd International Symposium on Computer and Information Sciences, pp. 1–6. IEEE(2008)
5. Barbic, J., Safonova, A., Pan, J.Y., et al.: Segmenting motion capture data into distinct behaviors. In: Proceedings of Graphics Interface, pp. 185–194. Canadian Human-Computer Communications Society, Canada (2004)
6. Qu, S., Yu, R.H., Wu, L.D., et al.: Automatic segmentation of motion capture data based on latent space. Appl. Res. Comput. **28**(8), 3128–3130 (2011)
7. Jing, S.X., Wu, Z., Huang, Z.Y.: Segmenting single actions from continuous capture motion sequence. In: 2009 WRI World Congress on Computer Science and Information Engineering, pp. 85–89. IEEE, Los Angeles (2009)
8. Peng, S.J.: Motion segmentation using central distance features and low-pass filter. In: International Conference on Computational Intelligence and Security, pp. 223–226. USA (2010)
9. Xiao, J., Zhang, Y.T., Wu, F.: Feature visualization and ineranctive segmentation of 3D human motion. J. Softw. **8**(19), 1995–2003 (2008)
10. Xiao, J., Zhang, Y.T., Wu, F., et al.: A group of novel approaches and a toolkit for motion capture data reusing. Multimedia Tools Appl. **47**(30), 379–408 (2010)
11. Chattopadhyay, S., Bhandarkar, S.M., Li, K.: Human motion capture data compression by model-based indexing: a power aware approach. Vis. Comput. Graph. **13**(1), 5–14 (2007)
12. Salamah, S., Zhang, L., Brunnett, G.: Hierarchical method for segmentation by classification of motion capture data. In: Brunnett, G., Coquillart, S., van Liere, R., Welch, G., Váša, L. (eds.) Virtual Realities. Lecture Notes in Computer Science, vol. 8844, pp. 169–186. Springer International Publishing, Switzerland (2015)
13. Lin, J.F.S, Kulic, D.: Segmenting human motion for automated rehabilitation exercise analysis. In: 2012 Annual International Conference of the IEEE Engineering in Medicine and Biology Society (EMBC), pp. 2881–2884. IEEE (2012)
14. Orrite, C., Rodriguez, M., et al.: Automatic segmentation and recognition of human actions in monocular sequences. In: 22nd International Conference on Pattern Recognition of the IEEE, pp. 4218–4223. IEEE (2014)
15. Wu, X.T., Yan, D.Q.: Analysis and research on method of data dimensionality reduction. Appl. Res. Comput. **26**(8), 2832–2835 (2009)

16. Cox, T.F., Cox, M.A.A.: Multidimensional Scaling. CRC Press, New York (2000)
17. Vander Maaten, L.J., Postma, E.O., van den Herik, H.J.: Dimensionality reduction: a comparative review. J. Mach. Learn. Res. **10**(1–41), 66–71 (2009)
18. Peng, S.J., Liu, X.: Double-feature combination based approach to motion capture data behavior segmentation. Comput. Sci. **40**(8), 303–308 (2013)

Single-Sample Face Recognition Based on WSSRC and Expanding Sample

Zhijing Xu[1($\boxtimes$)], Li Ye[1], and Xiangjian He[2]

[1] Shanghai Maritime University, Shanghai, China
xzjsmu@163.com, 412571337@qq.com
[2] University of Technology Sydney, Ultimo, Australia
Xiangjian.He@uts.edu.au

Abstract. This paper proposes a face recognition method with one training image per person, and it is based on compressed sensing. We apply nonlinear dimensionality reduction through locally linear embedding and sparse coefficients to generate multiple samples of each person. These generated samples have multi-expressions and multi-gestures are added to the original sample set for training. Then, a super sparse random projection and weighted optimization are applied to improve the SRC. This proposed method is named weighted super sparse representation classification (WSSRC) and is used for face recognition in this paper. Experiments on the well-known ORL face dataset and FERET face dataset show that WSSRC is about 15.53 % and 7.67 %, respectively, more accurate than the original SRC method in the context of single sample face recognition problem. In addition, extensive experimental results reported in this paper show that WSSRC also achieve higher recognition rates than RSRC, SSRC DMMA, and DCT-based DMMA.

Keywords: Sparse representation classification · Single sample · Nonlinear dimensionality reduction · WSSRC · Local neighborhood embedding

1 Introduction

Face recognition technology is a computer technology for identity authentication by analyzing and comparing facial feature information [1]. Because the two-dimensional image of face is greatly impacted by light face, facial expressions and gestures, face recognition is still an unresolved problem. At the same time, due to difficulties in collecting the sample and the limited storage capacity of the system cause the single-sample problem, which makes the extraction of intrinsic features of training objects more difficult [2]. In the case that per person has only one image in the face training dataset, the recognition performance of the majority of the conventional methods will be severely degraded. Compressed Sensing [3] (CS) has recently been emerging as a new signal processing method. By use of compressed sensing theory, Wright et al. presented a sparse representation classification algorithm (SRC) [3]. SRC method can effectively respond to changes of lighting, facial expression and partial occlusions, so it has a good robustness. In order to improve the recognition rate, we use a super sparse

© Springer International Publishing Switzerland 2015
A. Bikakis and X. Zheng (Eds.): MIWAI 2015, LNAI 9426, pp. 197–206, 2015.
DOI: 10.1007/978-3-319-26181-2_18

random projection and weighted optimization to improve the SRC. For single-sample problem, we propose a single-sample face recognition method based on the improved SRC. Firstly, it makes a face image become pieces by use of local neighborhood embedding nonlinear dimension reduction and sparse coefficients, and then considers these generated new multi-expression and multi-gesture samples as training samples, and finally uses the improved SRC to identify and classify.

2   CS and WSSRC

In this section, related work and our new method is proposed. One is CS (Compressed Sensing). Compressed sensing presented by Donoho and Candes is a new theoretical framework. Its main model is linear. The basic idea is that high-dimensional original signal can be effectively restored by taking advantage of the low-dimensional observation signal as long as the signal meets the "sparse" feature or the signal is compressible. The detail of CS theory is in Sect. 2.1.

SRC (sparse representation-based classifier) has been widely used in face recognition, but it also has many drawbacks such as non-real time, poor recognition performance in case of single sample. So a weighted super sparse representation classification is proposed to solve the problem, the detail is in Sect. 2.3.

2.1   CS Theory

Compressed sensing theory mainly covers the following aspects:

(1)   For the signal $x \in \mathbf{R}^n$, how to find an orthogonal basis or tight frame Ψ to make it sparse in Ψ, that is to say, the sparse representation problem of a signal: $x = \Psi \alpha$.

(2)   How to design a stable M × N-dimensional measurement matrix Φ unrelated to Ψ, which can ensure that the important information of the signal is not damaged from the N-dimension to M-dimensional, that is to say, the low-speed signal sampling problem: $y = \Phi x$.

(3)   How to design fast reconstruction algorithm to recover signal from linear observation $y = A\alpha$ (A is called CS information operator, $A = \Phi\Psi$), that is to say, the reconstruction problems of the signal. It is based on the rigorous mathematical optimization question:

$$\min_{\alpha} \|\hat{\alpha}\|_0 \quad \text{s.t.} \ A\hat{\alpha} = y \tag{1}$$

L0 norm optimization problem is a NP problem, that is to say, it is difficult to solve in polynomial time, or even impossible to verify the reliability of the solution of the problem. So, we have to change L0 norm to L1 norm:

$$\min_{\alpha} \|\hat{\alpha}\|_1 \quad \text{s.t.} \ A\hat{\alpha} = y \tag{2}$$

2.2   SRC

The basic idea of SRC is: the original data set which contains the all training samples can be seen as over-complete dictionary, so the test samples can be characterized sparsely by the over-complete dictionary. If each class has enough training samples, the test samples can only be represented as a linear combination of the training samples from the same class. For the over-complete dictionary, it is the best sparse representation of test samples, and can be obtained by the L1 norm minimization. Therefore, finding the best sparse representation can naturally distinguish different types of training samples.

On the face recognition problem, if there are enough training samples in class i: $A_i = [v_{i,1}, v_{i,2}, \cdots, v_{i,N_i}]$ a new face image (test sample) $x \in \mathbf{R}^n$ can be represented as a linear combination of these given training samples: $x = A_i \alpha_i$, $\alpha_i \in R^{N_i}$ is the coefficient vector. If the test sample's category is unknown, cascade all C categories of training samples whose number is $N = \sum_{i=1}^{C} N_i$ together to form a new matrix Ψ to represent the entire training set:

$$\Psi = [A_1, A_2, \cdots, A_c] = [v_{i,1}, v_{i,2}, \cdots, v_{c,N_c}]$$

So the x can be expressed linearly by all training samples:

$$x = \Psi \alpha \in R^n \tag{3}$$

Ideally $\alpha = [0, 0, \cdots, 0, \alpha_i^T, 0, 0, \cdots, 0]^T$ is a coefficient vector. The vector's other elements are 0 in addition to those elements related to class i.

In actual recognition scenarios, the test sample x may be partially broken or occluded. So the formula (3) can be amended as follows:

$$x = x_0 + e_0 = \Psi \alpha + e_0 \tag{4}$$

In it, $e_0 \in R^n$ is an error vector, due to the number of unknowns exceeds the number of equations n the formula (4), it is not direct to solve α. However, under a certain condition, the desired solution (α, e_0) is the best sparse solution of the formula (4):

$$(\alpha, e_0) = \arg \min \|\alpha\|_0$$
$$\text{s.t. } x = \Psi \alpha + e_0$$

In it, L0 norm refers to the number of non-zero elements in a vector. The literature [3] found α by solving the following convex question because of the theory of L0 and L1 minimization equivalence:

$$\min \|\alpha\|_1 + \|e\|_1$$
$$\text{s.t. } x = \Psi \alpha + e_0 \tag{5}$$

Once the L1 minimization problem is solved, use the residual $r_i = \|x - A_i \alpha_i\|$ to decide the category of the test sample.

2.3   WSSRC

This paper presents the algorithm by using a super sparse random projection matrix as the measurement matrix Φ to reduce the image dimension. The literature [4] has proved that the super sparse random projection matrix not only meets the necessary condition of CS measurement matrix but also has better measurement result than the Gaussian measurement matrix. Find the optimal sparse solution by use of dual augmented Lagrange multiplier (DALM) on L1 minimization problem. Its principle is to translate the formula (1) into the dual problem:

$$\max_t y^T t \text{ s.t. } A^T t \in B_1^{\infty}$$

In it, $B_1^{\infty} = \left\{ \alpha \in R^n : \|\alpha\|_{\infty} \leq 1 \right\}$ The dual problem's corresponding Lagrange function is:

$$\min_{t,z} -y^T t - \alpha^T \left(z - A^T t \right) + \\ \frac{\beta}{2} \left\| z - A^T t \right\|_2^2 \quad \text{s.t. } z \in B_1^{\infty} \tag{6}$$

In it, α is the Lagrange multiplier. Because it is impossible to change α, t, z at the same time, we can use the alternating iterative method to find the extreme value of the formula (6), that is to say, update the third variable continuously by minimizing the cost function while maintaining the two other variables held constant. This will not only ensure that the reconstruction accuracy but also reduce the algorithm's calculating velocity. Finally, use weighted residual to determine the category of test data for more precise classification. This is the improved SRC method which we named WSSRC (weighted super sparse representation classification) algorithm

Algorithm is as follows:

(1) Input the N training samples which have C classes to constitute the dictionary matrix $\Psi = \left[A_1, A_2, \cdots, A_c \right] \in R^{n \times N}$.

(2) Generate column-normalized very sparse random projection matrix $\Phi \in R^{M \times n}$, in it, $\Phi(i,j) = \frac{1}{\sqrt{M}} h_{ij}$, while h_{ij} obeys very sparse projection distribution:

$$h_{ij} \sim \begin{pmatrix} \sqrt[4]{n} & O & -\sqrt[4]{n} \\ \frac{1}{2\sqrt{n}} & 1 - \frac{1}{\sqrt{n}} & \frac{1}{2\sqrt{n}} \end{pmatrix}$$

(3) Give a test image $x \in R^n$.

(4) Calculate $Y_i = \Phi A_i, (i = 1, \cdots, C)$, and calculate the measurement matrix $Y = \left[Y_1, Y_2, \cdots, Y_c \right] = \Phi \Psi$, in it $Y \in R^{M \times N}$.

(5) Calculate the new projection sample $y = \Phi x$, and find the sparse vector $\hat{\alpha}$ by using DALM to make $Y\alpha = y$.

(6) Let $\hat{\alpha} = [r_1^T, r_2^T, \cdots, r_N^T]$ calculate the reconstruction error by use of weighted residual $e_i = \frac{\|y - Y_i r_i\|}{\|r_i\|}$ in each class.

(7) If $k = \arg\min_i(e_i)$, then x belongs to the class k.

3   Generation of Redundant Sample

The difficulty of single sample problem is that each object has only a sample. Adding redundant information of samples is an effective way to solve the single sample problem. This article firstly single training sample use the method of local neighborhood embedding nonlinear dimension reduction [6] to make single training sample become pose-varied images. Then generate multi-expression images by use of sparse coefficient. Finally, make these multi-expression and multi-gesture new samples as training samples for training. Its purpose is to get more useful information from only one training sample in order to identify further classification.

3.1   Generation of Multi-gesture Samples

The method of generating multi-gesture samples is based on the theory of local neighborhood embedding nonlinear dimension reduction. The specific implementation is:

(1) Let all face images represent in form of a column vector.
(2) Set the input face image which has one gesture as Ii. Regard Ii as one point of high-dimensional space, and the same gesture training set of face images T_i^n (including N images) is regarded as near point of Ii. Then find the weight value of near points by local neighborhood embedding nonlinear dimension reduction.
(3) Set the target gesture training set of face images as T_0^n (including N images), and the unknown face image which has the target gesture is Io. Then use the weight value solved in the step (2) to calculate a point of that high-dimensional space, that is to say, find the Io.
(4) Let the target pose face image vector represent in form of matrix.

This method can be easily and quickly to generate multi-pose samples, and can overcome the problem of complexity, large amount of computation, and non-ideal result of those existing similar algorithms. Figure 1 is the original face images and generated new samples.

3.2   Generation of Multi-expression Samples

The method of generating multi-expression samples is based on sparse representation. Make sure that all images' size uniform and the eye, nose, mouth of every image are fixed in the same position of them. For a single frame face test image which has a kind of expressive, use a linear combination of face images having the same expression to approximate the test image, so we can get the sparse reconstruction coefficients of the

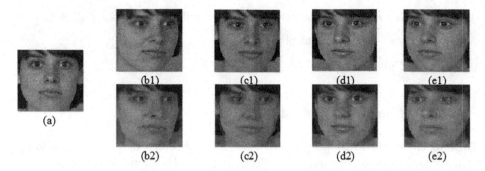

Fig. 1. The original face images and generated new samples

approaching linear combination. Then take advantage of the sparse reconstruction coefficients and target expression training set of face images to reconstruct the face image which has the target expression.

This method is simple and does not contain troublesome facial feature extraction step, so it is easy to implement. Besides, its algorithm complexity is relatively reduced. Figure 2 is the original face images and generated multi-expression samples, (a) the input happy face test image; (b1), (c1), (d1), (e1), (f1) respectively is the generated anger, disgust, fear, sadness, surprise face image, and (b2), (c2), (d2), (e2), (f2) is the corresponding real images.

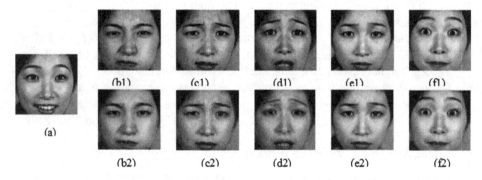

Fig. 2. The original face images and generated new samples

4   Experiment Results

The paper uses ORL, FERET face dataset for experimental verification. Select a face image for training, and the remaining images are regarded as test samples. Readjust all face images' size as 32×32 before the experiment. In the case of a single sample, this paper first takes the presented method to generate multi-expression and multi-pose new samples, which is based on gesture and expression library. Then the original training sample and new samples are regarded as training samples. Finally, use the WSSRC for classification.

4.1 ORL Experiment

ORL dataset has total 40 individuals' 400 pictures, that is to say, each person has 10 images. Some images are taken at different times, and human facial expressions and facial details have varying degrees of change. In the experiment, take a sample of each person to expand. Then let the generated redundant samples and the original sample together as training samples and samples the remaining images as test samples. Figure 3 depicts the recognition rate while using a different number of redundant samples.

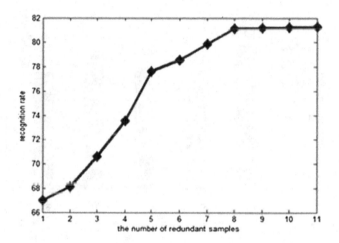

Fig. 3. Recognition rate under different number of redundant sample

As can be seen from Fig. 3, the recognition rate is low when the number of redundant sample is not enough. With the increase of the number of redundant sample, the recognition rate significantly increased. But when the redundant sample reaches 8, the recognition rate grows very slowly, almost gently. The reason is that the number of samples is the main cause of influencing the identification effect when the intrinsic characteristics of these samples are too few because of the too few training samples. But with the number of redundant sample continues to improve, it will be bound to cause a lot of redundant information, and the data distribution of the generated redundant samples has some inconsistency with the data distribution of the original samples, which can indicate that increasing the number of redundant sample has a limited influence on the identification rate.

Compare the algorithm of this paper with SRC and the algorithms in references 7, 8. As can be seen from Fig. 4, the method mentioned in this paper improves 15.53 % than the original SRC method, and has better recognition performance than those algorithms in references 7, 8.

4.2   FERET Experiment

Take some images from FERET face dataset, including 200 individuals' 1400 pictures. They are taken under conditions that expression, perspective and light intensity are different. Similarly, take a sample of each person to expand, and the remaining images as test samples. Figure 5 depicts the recognition rate while using a different number of redundant samples.

Some researches about single-sample face recognition were also done in literatures [8–12]. It can be seen from Fig. 6 that the presented method increases 7.63 % than the

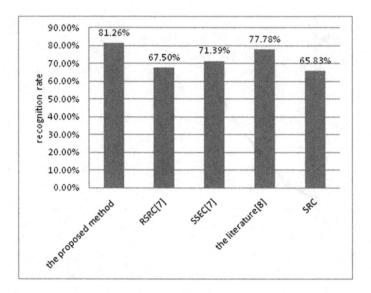

Fig. 4. Recognition rate of the proposed method and other algorithms

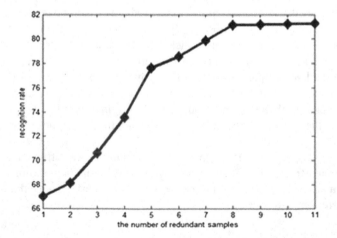

Fig. 5. Recognition rate under different number of redundant sample

original SRC on FERET face dataset. The presented method also has better recognition performance than DMMA, DCT-based DMMA, I-DMMA, etc. Figure 6 shows the single-sample recognition rate of these methods applied on the FERET.

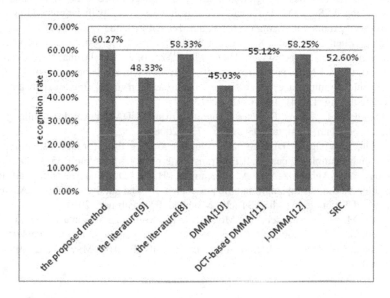

Fig. 6. Recognition rate of the presented method and other algorithms

5   Conclusion

For single-sample problem of face recognition research, we propose a CS-based single-sample face recognition algorithm. These comparison experiments show that the method not only takes advantage of the feature information of a single sample to generate some new images, but also greatly improves the recognition rate in case of single sample, which provides a new method for single-sample face recognition technology.

Acknowledgements. This work was supported by the National Natural Science Foundation of China under Grant No. 61404083 and Foundation of Shanghai Maritime University Grant No. 20120108.

References

1. Wang, M.: Study of Face Recognition Algorithm Based on Compressive Sensing. MS Thesis of Xi'an Electronic and Engineering University (2012)
2. Jie, W., Plataniotis, K.N., Lu, J., et al.: On solving the face recognition problem with one training sample per subject. Pattern Recogn. **39**(9), 1746–1762 (2006)
3. Wright, J., Yang, A.Y., Ganesh, A., et al.: Robust face recognition via sparse representation. EEEE Trans. Pattern Anal. Mach. Intell. **31**(2), 210–227 (2009)

4. Fang, H., Zhang, Q.-B., Wei, S.: Method of image reconstruction based on very sparse random projection. Comput. Eng. Appl. **43**(22), 25–27 (2007)
5. Yang, A.Y., Zhou, Z., Balasubramanian, A.G., Sastry, S.S., Ma, Y.: Fast l1-minimization algorithms for robust face recognition. IEEE Trans. Image Process. **20**(3), 681–695 (2011)
6. Ma, R., Wang, J., Song, Y.: Multimanifold learning using locally linear embedding nonliner dimensionality reduction. Tsinghua Univ. Sci. Tech. **48**(4), 583–586 (2008)
7. Chang, X., Zheng, Z., Duan, X., Xie, C.: Sparse representation-based face recognition for one training image per person. In: Huang, D.-S., Zhao, Z., Bevilacqua, V., Figueroa, J.C. (eds.) ICIC 2010. LNCS, vol. 6215, pp. 407–414. Springer, Heidelberg (2010)
8. Dan, G.: Single-sample face recognition algorithm based on virtual sample expansion method. Sci. Technol. Eng. **13**(14), 3908–3911 (2013)
9. Gao, Q., Zhang, L., Zhang, D.: Face recognition using FLDA with single training image per person. Appl. Math. Comput. **205**(2), 726–734 (2008)
10. Lu, J., Tan, Y., Vang, G.: Discriminative multimanifold analysis for face recognition from a single training sample per person. IEEE Trans. Pattern Anal. Mach. Intell. **35**(1), 39–51 (2013)
11. Nabipour, M., Aghagolzadeh, A., Motameni, H.: A DCT-based multimanifold face recognition method using single sample per person. In: The 8th Symposium on Advances in Science and Technology (8thSASTech), Mashhad, Iran, February 2014
12. Nabipour, M., Aghagolzadeh, A., Motameni, H.: Multimanifold analysis with adaptive neighborhood in DCT domain for face recognition using single sample per person. In: The 22nd Iranian Conference on Electrical Engineering (ICEE 2014), May 20–22, 2014. Shahid Beheshti University (2014)

Flotation Surface Bubble Displacement Motion Estimation Based on Phase Correlation Method

Liangqin Chen[✉] and Weixing Wang

College of Physics and Information Engineering,
Fuzhou University, Fuzhou, China
odiechen@fzu.edu.cn, znn525d@qq.com

Abstract. Phase correlation technique combined of bubble tracking algorithm is investigated to estimate the flotation surface bubble displacement movement in this work. Image segmentation is used to extract the high gray value area of each bubble after the two continuous images in a sequence are preprocessed by zooming out on minimum. Because of the bubble motion varying at different parts of the cell surface, block phase correlation is employed to obtain the detailed displacement feature for each block. A lead zinc flotation plant is used to carry out experiments for the estimation of the bubble displacement motion. Experimental results show that the bubble displacement motion of each flotation cell is in a certain cycle. The displacement motion curve distribution and the turbulence degree curve distribution of the same level flotation cell are the similar.

Keywords: Flotation · Bubble displacement movement · Motion estimation · Phase correlation · Bubble tracking

1 Introduction

Froth flotation is a common process used for extracting a desired mineral from its ore. The shape, size, stability and mobility of the bubbles, and other bubble features, have a significant effect on both grade and recovery of the valuable minerals achieved at the cell. These characters can help the operators infer the yield and adjust the process if required. However, due to natural complexity of the flotation process, achieving optimal control is often not possible for the human operators [1, 2]. The need to overcome these problems, coupled with rapid advances in computer technology has led to the development of computer vision systems for monitoring and intelligent control of flotation circuit [3].

In recent years, more and more works are done to measure froth stability quantitatively, and it has been found that there is a clear relationship between froth stability and flotation performance [4–9]. In their researches the bubble velocity on the top surface of the froth is always an essential and important parameter on the measurement and calculation formulations about air recovery and other froth stability parameters. However, Aldrich et al. [10] mentioned that motion estimation in froths is difficult, owing to the smoothness of the images, as well as the effects of bubbles bursting and

© Springer International Publishing Switzerland 2015
A. Bikakis and X. Zheng (Eds.): MIWAI 2015, LNAI 9426, pp. 207–216, 2015.
DOI: 10.1007/978-3-319-26181-2_19

merging. Ross [11] divided the total volume of the froth into four different stages. Holtham and Nguyen [12] argued that froth mobility means how fast froth moves for the plant operators, and therefore, surface froth displacement velocity at stage 4 is one of the common parameters that plant operators frequently use to evaluate the flotation process and to make appropriate changes to optimize the flotation production. Some algorithms have been proposed to measure the surface froth velocity including bubble tracking algorithms [1, 13], block matching algorithms [14–16] and pixel tracing algorithms [12, 17, 18]. However, motion estimation accuracy and computational complexity of these algorithms is difficult to balance.

In this work, phase correlation method is employed to estimate the surface bubbles displacement movement. Froth videos obtained from a lead-zinc flotation plant is used to investigate the proposed method. The study provides a foundation for the realization of the intelligent control system of flotation production based on computer vision.

2 Phase Correlation Method

Phase correlation method is a method based on Fourier transform. Let $f_1(x, y)$ and $f_2(x, y)$ are the two images that differ only by a displacement (x_0, y_0), i.e.

$$f_2(x, y) = f_1(x - x_0, y - y_0) \tag{1}$$

Their corresponding Fourier transform $F_1(u, v)$ and $F_2(u, v)$ will be related by:

$$F_2(u, v) = e^{-j2\pi(ux_0 + vy_0)} F_1(u, v) \tag{2}$$

The cross-power spectrum of the two images $f_1(x, y)$ and $f_2(x, y)$ with Fourier transform $F_1(u, v)$ and $F_2(u, v)$ is defined as

$$\frac{F_1(u, v)F_2^*}{|F_1(u, v)F_2^*|} = e^{j2\pi(ux_0 + vy_0)} \tag{3}$$

Where F_2^* is the complex conjugate of F_2. The shift theorem guarantees that the phase of the cross-power spectrum is equivalent to the phase difference between the two images. The result of the inverse Fourier transformation of the cross-power spectrum is an impulse function; that is, it is approximately zero everywhere except at the displacement. There is an obvious sharp peak at the displacement position. So, the shift theory indicates that the displacement movement between the two images can be calculated by searching the peak position of the impulse function.

3 Surface Bubble Displacement Motion Estimation Method Based on Phase Correlation

3.1 Phase Correlation Method Based on Bubble High Gray Value Area Tracking

Compared with a general scene image, froth image has its particularity. Many gray and black bubbles, sticking together and with one or more high gray value area on almost each bubble, constitute the foreground of froth image, with no background [19]. For froth video, the gray value changes little between frames. Because of the scraper and the bubble growth process of itself, bubble movement in froth video is a very complex movement containing translation, rotation, stretching and other movements. In order to obtain the translation feature, this paper designs a method combined of phase correlation and bubble tracking.

The current and reference images are segmented to extract the high gray value region of each bubble, and then phase correlation method is used to obtain the bubble translational motion parameters. Each bubble has one or more high gray value area, that is to say, the high gray value area is representatives of each bubble. The high gray value area is also made to do the corresponding displacement movement when the bubble is in the process of movement. By means of image segmentation processing, it can greatly reduce the effects of noise caused by the rotation and other non-translational movement.

It should be noted that the accuracy of motion estimation using phase correlation can not be guaranteed when the amplitude of motion between the two images is too large. Figure 1 shows the motion estimation results based on phase correlation method.

Figure 1f, g, h, and i are the cross-power spectrums calculated based on No. 101 frame as a current moving image, respectively using No. 100, 99, 97 and 90 as a reference image. The amplitude of the peak of the cross-power spectrum and its location, and the noise data are shown in Table 1.

Figure 1 and Table 1 show that, when dealing with the two consecutive frames, there is a distinct peak in the cross-power spectrum whiles the noise elsewhere are close to zero. The distinct peak indicates that phase correlation method can effectively obtain the motion information on small amplitude movement. With the interval frames between the current image and the reference image increasing, that is to say, the movements between the two images are getting greater, and the contraction, rotation and other non-translation movement in the bubbles are increasing, the peak value of the cross-power spectrum are getting decreasing, and it is equivalent to that the noise value are becoming increasing. When the mean value of noise is closer to the peak value, it means that the peak is submerged in the noise. For example, in Fig. 1i, the peak value is 0.0159, and the mean value of noise reaches to 0.0101, so it is difficult to obtain the correct velocity information by the cross-power spectrum. Therefore, in this work a consecutive previous frame is used as a reference frame for motion estimation to ensure the accuracy of motion estimation.

Furthermore, in order to improve the processing speed of the sequence, a preprocessing named zooming out using the minimum is also carried out on the two original images before segmentation. Thinking of the weak boundaries with low gray value

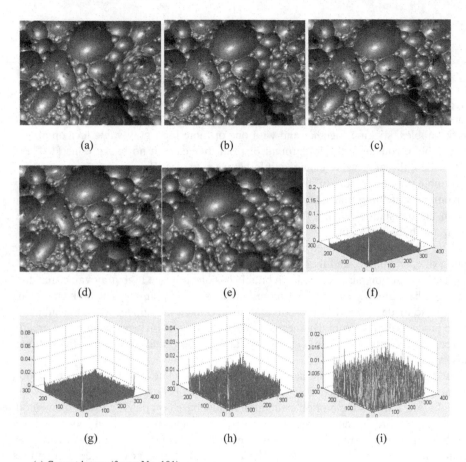

(a) Current image (frame No. 101)

(b) (c) (d) (e) Reference images (frame No. 100, 99, 97 and 90)

(f) (g) (h) (i) Cross-power spectrums corresponding to the reference image (b) (c) (d) (e) separately

Fig. 1. Motion estimation on froth image sequence

Table 1. The peak of the cross-power spectrum and its position (corresponding to Fig. 1)

Current image	Reference image	Peak value	Position	Noise mean	Ratio value
Frame 101	Frame 100	0.1686	(2,1)	0.0103	16.37
	Frame 99	0.0696	(4,1)	0.0103	6.76
	Frame 97	0.0373	(4,1)	0.0103	3.62
	Frame 90	0.0159	(233,64)	0.0101	1.57

between bubbles [20], the minimum gray value of the 4 neighborhood of the current pixel is selected as the gray value in the output image. It can protect the boundary characteristics of each bubble to ensure the accuracy of the extraction of the bubble

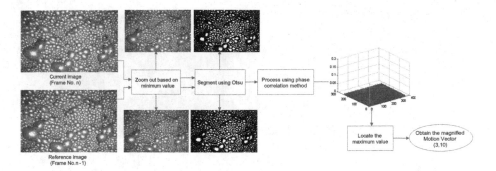

Fig. 2. Demo and workflow of the proposed method

displacement speed. In the last step the displacement velocity would be multiplied appropriately according to the zooming out parameter. A demo and workflow based on the above method is shown in the Fig. 2.

3.2 Block Phase Correlation Method

Only a motion vector is extracted when using phase correlation method for an image. However, detailed mobility information is sometimes required for some sub-regions of the froth surface such as a region of bubbles bursting and merging. In order to further obtain the movement velocity of each region of an image, and assess the consistency and disturbance of the bubbles movement, a method of block and phase correlation is used in the paper.

Phase correlation method is often used for the condition that the two images differ only by a displacement, not including rotation and scaling. However, the bubbles are often with the rotation and scaling change during the movement. Dividing image into sub-block in small size can weaken the non-translation movement ration of the whole image, and make it to be close to the translation movement. Therefore, the method based on block phase correlation, can well avoid the inherent shortcomings of the global phase correlation when applied on the forth videos, and can obtain an accurate velocity field for each block as well as retain its excellent anti-noise characteristics.

The steps of motion estimation based on block phase correlation are described as the following:

(1) Firstly, the current image and the reference image are preprocessed by zooming out on minimum as described in Sect. 3.1;
(2) Secondly, divide the two images into a series of non-overlapping blocks, respectively. Here, the block size can be set according to the mean size of bubbles, and then a block can be seen as the corresponding single bubble.
(3) Finally, calculate the Fourier transform and the cross-power spectrum for each block, and obtain the motion vector for each block.

Figure 3 gives the results based on the block phase correlation method.

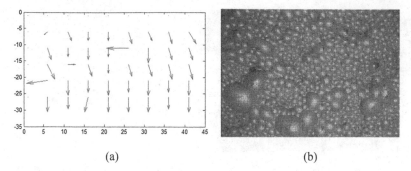

(a)   (b)

(a)Motion vector filed showed by display density 5, block size is 8×8;

(b) Result of overlap (a) and the current image;

Fig. 3. Motion vector fields obtained from block phase correlation method

4 Experiment and Result

4.1 Lead Zinc Flotation Procedure

The proposed motion estimation algorithm is been tested on froth videos from lead and zinc mines in Jin Dong flotation plant in China, one of our cooperation partners. The rude ore contains lead and zinc. Lead is extracted firstly, and then zinc is extracted from the tailings. Figure 4 shows the lead flotation procedure in their flotation plant, and the zinc flotation procedure is similar to this. We have set up an image acquisition system to grab the high quality froth images and videos (see Fig. 5). According to the flotation procedure, lead rougher cell A, lead concentration cell A, zinc rougher cell A, and zinc concentration cell A are chosen to carry out the experiment. Phase correlation method based on bubble high gray value area tracking is used to obtain the displacement velocity. The images are zoomed out to a quarter of the original sizes, and Otsu is used to extract the bubble high gray value area.

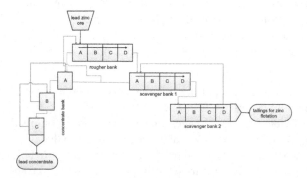

Fig. 4. Lead flotation procedure

Fig. 5. Image acquisition system

4.2 Surface Bubble Displacement Movement Data and Algorithm Analysis

The experiment results are shown in Fig. 6. In order to evaluate the bubble motion turbulence degree, we employ the turbulence degree parameter [15], which is defined as follows:

$$\Delta D_v = D_v - \overline{D_v} \tag{4}$$

$$D_T = \frac{\sqrt{\Delta D_v^2}}{\overline{D_v}} \tag{5}$$

Where D_v is the displacement of each frame, $\overline{D_v}$ is the average displacement of a continuous period of time, and D_T is the displacement turbulence degree of each frame.

The bubble motion seems irregular if seen from the data of an individual flotation cell. However, from a macro point of view, it is shown that the bubble displacement motion of each flotation cell has its own cycle. And the displacement motion distribution curve and the turbulence degree distribution curve of the lead rougher cell and the lead concentration cell are similar to the corresponding ones of zinc separately. The bubble movements always are in a certain turbulence degree as shown in Fig. 6b, d, f and h. In the heavy metal flotation production such as lead and zinc, maintaining a certain degree of bubble movement turbulence can effectively promote the desorption of the unwanted minerals from the mineralized bubbles, which can improve the grade of concentrate.

The proposed algorithm can obtain the features of bubbles translational motion over a period of time. In the case of the collapse of the bubbles, block phase correlation method can be used to provide the detailed motion features of each block. Compared with the algorithms of references [15, 18], motion estimation is based on the tracking of the extracted bubble high gray value area, and therefore the algorithm has its good anti-noise performance. However, the deviation of the position of the extraction of the high gray value also may lead into motion estimation error.

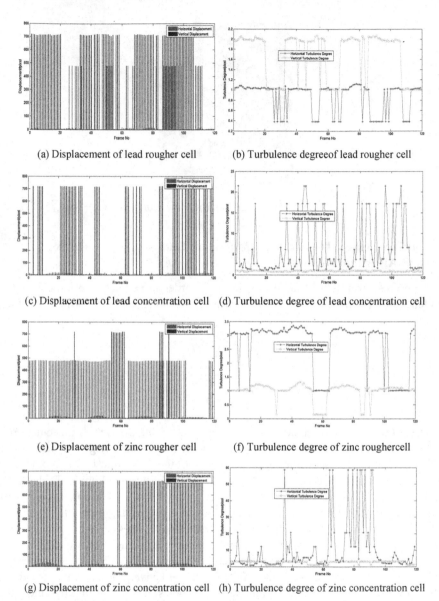

(a) Displacement of lead rougher cell   (b) Turbulence degreeof lead rougher cell

(c) Displacement of lead concentration cell   (d) Turbulence degree of lead concentration cell

(e) Displacement of zinc rougher cell   (f) Turbulence degree of zinc roughercell

(g) Displacement of zinc concentration cell   (h) Turbulence degree of zinc concentration cell

Fig. 6. Bubble displacement movement and turbulence degree curves

5   Conclusion

(1) Flotation surface bubble displacement movement features can be obtained by phase correlation method. The method can also be extended to other mineral flotation production.

(2) The bubble displacement motion of each flotation cell has a certain cycle. The displacement motion distribution curve and the turbulence degree distribution

curve of the same level flotation cell are the similar. Maintaining a certain degree of motion turbulence would help to improve the grade of concentrate.

(3) The further work is about to analysis and model the relationship between the bubble displacement motion and the flotation performance. It would provide a foundation for the establishment of flotation intelligent optimization control system on computer vision.

Acknowledgements. This research is financially supported by the National Natural Science Fund in China (grant no. 61170147) and the Science and Technology Development Fund of Fuzhou University (grant no. 2014-XY-31).

References

1. Mehrabi, A., Mehrshad, N., Massinaei, M.: Machine vision based monitoring of an industrial flotation cell in an iron flotation plant. Int. J. Miner. Process. **133**, 60–66 (2014)
2. Wang, W.X., Chen, L.Q.: Flotation bubble tracing based on harris corner detection and local gray value minima. Minerals **5**(2), 142–163 (2015)
3. Morar, S.H., Harris, M.C., Bradshaw, D.J.: The use of machine vision to predict flotation performance. Miner. Eng. **36–38**, 31–36 (2012)
4. Barbian, N., Hadler, K., Cilliers, J.J.: The froth stability column: measuring froth stability at an industrial scale. Miner. Eng. **19**, 713–718 (2006)
5. Barbian, N., Cilliers, J.J., Morar, S.H., Bradshaw, D.J.: Froth imaging, air recovery and bubble loading to describe flotation bank performance. Int. J. Miner. Process. **84**, 81–88 (2007)
6. Hadler, K., Cilliers, J.J.: The relationship between the peak in air recovery and flotation bank performance. Miner. Eng. **22**, 451–455 (2009)
7. Hadler, K., Greyling, M., Plint, N., Cilliers, J.J.: The effect of froth depth on air recovery and flotation performance. Miner. Eng. **36–38**, 248–253 (2012)
8. Qu, X., Wang, L.G., Nguyen, A.V.: Correlation of air recovery weith froth stability and separation efficiency in coal flotation. Miner. Eng. **41**, 25–30 (2013)
9. Rojas, I., Vinnett, L., Yianatos, J., Iriarte, V.: Froth transport characterization in a two-dimensional flotation cell. Miner. Eng. **66–68**, 40–46 (2014)
10. Aldrich, C., Marais, C., Shean, B.J., Cilliers, J.J.: Online monitoring and control of froth flotation systems with machine vision: A review. Int. J. Miner. Process. **96**(1–4), 1–13 (2010)
11. Ross, V.E.: A study of the froth phase in large scale pyrite flotation cells. Int. J. Miner. Process. **30**, 143–157 (1990)
12. Holtham, P.N., Nguyen, K.K.: On-line analysis of froth surface in coal and mineral flotation using JKFrothCam. Int. J. Miner. Process. **64**, 163–180 (2002)
13. Francis J.J., De Jager G.: An investigation into the suitability of various motion estimation algorithms for froth imaging. In: Proceedings of the 1998 South African Symposium on Communications and Signal Processing (COMSIG 1998), pp. 139–142 (2001)
14. Forbes, G., De Jager, G.: Unsupervised classification of dynamic froths. SAIEE Afr. Res. J. **98**(2), 38–44 (2007)

15. Tang, Z.H., Liu, J.P., Gui, W.H., Yang, C.H.: Froth bubbles speed characteristic extraction and analysis based on digital image processing. J. Cent. S. Univ. (Sci. Technol.) **40**(6), 1616–1622 (2009)
16. Jahedsaravani, A., Marhaban, M.H., Massinaei, M.: Prediction of the metallurgical performances of a batch flotation system by image analysis and neural networks. Miner. Eng. **69**, 137–145 (2014)
17. Nguyen K.K., Holtham P.N.: The application of pixel tracing techniques in the flotation process. In: Proceedings of the First Joint Australian and New Zealand Biennial Conference on Digital Imaging and Vision Computing and Applications, pp. 207–212 (1997)
18. Mu, X.M., Liu, J.P., Gui, W.H., Tang, Z.H., Li, J.Q.: Flotation froth motion velocity extraction and analysis based on SIFT features registration. Inf. Control **40**(4), 525–531 (2011)
19. Wang, W.X., Li, Y.Y., Chen, L.Q.: Bubble delineation on valley edge detection and region merge. J. China Univ. Min. Technol. **42**(6), 1060–1065 (2013)
20. Wang, W.X., Bergholm, F., Yang, B.: Froth delineation based on image classification. Miner. Eng. **16**(3), 1183–1192 (2003)

An Improved DCT-Based JND Model
Based on Textural Feature

Bo Huang$^{(\boxtimes)}$, XiuZhi Yang, KaiXiong Su, and MingKui Zheng

College of Physics and Information Engineering, Fuzhou University,
Fuzhou 350002, China
{nl31120053,yangxz,skx,zhengmk}@fzu.edu.cn

Abstract. In this paper, we propose an improved DCT-based just noticeable difference (JND) model incorporating the spatial contrast sensitivity function (CSF), the luminance adaptation (LA) effect and the contrast masking (CM) effect. For the CM model, we propose a novel image block texture classification method based on mean value and variance of AC coefficients which better represent the texture characteristics than conventional model. Experimental results confirm that the proposed model yields invisible distortions for test images with average PSNR of 28.24 dB, which is almost 3 dB lower than the other models for comparison. It is proved that the proposed model inject more redundancy into images in unperceived way and thus more consistent with human visual features.

Keywords: JND model · DCT · Mean value · Variance

1 Introduction

The human eye is the ultimate recipient of many video processing, Distortion in images or video may not attract the attention of the human eye because Information is usually processed and perceived in nonlinear way. Just Noticeable Distortion (JND) is the minimal distortion visibility threshold of the human visual system [1]. For JND model reflects the perception of HVS directly, it is widely used in various image and video processing field [2, 3].

JND model is generally divided into two categories: pixel-based and DCT-based model. The pixel-based model only considers luminance adaptation (LA) and contrast masking (CM) effects. and the other one, DCT-based model is based on the contrast sensitivity function (CSF) and modulated by the LA and CM effects. Currently the DCT transform is widely used in image and video compression coding, using DCT-based JND model is a good solution to remove the human psychological visual redundancy and thereby reducing more bits for higher compression efficiency [4, 5].

Wei et al. [6] proposed the classical JND model in DCT domain. There are four factors that affect the JND value in the model: CSF, LA, CM, and overall adjustment factor. Kim et al. [7] proposed an elaborate model as a continuous function of texture complexity which does not require image block classification or DCT computation. Bae et al. [8] proposed a new texture complexity metric that considers not only contrast intensity, but also structure of image patterns. Wu et al. [9] uses free-energy principle to

© Springer International Publishing Switzerland 2015
A. Bikakis and X. Zheng (Eds.): MIWAI 2015, LNAI 9426, pp. 217–225, 2015.
DOI: 10.1007/978-3-319-26181-2_20

predict the texture. Bae et al. [10] found that LA depends on Spatial frequency also, and the piecewise linear function of background luminance established in [6] can't fit the model in DCT domain.

The piecewise linear function of background luminance established in [6] increase the JND threshold where the image background brightness is too bright or too dark, but it ignores the correlation between the background luminance and the DCT coefficient frequency. The texture masking factor mainly uses the Canny operator to count the edge pixel density to distinguish different texture regions, but the effect of this algorithm is futile when block size is too small. Experiments show that this method unable to distinguish the texture and edges in human visual experience. As a result, JND value is overestimated in edge parts, while underestimated in texture parts. In order to improve the matching degree between JND model and human visual system (HVS), the method of texture classification which is based on coefficient variance and mean value in DCT domain is proposed in this paper.

2   Basic DCT-Based JND Model

Basic JND model in DCT domain generate the JND value of all coefficients at the same time for a DCT block and is formulated by three factors: CSF, LA and CM. which is expressed as [11]

$$JND = H_{CSF} \cdot MF_{LA} \cdot MF_{CM} \tag{1}$$

Where H_{CSF} is the base threshold which is generated by the spatial contrast sensitivity function. MF_{LA} and MF_{CM} are modulation factors of luminance masking and masking factor respectively.

All factors depend on frequency in DCT domain and $\omega_{i,j}$ is defined as the cycles per degree in spatial frequency for the (i,j)-th DCT coefficient and is calculated by [6]:

$$\omega_{i,j} = \frac{1}{2N} \sqrt{(i/\theta_x)^2 + (i/\theta_y)^2} \tag{2}$$

N is the dimension of DCT transform block and in this paper we set $N = 8$.

2.1   Contrast Sensitivity Function (CSF)

Human eyes show a low-pass property in the spatial frequency domain. H_{CSF} is the basic JND threshold in DCT domain and its value increases with the increase of the spatial frequency, which is consistent with the characteristics of human visual perception. The calculation formula is as follows:

$$H_{CSF}(i,j) = \frac{s}{\varphi_i \varphi_j} \times \frac{\exp(c\omega_{ij})/(a+b\omega_{ij})}{r+(1-r\cos\phi_{ij})} \tag{3}$$

φ_i and φ_j are DCT normalization factors, s is to account for the spatial summation effect and we set the value to 0.25 [12]. $r + (1 - r\cos\phi_{ij})$ represents the oblique effect and ϕ_{ij} stands for the directional angle of the corresponding DCT component. In this paper, we take $a = 1.33$, $b = 0.11$, $c = 0.18$, $r = 0.6$ [6].

2.2 Luminance Masking Factor (LM) Model

Human eyes show a band-pass property in the Luminance sensitivity. It reflects different sensitivity under different intensity of light. The HVS is known to be more insensitive in a dark or bright area than in the middle ranges of pixel intensities. Though the piecewise linear function of background luminance established in [6] is easy to calculate, different frequency coefficient using the same weighting is not desired in HVS. Model in [10] takes its frequency characteristics into account.

$$M_{LA}(\omega, \bar{I}) = \begin{cases} 1 + (M_{0.1}(\omega) - 1) \cdot \left|\frac{\bar{I}-0.3}{0.2}\right|^{0.8}, & \bar{I} \leq 0.3 \\ 1 + (M_{0.9}(\omega) - 1) \cdot \left|\frac{\bar{I}-0.3}{0.6}\right|^{0.6}, & \bar{I} > 0.3 \end{cases} \quad (4)$$

$\bar{I}$ is the normalized average pixel intensity which is defined as:

$$\bar{I} = \left(\frac{1}{255 \cdot N^2}\right) \sum_{x=0}^{N-1} \sum_{y=0}^{N-1} I(x,y) \quad (5)$$

$M_k(\omega)$ are quadratic polynomial functions with different parameters found in [10].

2.3 Texture Masking Factor (CM) Model

In the human visual characteristics, it shows a stronger tolerance in the texture areas than the plane areas when browsing image or video. In fact, there is no accurate physical theory supporting the relationship between the texture and its complexity. Research shows that distortion tolerance increases as the background texture patterns become more complex.

Wei et al. [6] use an edge pixel density metric to define the texture masking property which is based on the Canny operator. Firstly, the edge pixel density is calculated by the number of edge pixels in all pixels of a block by using Canny operator, then the block type is classified into either plain, edge or texture. Then different types are weighted by different weights. This method needs original image filtering before DCT transform. This processing will increase computational complexity if the images need DCT transform.

Kim et al. [7] use continuous functions associated with edge density statistics. Different from the Wei's model, Kim's model change classification into a continuous function and use Sobel operator to calculate the edge density in image block. Sobel operator has a better edge detection effect than Canny, but the detection efficiency is very low for the 8 × 8 block data have too little information to use. This method can hardly distinguish edges and textures because both Canny and Sobel operator can only reflect edge density in the whole image, but the authors in [7] did not consider this. Judging the texture types of image block by edge density always depends on the edge detection operator. Sobel operator is a differential operator, although the edges can be detected in a small block, its theory is not based on the image pixel processing and not matching the human visual features, so its ability to judge the texture is not good.

Normalized power spectral density used in [8] has good judgment ability but high computational complexity. Although the algorithm is better than above two approaches, it is necessary to estimate the pixel probability and calculate the power spectrum density of each frequency, which is complex in theory and practice.

The proposed method has a good outcome with lower complexity and is easier to understand. Calculating texture complexity after the DCT transform will be more flexible, easy to implement and not subject to the conditions of DCT transform block size. This paper argues that texture and edge detection is not directly related. For the human eye, besides the pixel brightness difference, factors that would affect human recognizing texture types including repeatability, orientation, structure, etc. The edge detection operators can only detect a very obvious mutation edge, which largely underestimated the JND threshold in regions where with a small structural or brightness step is not obvious. This paper presents a texture classification method that is based on the mean and variance of the block. The proposed model shows a better consistency with the HVS characteristics compared to the conventional models.

3 The Proposed CM Model

We know that disorderly texture can tolerate more distortion. According to that, we define image blocks of small structure or clutter pixel distribution as a texture type in this paper. We believe that the blocks with clear boundary have large structure and are defined as the edge. Figure 1-(a) shows three types of regions (plane, edge and texture, respectively) distinguished by HVS. In order to analyze the characteristics of the three blocks, we calculate the mean value of the AC coefficients after DCT transformation by using the following formula

$$\bar{C}_{avg} = \frac{1}{N} \left(\sum_{i=0}^{N-1} \sum_{j=0}^{N-1} |C(i,j)| - C(0,0) \right) \tag{6}$$

Most of the energy in the plane region is concentrated in the DC component, so the AC component is close to zero. Figure 1-(b) shows the mean value of plane region is much lower than other two, usually less than 5.

It is difficult to distinguish edge and texture by mean value, so we introduces the variance of the AC's coefficient absolute value into model, which is defined as

$$Var = \frac{1}{(N-1)^2} \left(\sum_{i=0}^{N-1} \sum_{j=0}^{N-1} \left(|C(i,j)| - \bar{C}_{avg} \right)^2 - \left(|C(0,0)| - \bar{C}_{avg} \right)^2 \right) \tag{7}$$

Fig. 1. The mean and variance value of AC coefficient in 8 × 8 block: (a) The original image; (b) mean values of AC coefficients (c)-(d) variances of AC coefficients in different scale.

Figure 1-(c) and (d) shows variance value with different scaling size. Values in Fig. 1-(c) range from 0 to 60. Values in Fig. 1-(d) range from 0 to 8000. It is shown in Fig. 1-(c) that variance values of plane areas are approximate to 10, which is much smaller than other two. Figure 1-(d) shows variance value of texture areas are approximate to 100, and variance value of edge areas are far more than 1000.

Corresponding to the characteristics of human visual judgment, we obtained the following block classification rules with AC coefficient variance and mean value in this paper: mean value and variance of plane areas are very small, close to 0; mean value of the texture or edge are larger, so use variance threshold to separate them two.

Figure 2 shows the image block classification results of the proposed method ((a)-(c)) and Kim's [6] method ((g)-(i)). Black, white and ash areas are judged as plane area, edge area and texture area respectively. The edge in Fig. 2 second row is clearer than third row. The outline of the object is clearly, the none-sensitive areas with small structure are better extracted. When the texture factor is calculated, the weight of these none sensitive areas will increase. It is proved that the proposed method is superior to Kim's method, and the classification method is more consistent with human visual features.

Fig. 2. Image block classification results of the proposed method and Kim et al.'s [6] method: (a)-(c) original image; (d)-(f) proposed method; (g)-(i) Wei et al.'s [6] method.

Texture masking factor MF_{CM} take into account the masking effect of the sub-band coefficients, which is calculated by [6]:

$$MF_{CM} = \begin{cases} \psi & (i^2+j^2) \leq 16 \ in \ Plane \ and \ Edge \\ \psi \times \min(4, \max(1, (\frac{C(i,j)}{H_{CSF}(i,j) \times M_{LA}(i,j)})^{0.36})) & others \end{cases}$$

$$(8)$$

We set weighted parameters as $\psi = 1$, $\psi = 2.25$, $\psi = 1.25$ in the area of plane, texture and edge regions respectively.

Program flowchart is shown as follows:

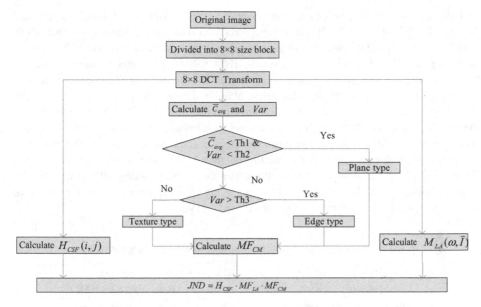

Fig. 3. Flowchart of the proposed image block classification method

Based on the mean values and variances obtained in image experiment, we have obtained the threshold estimates of $Th1 = 15$, $Th2 = 60$, $Th3 = 500$ (Fig. 3).

4 Experimental Results

JND model uses threshold value to reflect the amount of distortion that can be tolerated in insensitive area. For the performance measure, some amounts of noise are injected into all components based on their corresponding JND values by using following formula:

$$C_{JND}(n,i,j) = C(k,i,j) + random \times JND(n,i,j) \qquad (9)$$

Where $JND(n,i,j)$ the (i, j)-th DCT coefficient in the n-th block and is the coefficient with noise injected into $C_{JND}(i,j)$ and *random* are a bipolar random noise of ± 1.

Good performance for a JND profile means that the JND model yields distorted images with lower PSNR (more distortions) values with the same or higher MOS (perceptual quality) values against its reference models under comparison. Since JND is based on human visual characteristics, we use subjective evaluation and objective evaluation to test the proposed model.

Proposed model is compared with [6–8] when the objective evaluation is simulated. In this paper, the block size is set as 8 × 8 which is consistent with the other documents. Table 1 shows the test results for different content image. Half of them are high resolution, and the other half in low resolution. In Table 2, the negative MOS values mean that the distorted images are worse than reference ones. It can be seen in Table 1 that the average PSNR values of the proposed, Wei's, Bae's and Kim's model are 28.24, 31.03, 30.41 and 31.71 dB, respectively, and their corresponding average MOS values are 0.037, -0.018 and -0.018 compared to proposed model respectively. Weight in Wei's model for texture type is small, JND values are underestimated which is unable to remove redundancy very well. As can be seen in Fig. 2, The edge density calculated in Kim's model is not consistent with the human visual features, so the visual effect is poor. Bea's model underestimated the JND threshold in edge areas, and the brightness factor does not consider the relation between brightness and frequency, so the brightness distortion is too large to detect in dark places. That is, the proposed JND model produces better visual quality (higher MOS) and largest distortions (the lowest PSNR) value compared with the other three JND models. Illustrate that the imperceptible distortion added in test images is bigger than other models, having more redundancy means its effect would more close to HVS.

Table 1. Comparison of four JND models in PSNR and MOS values

Test picture	Proposed	Wei [6]		Kim [7]		Bae [8]	
	PSNR	PSNR	MOS	PSNR	MOS	PSNR	MOS
BQTerrace	27.53	30.89	0.1	32.25	0	29.65	-0.05
Cactus	28.56	31.24	0.05	30.97	0	30.53	0.1
PartyScene	28.31	30.72	0	31.98	-0.05	30.80	-0.05
RaceHorses	28.90	31.31	0.05	32.41	-0.1	31.02	0.05
Peppers	28.63	31.15	0	30.00	0	30.61	-0.05
Goldhill	28.64	31.26	0.05	31.81	-0.1	30.72	0
Baboon	27.61	30.48	0.1	32.88	0	30.21	-0.05
boats	27.77	31.25	-0.05	31.39	-0.05	29.80	-0.1
Average	28.24375	31.0375	0.0375	31.71125	-0.0375	30.4175	-0.01875

Table 2. Comparison scale

-3	Much worse
-2	Worse
-1	Slightly worse
0	The same
1	Slightly better
2	Better
3	Much better

5   Conclusion

In this paper, we propose JND model in DCT domain, consists of CSF function, LM model and CM model. To better reflect the perception characteristics of HVS and reduce computational complexity, we proposed Texture classification method in CM model based on AC coefficients' variance and mean values. JND map formed in image block, which is flexible and more suitable for DCT-based image coding and video coding. Compared with three conventional JND models in subjective quality assessment experiments, the proposed JND model yielded the average PSNR value of 28.24, much lower than others and MOS almost to zero. This is because JND values in edge areas are not overestimated, and the JND values of texture areas are not underestimated. Meanwhile, more redundancy injected means the model can tolerate more distortion. This implies that the proposed JND model shows a better consistency with the HVS characteristics compared to the conventional models.

Acknowledgments. This work was supported by the Science and Technology Major Project of Fujian Province (No. 2014HZ0003-3), Natural Science Foundation of Fujian Province (No. 2015J01251) and Project of Fujian Provincial Education Department (No. JA14065).

References

1. Jayant, N., Johnston, J., Safranek, R.: Signal compression based on models of human perception. Proc. IEEE **81**(10), 1385–1422 (1993)
2. Wu, W., Song, B.: Just-noticeable-distortion-based fast coding unit size decision algorithm for high efficiency video codin. Electron. Lett. **50**(6), 443–444 (2014)
3. Yang, K., Wan, S., Gong, Y., et al.: Perceptual based SAO rate-distortion optimization method with a simplified JND model for H. 265/HEVC. Sig. Process. Image Commun. **31**, 10–24 (2015)
4. Yang, X.K., Ling, W.S., Lu, Z.K., et al.: Just noticeable distortion model and its applications in video coding. Sig. Process. Image Commun. **20**, 662–680 (2005)
5. Ahumada, A.J, Peterson, H.A.: Luminance-model-based DCT quantization for color image compression. In: SPIE, Human Vision, Visual Processing, and Digital Display III (1992)
6. Wei, Z., Ngan, K.N.: Spatio-temporal just noticeable distortion profile for grey scale image/video in DCT domain. IEEE Trans. Circuits Syst. Video Technol. **19**(3), 337–346 (2009)
7. Kim, J., Bac, S., Kim, M.: An HEVC-Compliant Perceptual Video Coding Scheme based on JND Models for Variable Block-sized Transform Kernels. IEEE Trans. Circ. Syst. Video Technol. **1** (2015)
8. Bae, S.H., Kim, M.: A Novel Generalized DCT-Based JND Profile Based on an Elaborate CM-JND Model for Variable Block-Sized Transforms in Monochrome Images. IEEE Trans. Image Process. **23**(8), 3227–3240 (2014)
9. Wu, J., Shi, G., Lin, W., et al.: Just noticeable difference estimation for images with free-energy principle. IEEE Trans. Multimedia, **15**(7), 1705–1710 (2013)
10. Bae, S.H., Kim, M.: A novel DCT-based JND model for luminance adaptation effect in DCT frequency. IEEE Signal Process. Lett. **20**(9), 893–896 (2013)
11. Watson, A.B.: DCTune: a technique for visual optimization of DCT quantization matrices for individual images. In: Society of Information Display Digital Technical Papers XXIV, pp. 946–949 (1993)
12. Peterson, H.A., Ahumada, A.J., Watson, A.B.: Improved detection model for DCT coefficient quantization. In: Proceedings of the SPIE Human Vision, Visual Processing, Digit. Display IV, vol. 1913, pp. 191–201 (1993)

Scanned Document Images Skew Correction Based on Shearlet Transform

Fan Zhang[1,2(✉)], Yifan Zhang[1], Xingxing Qu[1], Bin Liu[1], and Ruoya Zhang[1]

[1] School of Computer and Information Engineering, Henan University,
Kaifeng 475001, China
zhangfan@henu.edu.cn
[2] Institute of Image Processing and Pattern Recognition, Henan University,
Kaifeng 475001, China

Abstract. During the documents scanning process, a few degrees of skew is unavoidable. Which will cause some adverse effects on the documents identification, such as Optical Character Recognition (ORC). This paper presents a scanned document images skew estimation and correction algorithm based on Shearlet transform. Shearlet transform offers very good time-frequency localization and direction selectivity. It is possible to detect the skew orientation of the document images accurately. Experimental results show that the proposed algorithm has a high accuracy rate of skew estimation even the scanned document images contain noise or include some pictures or diagrams.

Keywords: Shearlet transform · OCR · Skew correction

1 Introduction

Office automation has become an indispensable trend in information society. The documents are scanned and stored as PDF or images. Optical character recognition (OCR) is the most prominent and successful technique of pattern recognition to date. After many years of development, OCR can reached a high accuracy rate. When a document is fed to the scanner either mechanically or by a human operator, a few degrees of skew is unavoidable. Skew correction is a process which aims at detecting the deviation of the document orientation angle from the horizontal or vertical direction. Skew detection and correction are important preprocessing steps of document layout analysis and OCR approaches [1].

The commonly used algorithms of skew correction are as follows: (1) The skew correction algorithms based on Hough transform. Those algorithms determine the text direction through detecting the direction of lines, so the detection effect is very poor when there are many kinds of difference layouts in the scanned images. (2) The algorithms based on projection characteristics. If the images contain pictures or diagrams, the projector features is not obvious to detect the text direction. (3) The algorithms based on Fourier transform. Those algorithms worked in the frequency domain, but the direction estimation of the algorithm is rough, so the accuracy rate is not very good as well.

© Springer International Publishing Switzerland 2015
A. Bikakis and X. Zheng (Eds.): MIWAI 2015, LNAI 9426, pp. 226–232, 2015.
DOI: 10.1007/978-3-319-26181-2_21

In this paper, a skew estimation and correction algorithm based on Shearlet transform is proposed, which aimed at the complexity and variety of skew estimation and correction in document images. The proposed algorithm also works in the frequency domain. Shearlet transform offers very good time-frequency localization and direction selectivity. Based on Shearlet transform, the skewe direction and the angle of inclination of images can be detected more accurately, and the proposed algorithm has excellent anti-noise performance and robustness.

2 Shearlet Transform

Shearlcts transform is one of the most successful methods for the multiscale analysis. Shearlets are constructed by parabolic scaling, shearing and translation, which offer very good time-frequency localization and direction selectivity.

2.1 Shearlet System

Guo and Labate used a synthetic expanded affine system constructs a multidimensional Shearlet function nearly optimal representation [2, 3]. When the dimension n is 2, the affine systems with composite dilations are as the follows:

$$\Lambda_{AB}(\psi) = \{\psi_{j,l,k}(x) = |\det A|^{j/2} \psi(B^l A^j x - k) : j, l \in \mathbb{Z}, k \in \mathbb{Z}^2\}, \tag{1}$$

Where $\psi \in L^2(R^2)$, A and B are 2×2 invertible matrices, $|\det B| = 1$, The element of this system are called composite wavelets if $\Lambda_{AB}(\psi)$ forms a Parseval frame (also called tight frame) for $L^2(R^2)$ [4].
  Let

$$A_a = \begin{pmatrix} a & 0 \\ 0 & a^{1/2} \end{pmatrix}, \ S_s = \begin{pmatrix} 1 & s \\ 0 & 1 \end{pmatrix}, \tag{2}$$

are the (parabolic) scaling matrix and the shearing matrix with $a > 0$, then, assuming the existence of a generating function $\psi \in L^2(R^2)$ of shearlet, a two-dimensional continuous Shearlet system is as follows:

$$SH_{cout}(\psi) = \{\psi_{a,s,t} = a^{3/4} \psi(A_a^{-1} S_s^{-1}(\cdot - t)) | a > 0, s \in \mathbb{R}, t \in \mathbb{R}^2\}, \tag{3}$$

And the associated Shearlet transform is as follows:

$$f \to SH_\psi f(a, s, t) = \langle f, \psi_{a,s,t} \rangle, \tag{4}$$

Where $f \in L^2(R^2)$ and $(a,s,t) \in R_{>0} \times R \times R^2$. So SH_ψ maps a function $f \in L^2(R^2)$ to a set of coefficients where each coefficient is indexed by a scaling parameter a, a shearing parameter s and a translation parameter t.

2.2 Cone-Adapted Shearlet System

During the experiments of this paper, the cone-adaptive Shearlet system is used. The Fourier-domain is partitioned into four cones (two horizontal, two vertical) and a square-shaped low-pass region (see Fig. 1). As shown of Fig. 1, where in C1, C3 horizontal cone, corresponding to the generation function ψ, C2, C4 vertical cone, corresponding to the generation function $\tilde{\psi}$, LP is a low-pass section, corresponding to the scale function φ [5, 6].

Let

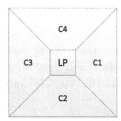

Fig. 1. The distribution of cone-adapted Shearlet system in frequency-domain.

$$\tilde{A}_a = \begin{pmatrix} a^{1/2} & 0 \\ 0 & a \end{pmatrix}, \tag{5}$$

is another scaling matrix, Given the Shearlet generators $\psi, \tilde{\psi} \in L^2(R^2)$ and the scaling function $\varphi \in L^2(R^2)$, the cone-adapted continuous Shearlet system $SH_{cout}(\phi, \psi, \tilde{\psi})$ is then given by the union of the following sets:

$$\Phi = \{\phi_t = \phi(\cdot - t) : t \in \mathbb{R}^2\}, \tag{6}$$

$$\Psi = \{\psi_{a,s,t} = a^{3/4}\psi(A_a^{-1}S_s^{-1}(\cdot - t)) : a \in (0,1], |s| \le 1 + a^{1/2}, t \in \mathbb{R}^2\}, \tag{7}$$

$$\tilde{\Psi} = \{\tilde{\psi}_{a,s,t} = a^{3/4}\tilde{\psi}(\tilde{A}_a^{-1}S_s^{-T}(\cdot - t)) : a \in (0,1], |s| \le 1 + a^{1/2}, t \in \mathbb{R}^2\} \tag{8}$$

3 The Feasibility of Stew Detection Based on Shearlet

The direction of the document image and the direction of text line is the same, and the horizontal strokes of Chinese characters can reflect the direction of the text, while the vertical strokes are perpendicular to the direction of the text. According to statistics, horizontal strokes and vertical strokes of Chinese characters occupy a dominant position in all the strokes. Jingxian Zhang made statistics of the strokes frequency on the 6196 general Chinese characters, and the statistics results show that the frequency of horizontal strokes is the highest. It is 27.65 %. The frequency of vertical strokes is 17.6 % [7]. The stroke frequency statistics form Cihai [8] by Zhiwei Feng showed the same results [9], the frequency of horizontal stroke is the highest, it's 30.66 %, and the frequency of vertical stroke is 19.17 %. In spite of the statistics from Jingxian Zhang and Zhiwei Feng not always be identical, half of the total number of strokes is horizontal and vertical strokes, which can demonstrate that, the horizontal and vertical strokes are the main components of Chinese characters [10].

Shearlet transform can offers very good time-frequency localization and direction selectivity. So it can detects the strokes of Chinese characters in all directions. Because the horizontal and vertical strokes are the main components of Chinese characters, therefore, the test results will be very prominent when the direction of detection is same as the horizontal or vertical strokes. So as to achieve the aim of detecting text direction.

The whole process of proposed algorithm determines the text direction by detecting the direction of horizontal strokes based on the time-frequency localization and direction selectivity of Shearlet. The skew detection results will not be affected even if there is some diagrams and noises in the image, or image quality is poor. Which means that the proposed algorithm is robust in the skew estimation and correction.

4   The Process of Skew Detection

The proposed algorithm is divided into three parts: (1) Image pre-processing, and the establishment of Shearlet system based on the images. (2) Computing coefficient of each scale one by one and find the component that has strongest energy in each scale. (3) Computing the angle of inclination according to the component that has the strongest energy. And then correcting the skew image. Algorithm flow is as follows.

(1)   Proper scaling the image and establishing the Shearlet system S with scale is n.
(2)   Hypothesis $S_{i,j}$ is the j coefficient or component in scale i. As shown in Fig. 2. There are $2^{n+2}+2$ coefficients in scale n. In the first scale, the component with the strongest energy are $S_{1,x1}$ and $S_{1,x2}$. These two components are corresponding to the direction of the vertical and horizontal direction of the text.

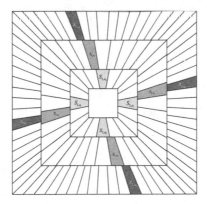

Fig. 2. The frequency domain distribution of Shearlet in each scale.

(3) In the first scale, we can measure the approximate scope of skew angle according to the strongest energy. In the next scale, only components related to $S_{1,x1}$ and $S_{1,x2}$ need to be calculated, that is $S_{2,y1}, S_{2,y2}, S_{2,y3}$ and $S_{2,y4}, S_{2,y5}, S_{2,y6}$. Then we can get the highest energy coefficient, $S_{2,y1}$ and $S_{2,y4}$.

(4) Keep doing decomposition in the next scale, until get the highest energy coefficients in the scale n.

(5) Computing the skew angle according to the highest energy coefficients obtained in the highest scale, then correcting the original image.

5   Experimental Results

The experiment is completed in Matlab2014a. The document image contains noise or a small amount of pictures and diagrams, as shown in Fig. 3.

1. Scaling the size of image to 800×800. The Shearlet system is established and the highest scale is 5.
2. The following five sub-figures in Fig. 4 show the energy of all coefficients at each scale, and there are two peaks in each scale. This shows all the coefficients, but in actual operation, it only computes the coefficients relate to the last scale. By comparison in the 5th scale, the 88th coefficient has the highest energy, and the corresponding angle is16.99 degrees.
3. Correcting image according to the skew angle corresponding to the highest energy.
4. Random noise is added to the original image (Fig. 5).
5. Comparing Figs. 6 and 4, it shows that proposed algorithm has an excellent robustness to noises.

Fig. 3. The original image.

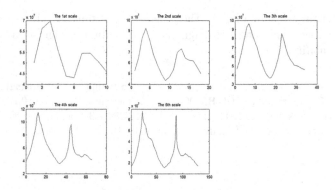

Fig. 4. The energy of all coefficients at each scale.

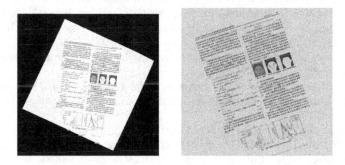

Fig. 5. Corrected image and the image with random noises.

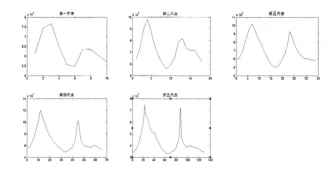

Fig. 6. The energy of all coefficients at each scale with random noises.

6   Conclusions

In this paper, we research and discuss scanned document image skew correction, and present a document image skew correction method based on Shearlet transform, which can correct the skew image successful. Experimental results show that the proposed algorithm based on Shearlet transform is more robust than former algorithms. The robustness is mainly reflected in: (1) The proposed algorithm has a high accuracy rate even the image contain pictures and diagrams. (2) The proposed algorithm has an excellent anti-noise ability.

Acknowledgments. This research was supported by the Foundation of Education Bureau of Henan Province, China grants No. 2010B520003, Key Science and Technology Program of Henan Province, China grants No. 132102210133 and 132102210034, and the Key Science and Technology Projects of Public Health Department of Henan Province, China grants No. 2011020114.

References

1. Sarfraz, M., Mahmoud, S.A., Rasheed, Z.: On skew estimation and correction of text. In: Computer Graphics, Imaging and Visualization, CGIV 2007, pp. 308–313. IEEE (2007)
2. Guo, K., Labate, D.: Analysis and detection of surface discontinuities using the 3D continuous shearlet transform. Appl. Comput. Harmonic Anal. **30**(2), 231–242 (2011)
3. Guo, K., Labate, D.: Optimally sparse multidimensional representation using shearlets. SIAM J. Math. Anal. **39**(1), 298–318 (2007)
4. Easley, G.R., Labate, D., Lim, W.Q.: Optimally sparse image representations using shearlets. In: Fortieth Asilomar Conference on Signals, Systems and Computers, ACSSC 2006, pp. 974–978. IEEE (2006)
5. http://shearlab.org/files/documents/ShearLab3Dv10_Manual.pdf
6. Easley, G., Labate, D., Lim, W.Q.: Sparse directional image representations using the discrete shearlet transform. Appl. Comput. Harmonic Anal. **25**(1), 25–46 (2008)
7. Zhangm, J.: The theory of modern Chinese stroke characters. In: The Second International Symposium of the Chinese Language Teaching (1988). (in Chinese)
8. The editorial board of ChiHai. ChiHai dictionary (1999). (in Chinese)
9. Feng, Z.: Natural Language Computing Processing. Shanghai Foreign Language Education Press, Shanghai (1996). (in Chinese)
10. Fan, L.,: Research on stroke characteristics and feature of Chinese characters, MS thesis. Southwestern University (2013). (in Chinese)

Classification of German Scripts by Adjacent Local Binary Pattern Analysis of the Coded Text

Darko Brodić[1]([✉]), Alessia Amelio[2], and Milena Jevtić[1]

[1] Technical Faculty in Bor, University of Belgrade, V.J. 12, 19210 Bor, Serbia
{dbrodic,mjevtic}@tf.bor.ac.rs
[2] Institute for High Performance Computing and Networking,
National Research Council of Italy, CNR-ICAR, Via P. Bucci 41C,
87036 Rende, CS, Italy
amelio@icar.cnr.it

Abstract. The paper proposes a script classification method which is based on textural analysis of the script types. In the first stage, each letter is coded by the equivalent script type, which is defined by its baseline position. Obtained coded text is subjected to the adjacent local binary pattern analysis to extract the features. The result shows the diversity of the extracted features between scripts, which makes the feature classification easier. It is the basis for decision-making process of the script identification by automatic classification. The proposed method is tested on an example of synthetic and historical German printed documents written in Antiqua and Fraktur scripts. The experiment shows very positive results, which proved the correctness of the proposed algorithm.

Keywords: Classification · Historical documents · Local binary pattern · Optical character recognition · Script recognition

1 Introduction

Differences between Fraktur and Antiqua scripts have a historical background. Fraktur script originates from the period of German emperor Maximilian (1493–1519). It is established as a "German answer" to widely accepted Antiqua script, which typically rages in Italy. During and after the reformation, Fraktur was widely accepted in Germany and in many Scandinavian countries as an opponent to books written in Antiqua representing Catholic church. Fraktur was used up to 1941, when it was surprisingly banned in a Schrifterlass (edict on script). Still, many historical books from Germany and Scandinavian countries (originated from 15th-20th century) are written in Fraktur. Hence, the interest in the script recognition of Fraktur and its discrimination compared to Antiqua is still present.

Script recognition is one of the most important steps in document image analysis [8]. Typically, the methods for the script recognition need a relatively large amount of data to evaluate some script [17]. Many methods for the script recognition have been proposed. They are usually classified as global or local techniques.

© Springer International Publishing Switzerland 2015
A. Bikakis and X. Zheng (Eds.): MIWAI 2015, LNAI 9426, pp. 233–244, 2015.
DOI: 10.1007/978-3-319-26181-2_22

Global techniques divide the document image into wider blocks. Obtained image blocks should be normalized and cleaned from the noise [7]. Then, they are subjected to the statistical or frequency-domain analysis [10].

In contrast, local techniques separate the document image into small blocks, typically small blocks of text called connected components. Then, these small blocks are subjected to the feature analysis like the black pixel runs, or similar ones [15]. This approach is much more computer-intensive than global approach. However, it is suitable for the low quality document images, which include noise.

Previously proposed methods receive a script identification accuracy in the range from 85 % to 95 %. However, in many cases the obtained results are misleading. In Ref. [8] German image given in Fraktur font was misclassified as Ethiopic or Thai, while in Ref. [10] 195 documents of 13 scripts are classified with an accuracy of 93.3 %. Still, these 13 scripts are too different to be easily overlapped in recognition. Furthermore, Ref. [7] receives classification error from 12.3 % to 15.9 % or at much complicated training level from 2.1 % to 12.5 %.

In this paper, we propose a document-level script classification method, which recalls the technique introduced by Brodić et al. [6]. The main idea of the method is the discrimination of the documents as written in different kinds of script, in this specific case distinguishing documents written in Fraktur from documents written in Antiqua scripts, by adopting an ad-hoc classification algorithm.

The proposed method includes the synergy of both local and global approaches. First, it extracts the characters from the text as connected components like in local methods. Then, it maps each character according to the established correlation with character shape codes [17], which we optimize in smaller set called script type. The obtained coded text is subjected to the adjacent local binary pattern analysis (ALBP) [13]. In contrast, the decision-making procedure of the script identification is established by an automatic classification algorithm Genetic Algorithms Image Clustering for Document Analysis (GA-ICDA) based on the feature vectors obtained from local binary pattern analysis of Antiqua and Fraktur scripts. It is a bottom-up genetic-based algorithm extending an image clustering approach of the state of the art which is successfully able to discriminate text documents written in Antiqua and Fraktur scripts.

Organization of the paper is as follows. Section 2 describes the proposed algorithm. Section 3 illustrates the experiment. Section 4 presents and discusses the results. Section 5 presents the conclusions.

2   The Algorithm

The proposed algorithm represents a multi-stage method, which performs script mapping, feature extraction by adjacent local binary pattern (ALBP) and automatic classification. At the end, it establishes the process of the script identification by automatic classification from the ALBP feature vectors.

The method is composed of three main steps:

1. Extraction of the coded text from each document by script mapping.

2. ALBP features computation from the coded text of each document corresponding to 1-D image, which corresponds to the initial text document.
3. Classification of the features representing the documents in order to discriminate the documents written in different scripts.

2.1 Script Mapping

We opt for grapheme compared to morpheme approach, because the script recognition in our case is based on the same language. In grapheme approach the smallest unit like letter in alphabet script represents the smallest unit under consideration. The first step is a script mapping. It is established on the center position and height of each letter in the text line. The text-line in a document can be divided into three vertical zones [22]: (i) the upper zone, (ii) the middle zone, and (iii) the lower zone. Each letter in text-line can be mapped taking into account these zones. Figure 1 illustrates the script type definition.

The partition of the letters according to the text-line zones can be established by the following script type classification [22]: (i) the short letter (S), (ii) the ascender letter (A), (iii) the descender (D), and (iv) the full letter (F). The letters can be classified depending on their position in vertical zones of the line, which reflects theirs energy profile [4].

Starting from the text images, a set of multidimensional data is extracted from the corresponding texts [3]. Accordingly, all letters can be classified into four sets. In this way, they are substituted with their equivalent script types. As a consequence, the text is mapped into the coded text. This kind of classification considerably reduces the number of variables under consideration. Furthermore, the following mapping is carried out to effectively perform statistical analysis:

$$S \rightarrow 1, \ A \rightarrow 2, \ D \rightarrow 3, \ F \rightarrow 4. \tag{1}$$

The application of the proposed concept to German text written in Antiqua and Fraktur script is illustrated below. First, the two lines of the text written in Fraktur are:

Füllest wieder Busch und Tal
Still mit Nebelglanz,

Then, the text coded according to the script type definition is given as (each blob represents different object):

Fig. 1. Definition of the script type classification.

4222142 111211 2144 112 212
22122 122 2121232113

Furthermore, the same text is written in Antiqua script such as:

Füllest wieder Busch und Tal
Still mit Nebelglanz,

At the end, it is coded by aforementioned script type model as:

2222111 111211 21112 112 212
22122 112 2121232111

It is obvious that Fraktur and Antiqua scripts modeled by script type analysis show a quite different combination of the coded text. Hence, it is a good starting point for the statistical analysis.

2.2   Script Type ALBP Analysis

Coded text represents a one dimensional image from which Adjacent Local Binary Pattern (ALBP) features are extracted. To explain the whole procedure, we need to start with a multidimensional local binary pattern (LBP).

LBP indicates a magnitude relation between a center pixel and its neighbor pixels in the window of interest (WOI), which is under consideration [14]. The initial WOI from the image is illustrated in Fig. 2(a).

34	77	91
244	128	222
0	111	134

0	0	0
1	X	1
0	0	1

(a)   (b)

Fig. 2. LBP calculation: (a) Initial WOI (3 × 3), (b) The result of the analyzed WOI.

LBP is calculated by thresholding the image intensity of the surrounding pixels with the vividness of the center pixel (gray cell from Fig. 2(b)). The pixels in this block are threaded by its center pixel value, multiplied by powers of two and then summed to obtain a label for the center pixel. Because the neighborhood consists of 8 pixels, then a total of $2^8 = 256$ different labels is obtained. The initial image is given by the vector $\mathbf{I}$, and $r = (x, y)^T$ represents a position vector. The LBP is calculated as:

$$b_i(r) = \begin{cases} 1, & I(r) < I(r + \Delta r_i) \\ 0, & \text{otherwise} \end{cases}, \tag{2}$$

where $i = 1, ..., N_n$, N_n is the number of neighbor pixels and Δr_i are displacement vectors from the position of center pixel r to neighbor pixels [9]. N_n is equal to 8 for 8-connected neighborhood, while displacement distance $d(\Delta r_i)$ is equal to 1 (pixel). The LBP $b(r)$ is converted into a decimal number. At the final stage, the LBP histogram is made by considering the decimals as labels. In this case, the coded text represents a 1-D image. Hence, the neighborhood consists of 2 pixels leading to a total of $2^2 = 4$ different labels [5].

The problem arises because LBP extracts a too small number of features for the statistical analysis. Hence, an extension of LBP method is needed. In Ref. [6] two LBP subsets called LBP(+) and LBP(×) are proposed. They are called adjacent local binary pattern (ALBP). LBP(+) considers two horizontal and two vertical pixels, while LBP(×) examines the four diagonal pixels. Because the coded text represents a 1-D image, the horizontal direction is the only possible. The ALBP is created with two adjacent LBP(+) establishing a 4-bit binary label. Hence, the total number of different labels is equal to $4 \times 4 = 16$, i.e. from '0000' to '1111'. The obtained labels represent the so-called co-occurrence LBP.

Figure 3 shows the comparison between ALBP distributions for the aforementioned coded text example given in Fraktur and Antiqua script.

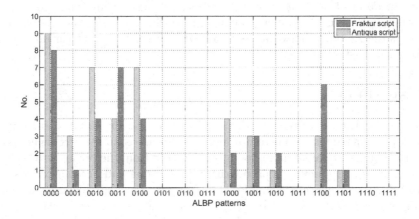

Fig. 3. Comparison of typical ALBP distributions obtained from the text written in Fraktur and Antiqua script.

2.3 Classification

Unsupervised classification according to the ALBP features of the coded text obtained from the text given in Fraktur and Antiqua scripts is performed by Genetic Algorithms Image Clustering for Document Analysis (GA-ICDA) framework, an extension of the Genetic Algorithms Image Clustering (GA-IC) approach [1]. GA-IC is an evolutionary method for image database clustering based on co-occurrence and color centiles features. The database is modeled as a weighted graph, where nodes represent images. Edges link each node with its most similar nodes, which define the h-neighborhood. Node similarity is between the images corresponding to nodes. h is a parameter controlling the neighborhood size. The edge weights quantify the similarity among the nodes. Then a genetic algorithm on this graph detects clusters of nodes which are the image classes.

Three main differences make GA-ICDA suitable for document classification. The first one consists in the feature representation. Each node of the graph is a text document defined as a feature vector of the 16 ALBP referent micro patterns from '0000' to '1111'. The second difference introduces a novelty derived from the concept of Matrix Bandwidth [12] in the graph construction. In particular, let f be a one-to-one function from graph nodes to integers $f : V \rightarrow \{1, 2, ...n\}$. It represents the node ordering induced by the graph adjacency matrix. For each node $v \in V$ with corresponding label $f(v)$, the difference between $f(v)$ and the labels F of the nodes nn_v^h in the h-neighborhood is computed. Then, for each node v, edges are inserted between v and the only nodes in nn_v^h with label difference less than a threshold value T. Figure 4 illustrates an example of graph construction.

The last difference realizes a refinement phase at the end of the genetic algorithm. Pairs of detected clusters are merged, having minimum distance to each other, until a fixed number of clusters is reached. The distance between two clusters is the L_1 norm between the two farthest text document feature vectors, one for each cluster.

3   Experiment

The experiment represents the process of testing the proposed algorithm. To perform the experiment, a custom-oriented database is created. It includes 100 synthetic and historical German documents mainly from the poems written by J. W. von Goethe. All documents are written in Antiqua and Fraktur scripts. The smallest document contains roughly 90 characters, while the biggest one counts around 1k characters. We can realize that the smallest number of characters is quite small for statistical analysis. If the algorithm is successful, then it can be used for scene recognition. The algorithm is applied to the database. Figure 5 shows an excerpt from the database.

4   Results and Discussion

The experiment is linked with the calculation of ALBP micro patterns obtained from the coded text written in Antiqua and Fraktur scripts in documents of the database. From obtained ALBP results, seven features show diversity between Antiqua and Fraktur, simplifying the classification process. It is illustrated in Fig. 6.

Obviously, the main point of discrimination will be connected to ALBP '0000'. The experiment is performed by using GA-ICDA on the ALBP feature

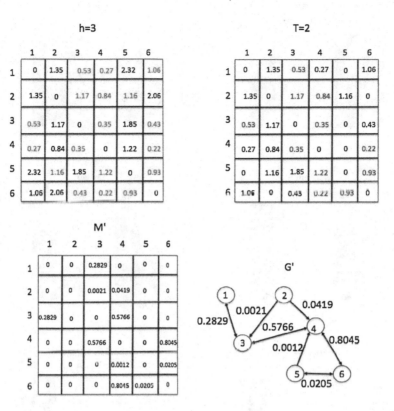

Fig. 4. Example of graph construction. From top to down for each node in the distance matrix, detection of the 3-nearest neighbors (in red), for each node, detection of the neighbors with label difference smaller or equal to $T = 2$ with respect to the label of that node (in blue), creation of the graph G' from the obtained adjacency matrix M'. Two or more clusters can be obtained by G' graph partitioning (Colour figure online).

Zueignung.

Ihr naht euch wieder, schwankend
Die früh sich einst dem trüben Bli
Versuch' ich wohl, euch diesmal fest
Fühl' ich mein Herz noch jenem A
Ihr drängt euch zu! nun gut, so v
Wie ihr aus Dunst und Nebel um
Mein Busen fühlt sich jugendlich e

Fig. 5. Text excerpt from the database of documents.

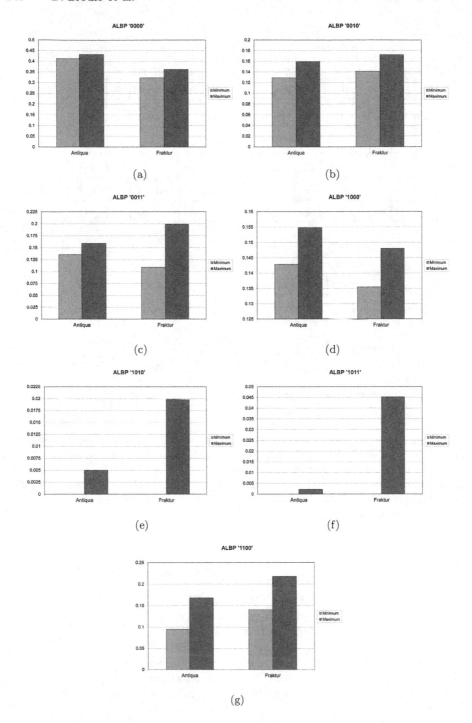

Fig. 6. ALBP micro patterns with values different from zero given in min-max manner: (a) '0000', (b) '0010', (c) '0011', (d) '1011', (e) '1000', (f) '1010', (g) '1100'.

vectors obtained from the text written in Fraktur and Antiqua scripts. The main goal is the evaluation of the classifier in correctly discriminate between the two different kinds of scripts in the database.

A trial and error procedure is adopted on benchmark documents, different from the documents of the database, for tuning the parameters of GA-ICDA. Then, parameter values giving the best results on the benchmark are also applied for clustering. Consequently, the h value of the node neighborhood is fixed to 10 and the T threshold value is fixed to 18.

To demonstrate the effectiveness of this technique, a comparison with other two unsupervised classifiers is also provided. We chose the Complete Linkage Hierarchical clustering and Self-Organizing-Map (SOM) methods, well-known for document classification [9, 11, 19–21]. Hierarchical and SOM classifiers use the same ALBP feature representation as GA-ICDA.

The measures chosen for performance evaluation of script classification are *Precision, Recall, F-Measure* (computed for each script class) and *Normalized Mutual Information (NMI)* [2, 18]. Because NMI is a similarity measure which is not dependent from the script class, a single value is reported for both Antiqua and Fraktur. Specifically, Precision, Recall and F-Measure are computed by considering the correspondence of each detected cluster with that true script class which is in majority inside the cluster.

Experiments have been performed on a Desktop computer quad-core 2.3 GHz with 8 Gbyte of RAM and Windows 7. Table 1 and Fig. 7 show the results of GA-ICDA classification on the database compared with hierarchical and SOM. In bold are the cases when GA-ICDA outperforms the other classifiers. Algorithms have been executed 100 times and the average values together with the standard deviation (in parenthesis) of the performance measures have been reported.

It is worth to observe that GA-ICDA has better performances than the other classifiers. In fact, hierarchical clustering obtains a F-Measure of 0.76 for Antiqua and 0.79 for Fraktur, while NMI is 0.25. SOM classifier reaches a F-Measure of 0.62 for Antiqua and 0.59 for Fraktur. NMI is 0.14. On the contrary, GA-ICDA obtains a very good script classification with a F-Measure of 0.9996 and 0.9994 for respectively Antiqua and Fraktur and NMI values of 0.9900. Standard deviation is zero, demonstrating the stability of the method.

If we compare the evaluation results on Antiqua and Fraktur recognition task given in Ref. [16] obtaining a very good 96.61 % accuracy, then it is obvious that our algorithm has an advantage. Furthermore, the method is primarily based on the bounding box approach, its height and center point position in the text line. Hence, the discrimination between characters is no needed in order to identify different script. It is an advantage when circumstances the historical documents are noise injected. It leads that only the allocation of various bounding boxes configuration creates differences between the scripts, which can be successfully discriminated with texture like analysis. At the end, a specific automatic classification tool like GA-ICDA is necessary to prove the full effectiveness of the proposed approach.

Table 1. Results of Precision, Recall, F-measure and Normalized Mutual Information (NMI) obtained for Antiqua and Fraktur classes on the script database from the unsupervised clustering algorithms: GA-ICDA, Hierarchical Clustering and SOM. *nc* is the number of clusters detected from the algorithm.

	nc	classes	Precision	Recall	F-Measure	NMI
GA-ICDA	2	Antiqua	**0.9993** (0.0000)	**1.0000** (0.0000)	**0.9996** (0.0000)	**0.99**(0.0000)
		Fraktur	**1.0000** (0.0000)	**0.9989** (0.0000)	**0.9994** (0.0000)	
Hierarch.	2	Antiqua	0.8286 (0.0000)	0.7073 (0.0000)	0.7632 (0.0000)	0.25 (0.0000)
		Fraktur	0.7447 (0.0000)	0.8537 (0.0000)	0.7955 (0.0000)	
SOM	4	Antiqua	0.8260 (0.0124)	0.4941 (0.0128)	0.6182 (0.0085)	0.14 (0.0085)
		Fraktur	0.8478 (0.0309)	0.4544 (0.0118)	0.5912 (0.0055)	

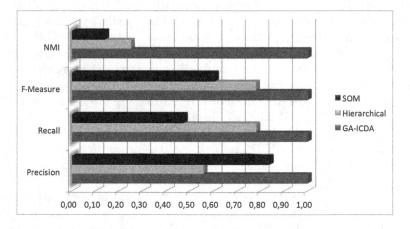

Fig. 7. Results of Precision, Recall, F-Measure and NMI averaged on Antiqua and Fraktur classes for GA-ICDA, Hierarchical and SOM classifiers.

5   Conclusion

The manuscript proposed a new approach for the script identification on the example of synthetic and historical German printed documents, which were written in Antiqua and Fraktur scripts. The algorithm performed the adjacent local binary pattern analysis of the coded text mapped from the text document. The coded text was subjected to the ALBP statistical analysis (ALBP). It showed significant diversity between both scripts. Hence, a solid script classification and identification method GA-ICDA was introduced and applied to the obtained statistical features. It is an extension of a state of the art image clustering method based on genetic algorithms successfully distinguishing among different text documents written in Antiqua and Fraktur scripts. The proposed algorithm was tested on a custom oriented database, which consists of historical documents written in Antiqua and Fraktur scripts. The experiments showed the ability of the algorithm in correctly discriminate between the scripts and the superiority of the results with respect to the state-of-the-art classification techniques. The presented concept is suitable for use in processing of OCR.

Future work will investigate the run-length coding for feature representation in script discrimination.

Acknowledgments. This work was partially supported by the Grant of the Ministry of Science of the Republic Serbia within the project TR33037.

References

1. Amelio, A., Pizzuti, C.: A new evolutionary-based clustering framework for image databases. In: Elmoataz, A., Lezoray, O., Nouboud, F., Mammass, D. (eds.) ICISP 2014. LNCS, vol. 8509, pp. 322–331. Springer, Heidelberg (2014)
2. Andrews, N.O., Fox, E.A.: Recent Developments in Document Clustering. Computer Science, Virginia Tech, Virginia (2009)
3. Bourennane, S., Marot, J., Fossati, C., Bouridane, A., Spinnler, K.: Multidimensional signal processing and applications. World Sci. J. **2014**(365126), 1–2 (2014)
4. Brodić, D., Milivojević, Z.N., Amelio, A.: Analysis of the South Slavic Scripts by Run-Length Features of the Image Texture. Elektronika Ir Elektrotechnika **21**(4), 60–64 (2015)
5. Brodić, D., Milivojević, Z.N., Maluckov, Č.A.: An approach to the script discrimination in the Slavic documents. Soft Comput. **19**(9), 2655–2665 (2015)
6. Brodić, D., Milivojević, Z.N., Maluckov, Č.A.: Multidimensional signal processing and applications. World Sci. J. **2014**(896328), 1–14 (2013)
7. Busch, A., Boles, W.W., Sridharan, S.: Texture for script identification. IEEE Trans. Pattern Anal. Mach. Intell. **27**(11), 1720–1732 (2005)
8. Hochberg, J., Kelly, P., Thomas, T., Kerns, L.: Automatic script identification from document images using cluster-based templates. IEEE Trans. Pattern Anal. Mach. Intell. **19**(2), 176–181 (1997)
9. Isa, D., Kallimani, V.P., Lee, L.H.: Using the self organizing map for clustering of text documents. Expert Syst. Appl. **36**(5), 9584–9591 (2009)
10. Joshi, G., Garg, S., Sivaswamy, J.: A generalised framework for script identification. Int. J. Doc. Anal. Recogn. **10**(2), 55–68 (2007)
11. Marinai, S., Marino, E., Soda, G.: Self-organizing maps for clustering in document image analysis. In: Marinai, S., Fujisawa, H. (eds.) Machine Learning in Document Analysis and Recognition. Studies in Computational Intelligence, vol. 90, pp. 193–219. Springer, Heidelberg (2008)
12. Marti, R., Laguna, M., Glover, F., Campos, V.: Reducing the bandwidth of a sparse matrix with Tabu search. Eur. J. Oper. Res. **135**(2), 450–459 (2001)
13. Nosaka, R., Ohkawa, Y., Fukui, K.: Feature extraction based on co-occurrence of adjacent local binary patterns. In: Ho, Y.-S. (ed.) PSIVT 2011, Part II. LNCS, vol. 7088, pp. 82–91. Springer, Heidelberg (2011)
14. Ojala, T., Pietikainen, M., Harwood, D.: A comparative study of texture measures with classification based on featured distributions. Pattern Recogn. **29**(1), 51–59 (1996)

15. Pal, U., Chaudhuri, B.B.: Identification of different script lines from multi-script documents. Image Vis. Comput. **20**(13–14), 945–954 (2002)
16. Rashid, S.F., Shafait, F., Breuel, T.M.: Discriminative learning for script recognition. In: Proceedings of 17th IEEE International Conference on Image Processing (ICIP), pp. 2145–2148. IEEE Press, New York (2010)
17. Sibun, P., Spitz, A.L.: Language determination: natural language processing from scanned document images. In: Proceedings of the Fourth Conference on Applied Natural Language Processing (ANLC 1994), pp. 15–21. Association for Computational Linguistics Stroudsburg (1994)
18. De Vries, C., Geva, S., Trotman, A.: Document Clustering Evaluation: Divergence from a Random Baseline. CoRR, abs/1208.5654 (2012)
19. Yen, G.G., Zheng, W.: A self-organizing map based approach for document clustering and visualization. In: Proceedings of International Joint Conference on Neural Networks (IJCNN 2006), pp. 3279–3286. IEEE Press, New York (2006)
20. Pu, Y., Shi, J., Guo, L.: A hierarchical method for clustering binary text image. In: Yuan, Y., Wu, X., Lu, Y. (eds.) ISCTCS 2012. CCIS, vol. 320, pp. 388–396. Springer, Heidelberg (2013)
21. Zhu, Q.-M., Li, J., Zhou, G., Li, P., Qian, P.: A novel hierarchical document clustering algorithm based on a kNN connection graph. In: Matsumoto, Y., Sproat, R.W., Wong, K.-F., Zhang, M. (eds.) ICCPOL 2006. LNCS (LNAI), vol. 4285, pp. 120–130. Springer, Heidelberg (2006)
22. Zramdini, A., Ingold, R.: Optical font recognition using typographical features. IEEE Trans. Pattern Anal. Mach. Intell. **20**(8), 877–882 (1998)

Tracking and Identifying a Changing Appearance Target

Somnuk Phon-Amnuaisuk[(⊠)] and Azhan Ahmad

Media Informatics Special Interest Group, School of Computing and Information
Technology, Institut Teknologi Brunei, Gadong, Brunei
{somnuk.phonamnuaisuk,azhan.ahmad}@itb.edu.bn

Abstract. Camshift algorithm has been popularly applied to tasks
such as pedestrian tracking and traffic tracking. Camshift employs a
histogram-based technique. The common hue-based histogram is robust
to minor changes in the shape of a tracked target but it is only suitable to track a target having, relatively, a constant appearance. In this
paper, we investigate the application of the Camshift algorithm and the
Speeded Up Robust Features (SURF) in tracking and identifying a target. The target is tracked using the Camshift algorithm and identified
using SURF. By combining the two techniques we show that a changing
appearance target can be tracked and identifed at the same time.

Keywords: Object tracking · Identifying a changing appearnce object ·
Camshift analysis · Speeded Up Robust Features

1 Introduction

The ability to correctly locate an object of interest in sequential video frames is
fundamental for many applications such as pedestrian tracking, motion analysis,
traffic monitoring, surveillance, etc. Different techniques have been developed
for various object tracking tasks. For example, in a pedestrian tracking task, if
we are only interested in tracking the number of anonymous pedestrians and
their positions in the scene, then it is not important to identify further details
such as articulated limbs movements. In such a case, features based on colour
histogram or movement information extracted from frame-difference technique
would be good features. However, if more detailed analyses such as gait analysis,
facial expression or hand gesture analyses are desired, then local points, object
silhouelettes, object skeletons and object contours will be better features.

Object tracking is a wide research area and different techniques have been
devised to handle different problems at hands. This paper reports our work in
progress on the Smartspaces project. In one of the subtasks, we hope to promote an expressive presentation by letting a presenter control the multimedia
contents associated with various symbols. In the current setup, the system tracks
and identifies 5 symbols, each printed on a 8 cm by 8 cm card. Imagine a weather
girl reporting the weather conditions and displaying various augmented reality

© Springer International Publishing Switzerland 2015
A. Bikakis and X. Zheng (Eds.): MIWAI 2015, LNAI 9426, pp. 245–252, 2015.
DOI: 10.1007/978-3-319-26181-2_23

content associated with different symbols; imagine a newscaster/presenter displaying various presentation materials associated with different symbols; these could provide a rich and informative presentation to the audience. Although the setup in this experiment aims at tracking and identifying printed symbols on a card, the symbols can be displayed on any device in practice.

The rest of the materials in the paper are organized into the following sections; Sect. 2: Related Works; Sect. 3: Problem Formulation; Sect. 4: Experimental Results and Discussion; and Sect. 5: Conclusion.

2   Related Works

Object tracking has been intensively investigated by researchers in computer vision community for decades. This task is computationally expensive and therefore, the popularity of research in this area is limited somewhat. Recently, with a decrease in computing cost, we see a lot of research activities in this area.

Majority of previous works formulate their problems as tracking desired features derived from the pixel information of an image. Features such as *colour histogram, intensity map, colour map, texture, intensity gradient, edge, contour, corner point*, etc., are popularly employed to describe an object of interest. These features are derived from 2D pixel information in a single image [1]. Combining temporal information from sequential 2D images, *spatio-temporal features* have been extended from static features obtained from each frame [2], for example, the optical flow method. The optical flow tracks an object by calculating the motion of each voxel between multiple frames in an image sequence. Apart from identifying moving objects, spatio-temporal features can also be used to identify the speed of moving objects [3] which can contribute to many useful applications such as traffic monitoring systems, virtual musical instruments [4], etc.

Object tracking is a hard problem since the representation of the region of interest (ROI) in an image using features listed earlier is sensitive to various factors such as illumination variations, occlusion, complexity of the shape and speed of an object. Various techniques and tactics have been devised by researchers to handle the challenges (see a survey by [5]). In providing the background to our work, we shall focus the review on the following areas: Camshift algorithm [6] and Speeded Up Robust Features (SURF) [7].

Camshift stands for *continuously adaptive Mean shift*. It is the application of the Mean shift algorithm [8,9] across video frames where the initialised candidate ROI on a frame f is the target ROI from the previous frame $f - 1$. The Mean shift locates the target ROI in a frame by iteratively moving the candidate ROI according to the gradient of the Mean shift vector. The target ROI is located if the similarity between the target ROI and the candidate ROI is within a desired threshold.

Camshift has been popularly employed to track arbitrary objects without prior training since the target model can be constructed from a user's specified ROI on the fly. It is also suitable to employ Camshift to track objects with changing shapes since describing a target model using histogram features constructed from hue information is quite robust to variations in shapes, for example, changes in a pedestrain's shape due to body movements.

It is possible to pre-construct a target model for generic object e.g., nose, eyes, etc. and then track it using Camshift. It is, however, not common to use Camshift to track and to identify particular objects e.g., John's nose, Jane's eyes, etc. This is because constructing a target model using histogram emphasises the global appearance and neglects local details that are crucial in discriminating objects sharing many similarities.

Local details are better detected using detectors such as Kanade-Lucas-Tomasi (KLT) corner detectors [10], and SIFT [11] or SURF [7] blob detectors. These intensity-based detectors locate points with peak intensity gradient changes [5]. Since intensity gradient peaks are more robust to illumination variations and rotations, these detected points have a good repeatability. Hence, they provide good local features in which two versions of an object can be successfully identified by matching their local points.

Therefore, employing Camshift to detect ROI based on global features and employing SURF to identify the class of ROI based on local features seem to be a good combination. It is the aim of this paper to investigate this point.

3   Problem Formulation

In our problem formulation, 5 symbols are printed on 5 cards, one for each card. Combining Camshift and SURF detector allows us to track a symbol using Camshift and identify the symbol using local descriptors obtained from SURF. Figure 1 gives an overview of our approach.

Let us formally define relevant vocabularies according to [12] that will be used in our discussion later.

Definition 1. *Target model: Let the target model be the region of interest (ROI) centered at (0,0) and let us define the target model using a feature vector* $\mathbf{q} = \{q_u\}_{u=1..m}$ *where u is the bin index number and C is the normalisation factor such that* $\sum_u q_u = 1$. *A vector* $\mathbf{q}$ *may be constructed from any type of features (e.g., hue, intensity). The probability of feature u in a model can be expressed as:*

$$q_u = C \sum_{b(x_i)=u} k(\|x_i\|^2) \qquad (1)$$

where $\{x_i\}_{i=1..n}$ *denotes target pixel locations;* $b(x_i)$ *denotes the bin index of pixel at* x_i; *and* $k(.)$ *is a differentiable, isotropic, convex and monotonically decreasing kernel (e.g., Epanechnikov kernel, Gaussian kernel). Hence* q_u *reflects a summation of desired feature (e.g., intensity) of the histogram bin u.*

Definition 2. *Target candidate model: Similarly, the candidate region of interest (candidate ROI) centered at y can be described using a feature vector* $\mathbf{p}$. *The probability of feature u in a model can be expressed as:*

$$p_u(y) = C_h \sum_{b(x_i)=u} k(\|\frac{x_i - y}{h}\|^2) \qquad (2)$$

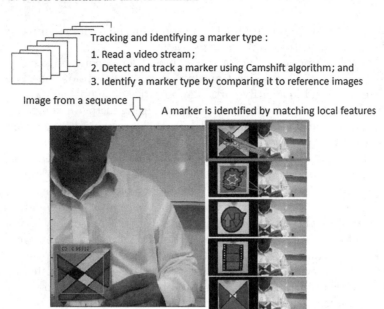

Fig. 1. The overall concept of our proposed approach. Left pane: the Camshift algorithm successfully tracks the ROI. Right pane: 5 small images showing matches in local features. Here, the tracked target is correctly matched to the correct reference symbol.

where C_h is the normalisation factor; y is the reference position of a candidate ROI; and h is the size of the candidate ROI.

Definition 3. *Similarity function: Let $f(y) = f[\mathbf{q}, \mathbf{p}(y)]$ be a similarity function based on the Bhattacharyya coefficient. The similarity of a candidate ROI centered at y is defined as*

$$\sum_{u=1}^{m} \sqrt{p_u(y)q_u} \tag{3}$$

Tracking a Marker Using Camshift. Camshift algorithm extends Mean-shift algorithm to handle object tracking in video frame sequences. Camshift applies Mean shift to detect the target ROI in each frame. For each frame, the Mean-shift algorithm locates the center of the density by calculating the center position y_c:

$$y_c = \frac{\sum_{y_i \in ROI_{candidate}} w_i y_i}{\sum_i w_i};$$

$$\text{where } \forall y_i \in ROI_{candidate}[b(y_i) = u] \rightarrow [w_i = \sqrt{\frac{q_u}{p_u(y)}}] \tag{4}$$

The Mean shift iteratively estimates the Mean shift vector from the y_c. For each iteration, the candidate ROI moves along the Mean-shift vector until it reaches the center of the density i.e., the target ROI.

Identifying Marker Types Using SURF. Speeded Up Robust Features (SURF) compute the points of interest using a blob detector based on the Hessian matrix. Given a point $p = (x, y)$ in a grayscale image I, if $L_{xx}(p, \sigma)$ denotes the second-order derivatives of the grayscale image along the x direction, then the Hessian matrix $H(p, \sigma)$ at point p and scale σ, is defined as follows:

$$H(p, \sigma) = \begin{bmatrix} L_{xx}(p, \sigma) & L_{xy}(p, \sigma) \\ L_{xy}(p, \sigma) & L_{yy}(p, \sigma) \end{bmatrix} \tag{5}$$

The local points at a scale σ are chosen when the Hessian matrix has the highest determinant.

We can match the local points identified on the tracked ROI with the local points on reference symbols. The tracked ROI is identified as the reference symbol if it has the maximum matched points. In our implementation, we smooth out the detection by delaying the classification result for 10 frames. The identification result on frame 11 is the result of the voting from frames 1–11, the result on frame 12 is the voting from frames 2–12, and so on. In other words, the result from frame n is determined from the voting of results from frames $n, n - 1, ..., n - 10$.

4   Experimental Results and Discussion

Since we are motivated by the need to provide an interactive and proactive environment that supports expressive presentation, we would like to see that the idea of tracking and identifying a symbol printed on a card could work. In this preliminary study, we design two experiments to evaluate our proposed approach.

Experimental Design. Two video clips were recorded at 15 frames per second under our typical classroom lighting condition: (i) each card was displayed for approximately 6 s. The card occupied approximately 60 % of the frame area so their movement was quite small; and (ii) each card moved toward the camera from the bottom-left to the upper-right corner of the frame and then back to the bottom-left of the frame. This took around 6 s for each card. Since the card was moved toward the camera, the size of the card appeared with, approximately, between 8 % to 40 % of the frame area. The motivation behinds this setup was to (i) verify that the local features could be used to identify symbols and (ii) evaluate the performance of the system with a moving target having various sizes of ROI. In each video clip, a presenter was instructed to move the card smoothly (Fig. 2).

Evaluation Criteria. For each clip, the symbol ID in a frame was manually labelled as a ground truth. The accuracy was then calculated based on this ground truth. Since the accuracy of tracking and identification was dependent

Fig. 2. Six snapshots from one of the experiments. This shows that task setup in the experiment: to track a card moving in 3D space that changes in its size and appearance.

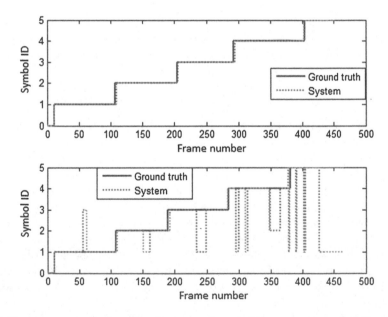

Fig. 3. Top pane: results from the first experiment. The system successfully identifies all 5 symbols. Bottom pane: results from the second experiment. Although the system can track and identify the symbols, from time to times, the system errornously identify the target ID.

on the ROI region set by a user, the accuracy value obtained from each video clip was averaged over 10 runs.

$$accuracy = \frac{\text{number of frames the markers were correctly identified}}{\text{total numbers of frames}} \quad (6)$$

Experimental Results. Figure 3 summarises the results from both experiments carried out in this work. The results from the first experiment show that the local features can be used to identify the correct target ID. The SURF points seems to be quite expressive for this task. The average accuracy was 98.4%. However, there were quite a number of misidentified targets in the second experiment with the average accuracy of 76.8%.

Upon inspection, we observe that misidentification occurred when the ROI was not properly located on the tracked-target. If the tracked ROI was correctly located, the local points were expressive enough to identify the target ID. This observation confirms the outcome of the first experiment. This implies that the accuarcy of the second experiment can be further improved if the performance of the tracking algorithm is improved.

From the experiment, we observe that the performance of the Camshift algorithm was sensitive to various parameters: the initialised ROI; the size of bins; and the similarity threshold controlling the iteration of the Meanshift processing for each frame. The setting of these parameters was dependent on the type of tracked-target. Here, the ROI was manually initialised; the numbers of hue bins were set at 36; and the similarity threshold was set at 0.8 (80%). We observe that the SURF points had a good repetability and it was suitable to employ them to identify a target with exact appearance.

5   Conclusion

In this work, we have shown a hybrid of two techniques: Camshift tracking and SURF interested-points matching. The Camshift performance is sensitive to parameters setting but it provides a good global feature of the ROI. The Camshift algorithm provides an inexpensive means to track a target and it is quite resistant to appearance variations since its histogram-based nature represents a global appearance of the target.

On the other hand, the SURF points reveal local features and it is possible to employ this information to identify different targets. The SURF is also sensitive to changes in appearance and it is relatively more expensive to track SURF interested-points (in comparison to the Camshift technique).

By combining these two techniques, we have successfully implemented a system that could track and identify the identity of the tracked target. There are many directions that may fork out from this point. In future work, we are interested in looking into users' attentions. We are also interested in creating various augmented reality content for each symbol. In the big picture, we would like to explore further on how to provide an interactive and proactive environment to the users.

Acknowledgments. We wish to thank anonymous reviewers for their comments, which help improve this paper. We would like to thank the GSR office for their financial support given to this research.

References

1. Gonzalez, R.C., Woods, R.E.: Digital Image Processing. Prentice Hall, Upper Saddle River (2001)
2. Laptev, I., Lindeberg, T.: Space-time interest points. In: Proceedings of the Ninth IEEE International Conference on Computer Vision (ICCV 2003), vol. 2, p. 432 (2003)
3. Kravchonok, A.: Detection of moving objects in video sequences by the computation of optical flow based on region growing. Pattern Recogn. Image Anal. **21**(2), 283–286 (2011)
4. Phon-Amnuaisuk, S., Rezahanjani, K., Momeni, H.R., Khor, K.C.: Virtual musical instruments: air drums. In: Proceedings of the Sixth International Conference on Information Technology in Asia (CITA 2009), Kuching, Sarawak, pp. 95–99 (2009)
5. Yilmaz, A., Javed, O.: Object tracking: a survey. ACM Comput. Surv. **38**(4), 1–45 (2006). Article 13
6. Bradski, G., Gary, G.R., Olga, L.M.: Computer vision face tracking for use in a perceptual user interface. Intel Technol. J. Q2-98 **10**, 1–15 (1998)
7. Bay, H., Ess, A., Tuytelaars, T., Van Gool, L.: SURF: speeded up robust features. Comput. Vis. Image Understand. (CVIU) **110**(3), 346–359 (2008)
8. Fukunaga, K., Hostetler, L.D.: The estimation of the gradient of a density function with applications in pattern recognition. IEEE Trans. Inf. Theory **21**(1), 32–40 (1975)
9. Cheng, Y.: Mean shift, mode seeking, and clustering. IEEE Trans. Pattern Anal. Mach. Intell. **17**(8), 790–799 (1995)
10. Shi, J., Tomasi, C.: Good features to track. In: Proceedings of the IEEE Conference on Computer Vision and Pattern Recognition, pp. 593–600 (1994)
11. Lowe, D.G.: Object recognition from local scale-invariant features. In: Proceedings of the Seventh IEEE International Conference on Computer Vision, Kerkyra, vol. 2, pp. 1150–1157 (1999)
12. Comaniciu, D., Ramesh, V., Meer, P.: Kernel based object tracking. IEEE Trans. Pattern Anal. Mach. Intell. **25**(5), 564–577 (2003)

Motion Detection System Based on Improved LBP Operator

Peijie Lin, Bochun Zheng, Zhicong Chen, Lijun Wu$^{(\boxtimes)}$, and Shuying Cheng$^{(\boxtimes)}$

School of Physics & Information Engineering, Fuzhou University,
Fuzhou 350116, China
{lijun.wu,sycheng}@fzu.edu.cn

Abstract. A fast and reliable motion detection algorithm is very important to an intelligent surveillance system. Local Binary Pattern (LBP) is one of powerful texture description and comparison mechanisms, but in contrast consumes a large portion of computational time in a CPU based system. In this paper, we propose a moving object detection algorithm based on the improved LBP operator which is tolerant against pixel noise. Combining the background subtraction algorithm and the frame difference algorithm, the automatic refreshing of the background is realized. The moving object detection system which can achieve real time processing of a 1024 × 768/60 Hz VGA signal is designed on a PFGA chip and all the algorithms are mapped to hardware logic. ROC curves of the experiments demonstrate that in the condition with shifty illumination, the algorithm based on LBP operator has a better performance than the algorithm based on grayscales.

Keywords: Local Binary Pattern · Frame difference · Background removal · FPGA · Real-time video process

1 Introduction

Motion detection algorithm is widely used in the area of intelligent video surveillance. The performance of moving object detection algorithm directly impacts on the performance of subsequence machine vision algorithm such as tracking and recognition. Currently, Optic Flow, Background Subtraction and Frame Difference are regard as the three mainstream algorithms to detect moving object. Optic flow reflects the velocity vector of each pixel, but this method has high computational complexity and big resource consumption [1, 2] and is usually used in the scenarios with moving camera. The basic principle of background removal method is building a background model and providing a classification of the pixels into either foreground or background [3–5]. In a complex and dynamic environment, it is difficult to build a robust background model. Frame difference based method need not model the background. It detects moving objects based on the frame difference between two continuous frames. Frame difference based method is easy to be implemented and it can realize real-time detection, but this method cannot extract the full shape of the moving objects [6].

There are three problems of the current video surveillance systems based on above algorithms:

© Springer International Publishing Switzerland 2015
A. Bikakis and X. Zheng (Eds.): MIWAI 2015, LNAI 9426, pp. 253–261, 2015.
DOI: 10.1007/978-3-319-26181-2_24

(1) The algorithms we have discussed above are all based on grayscale. In practical applications especially outdoor environment, the grayscales of each pixel are unpredictably shifty because of the variations in the intensity and angle of illumination.

(2) Besides the moving foreground, the background of the outdoor environment is always changing randomly, which makes building a robust background model a hard work.

(3) Software algorithms consume a large portion of computational time in a CPU based system [7, 8].

In order to solve the above problems, we proposed our LBP based algorithm which is resistance to the variations of illumination. Our algorithm can also auto refresh the background model considering of the moving foreground. To realize real-time processing, the algorithm is mapped to hardware logic and the whole system is implemented on a FPGA chip.

2 Paper Preparation

Local binary pattern (LBP) is widely used in machine vision applications such as face detection, face recognition and moving object detection [9–11]. LBP represents a relatively simple yet powerful texture descriptor which can describe the relationship of a pixel with its immediate neighborhood. The fundamental of LBP operator is showed in Fig. 1. The basic version of LBP produces 2^8 texture patterns based on a 3×3 neighborhood. The neighboring pixel is set to 1 or 0 according to the grayscale value of the pixel is larger than the value of centric pixel or not. For example, in Fig. 1 7 is larger than 6, so the pixel in first row first column is set to 1. Arranging the 8 binary numbers in certain order, we get an 8 bits binary number, which is the LBP pattern we need. For example in Fig. 1, the LBP pattern is 10001111. LBP pattern is tolerant against illumination changing. When the grayscales of pixels in a 3×3 window are shifted due to illumination changing, the LBP value will keep unchanged.

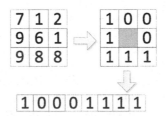

Fig. 1. Basic principle of LBP operator

LBP is not numerical value, the distance between two LBP patterns is normally measured by Hamming Distance $d_{ham}(\cdot, \cdot)$. In the information theory, Hamming Distance between two strings of equal length is the number of positions at which the corresponding symbols are different. For instance, the second and third position of string

"11110011" and "11111111" are different, so the Hamming Distance of these two strings is 2. For two LBP patterns, Hamming Distance can be computed by bitwise XOR operation.

The above method of calculating LBP patterns cannot meet the need of practical application of video surveillance for two reasons: Firstly, a "window" which only contains 9 pixels is a small area in which the grayscales of pixels are approximate to each other, and the texture feature in such a small area is too weak to be reflected by a LBP pattern. Secondly, pixel noise will immediately cause the noise of LBP pattern, which may lead to a large number of wrong detection. In order to obtain a better performance, we proposed a LBP algorithm based on the mean value of "block". In our design, one block contains 3 × 3 pixels. Compared with original LBP pattern calculated in a local 3 × 3 neighborhood between pixels, the improved LBP operator is defined by comparing the mean grayscale value of central block with those of its neighborhood blocks (see Fig. 2). By replacing the grayscales of pixels with the mean value of blocks, the effect of the pixel noise is reduced. The texture feature in such a bigger area is more significant to be described by LBP pattern.

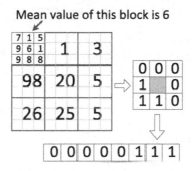

Fig. 2. Principle of LBP operator based on blocks

3 Background Updating

In practical application, the background is changing randomly. For traditional background subtraction algorithm the incapability of updating background timely will cause wrong detection. In order to solve this problem, we propose an algorithm with dynamic self updating background model. As we know, frame difference method can detect moving object without a background model, but this method cannot extract the full shape. Background subtraction method can extract the full shape but needs a background model. The basic principle of our algorithm is running a frame difference moving object detection process concurrently with the background subtraction process. What's time to update the background is according to the result of frame difference detection.

In our algorithm, the number of static frames is counted. If for successive N frames the frame difference process doesn't detect a moving block, our algorithm determines that the current surveillance vision is static, and replaces the old background model with the new one. So if there is no moving object in the surveillance vision, the background will be updated in each N frames time. The moving foreground will not be mistaken for background because the static frame counter will be clear to zero by frame difference moving object detector. Denote LBP(t, k) the LBP pattern of the block k in the frame t. Denote LBP(b, k) the LBP pattern of the block k in the background. The algorithm pseudo code is showed below:

```
while (1)
{
number of moving blocks = 0;
for (k=0 : k< number of all blocks in one frame : k++ )
    {
    if( Dham( LBP(t,k), LBP(b,k) ) > Dham threshold )
                { number of moving blocks ++ ;}
    }
    if( number of moving blocks > moving blocks number
        threshold )
            {number of moving frames ++;}
            if (number of moving frames > N )
            {
              update background;
              number of moving frames = 0;
            }
            t++ ;
    }
```

4   System Implementation

FPGA have features of concurrent computation, reconfiguration and large data throughput [7, 8]. It is suitable to be built an embedded surveillance system on. The algorithm introduced above is implemented on a Xilinx Virtex5 FPGA development board.

To make this system more practical, we add a function which allows security personnel to draw several rectangles on the screen. The system will detect moving object in each rectangle respectively. If there is moving object in a certain rectangle, the framework of this rectangle will be shining to alarm.

The structure of the system is showed in Fig. 3. In this system, a $1024 \times 768/60$ Hz VGA signal is inputted to the development board on which the analogy VGA signal is converted to a digital RGB signal by an AD9880 ADC chip. The output RGB signal of this system which combines the raw picture and the rectangles is converted to an analogy VGA signal by a CH7301 chip, and the VGA signal will be displayed on a LCD monitor finally. For hardware logic inside FPGA chip, LBP patterns of current frame is calculated by the LBP calculating module. The previous frame buffer and background buffer stores the LBP patterns of previous frame and background respectively. Through the calculation of the LBP patterns of current frame and background, we obtain the Hamming Distance of those two patterns by the background difference module. If the Hamming Distance is bigger than the threshold, this area will be identified as a moving object.

At the same time, the Hamming Distance between the LBP patterns of the current frame and previous frame are calculated by the frame difference module. Frame difference module also compares the current frame and the previous frame to determine whether there is a moving object in the surveillance vision. After the comparing, the LBP patterns of current frame will be stored in the former frame buffer, those patterns will be used in the next contiguous frames comparing. If the surveillance vision is static for a certain amount of frame, the write enable control signal of the background buffer RAM will be set to high level and the background model will be updated. After the updating the write enable control signal will be set to low level and wait for next updating. All the buffers we talk about are all implemented on the block RAM resource of FPGA.

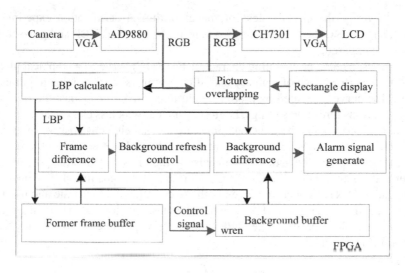

Fig. 3. Structure of motion detection system

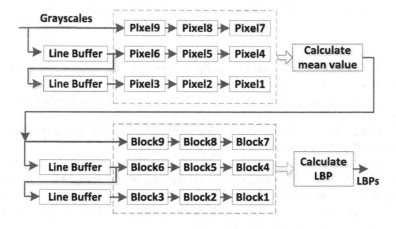

Fig. 4. Structure of LBP calculate module

To achieve real time computation of the LBP pattern, a circuit structure is put forward as showed in Fig. 4. Two line buffers and nine resisters are connected in the way showed in the figure. Nine neighbor pixels are extracted with minimum delay, and the mean value of this block is calculated by the mean value calculate module which contains some adders and shifters. The mean values of the blocks are inputted to a similar structure and extracted in a similar way, and the LBP pattern is calculated by the consequence LBP calculate module.

5   Experimental Results

In order to verify the effectiveness of the algorithm, a certification experiment is designed, and the ROC curves of the two algorithms based on LBP and grayscale are plotted and compared.

In statistics, a receiver operating characteristic (ROC) curve, is a graphical plot that illustrates the performance of a binary classifier system as its discrimination threshold is varied. The curve is created by plotting the true positive rate against the false positive rate at various threshold settings. Let us consider a two-class classification problem (binary classification), in which the outcomes are labeled either as positive (p) or negative (n). There are four possible outcomes from a binary classifier. If the outcome from a classification is p and the actual value is also p, then it is called a true positive (TP); however if the actual value is n then it is said to be a false positive (FP). Conversely, a true negative (TN) has occurred when both the classification outcome and the actual value are n, and false negative (FN) is when the classification outcome is n while the actual value is p. True positive rate (TPR) and the false positive rate (FPR) are defined as the following formula:

$$TPR = \frac{TP}{TP + FN} \tag{1}$$

$$FPR = \frac{FP}{FP + TN} \tag{2}$$

In the experiment, positive samples are the video clips with moving objects. On the contrary, negative samples are the video clips without moving objects. Two test collections are prepared for the experiment:

Collection 1: 40 short video clips with fixed illumination, including 20 positive samples and 20 negative samples.

Collection 2: 40 short video clips with shifty illumination, including 20 positive samples and 20 negative samples.

For a ROC curve, one threshold value corresponds to one point. In the LBP algorithm, we classify areas into background and moving object according to whether the hamming distance is bigger than the threshold or not. For the 8 bits patterns, the threshold can be an integer from 1 to 8. For the traditional algorithm based on 256 levels grayscale, the threshold can be an integer from 0 to 255.

The picture of the certification experiment is showed in Fig. 5. A PC acts as the source of the test signal which is input to the FPGA in the form of VGA. Passing through the

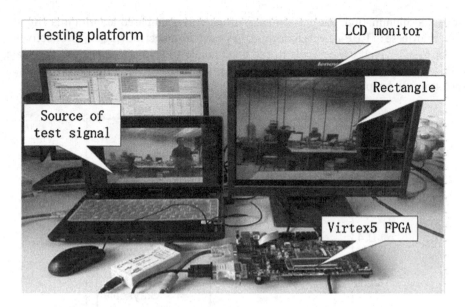

Fig. 5. Experiment platform

FPGA board, video signal is displayed on a LCD screen. This system supports a VGA video processing up to 1024 × 768/60 fps, and can output a clear and stable video signal. On the screen, we can see 4 rectangles whose boundary will be shining when something is moving in it.

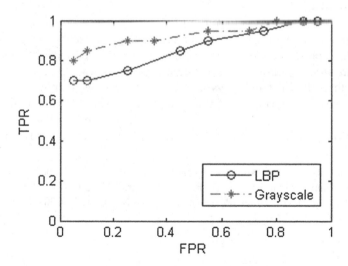

Fig. 6. ROC curves of the test collection 1

Figures 6 and 7 are the ROC curves of two experiments using test collection 1 and test collection 2 respectively. We can see in the Fig. 6 which corresponds to the condition with fixed illumination, the performance of the grayscale-based algorithm is slightly better than these of LBP-based algorithm, they can both detect moving object effectively. But in Fig. 7 which corresponds to the condition with shifty illumination, grayscale based algorithm deteriorates drastically and nearly lose efficacy. But the LBP based algorithm still keeps a good performance.

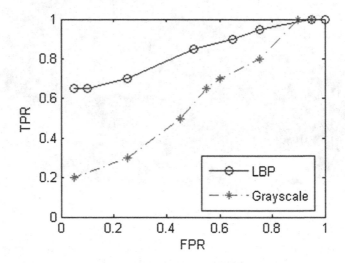

Fig. 7. ROC curves of the test collection 2

6 Conclusion

To overcome or at least alleviate the problems existing in current moving object detect system, we have presented an efficient and effective LBP based and self-updating background subtraction algorithm. A verification experiment was designed and the ROC curves of both of the LBP and grayscale-based algorithm were drawn and compared. One find that the algorithm based on improved LBP operator is more robust to the change of illumination. According to the theoretical analysis and verification experiment, our algorithm can improve the reliability of a surveillance system and is easily to implement.

Acknowledgements. This research is supported by the grant No. JK2014003 and No. JA14038 from the Educational Department of Fujian Province, the grant No. 2015J05124 from Science and Technology Department of Fujian Province, the grant No. LXKQ201504 from ministry of education of China.

References

1. Horn, B.K., Schunck, B.G.: Determining optical flow: a retrospective. Artif. Intell. **59**, 81–87 (1993)
2. Paul, J., Laika, A., Claus, C., et al.: Real-time motion detection based on SW/HW-codesign for walking rescue robots. J. Real-time Image Pr. **8**, 353–368 (2013)

3. Elhabian, S.Y., El-Sayed, K.M., Ahmed, S.H.: Moving Object Detection in Spatial Domain Using Background Removal Techniques-State-of-Art. Recent Pat. Comput. Sci. 1, 32–54 (2008)

4. Heikkilä, M., Pietikäinen, M.: A texture-based method for modeling the background and detecting moving objects. IEEE T. Pattern Anal. 28, 657–662 (2006)

5. Wenbin, L., Xiaomin, Z., Changsong, W.: Detection algorithm of moving objects based on background subtraction method. J. Univ. Sci. Technol. Beijing 2, 212–216 (2008)

6. Wang, K., Xu, L., Fang, Y., et al.: One-against-all frame differences based hand detection for human and mobile interaction. Neurocomputing 120, 185–191 (2013)

7. Cetin, M., Hamzaoglu, I.: An adaptive true motion estimation algorithm for frame rate conversion of high definition video and its hardware implementations. IEEE T. Consum. Electr. 57, 923–931 (2011)

8. Ho, H., Klepko, R., Ninh, N., et al.: A high performance hardware architecture for multi-frame hierarchical motion estimation. IEEE T. Consum. Electr. 57, 794–801 (2011)

9. Ahonen, T., Hadid, A., Pietikäinen, M.: Face Recognition with Local Binary Patterns. In: Pajdla, T., Matas, J(. (eds.) ECCV 2004. LNCS, vol. 3021, pp. 469–481. Springer, Heidelberg (2004)

10. Hadid, A., Pietikäinen, M., Ahonen, T.: A discriminative feature space for detecting and recognizing faces. In: Proceedings of the 2004 IEEE Computer Society Conference on Computer Vision and Pattern Recognition, CVPR 2004, vol. 2, pp. II-797. IEEE (2004)

11. Ojala, T., Pietikäinen, M., Mäenpää, T.: Multiresolution gray-scale and rotation invariant texture classification with local binary patterns. IEEE T. Pattern Anal. 24, 971–987 (2002)

Design History Retrieval Based Structural Topology Optimization

Jikai Liu and Yongsheng Ma[✉]

Department of Mechanical Engineering,
University of Alberta, Edmonton, Canada
yongsheng.ma@ualberta.ca

Abstract. This paper presents a novel topology optimization (TO) method which relies on design history retrieval and surrogate modeling. With this method, a new design case starts by retrieving the design history to find similar cases in both design domain geometry and boundary condition (BC), for which an innovative BC similarity evaluation has been developed. For the best-match history case, feature based topological design was available in database and is predictably similar to that of the new design case. Therefore, it can be used as the feature model input of the new design case, and the TO problem is simplified into a sizing optimization problem to find the optimal feature parameter set. Surrogate model based method has been employed to solve the sizing optimization problem. Overall, this new TO method characterizes as: first, the efficiency is much higher than the conventional TO methods; second, it obtains feature-based topological design without post-treatment effort.

Keywords: Topology optimization · Design history retrieval · Surrogate modeling · Sizing optimization

1 Introduction

Topology optimization (TO) is an effective structural design method. It addresses both shape and topology changes, and generates light-weight and high-performance conceptual designs. Even for already highly-engineered products, it helps the engineers think out of the box to create further innovations. Currently, there are mainly three TO methods: the density based method [1], the evolutionary structural optimization method [2], and the level set method [3]. Comprehensive reviews about these methods can be found in [4, 5].

Even though TO has gained great development in the past decades, there are still remaining issues to be addressed. First, TO methods rely on the rigorously-derived sensitivity result and the carefully-customized design update scheme, with which the convergence process tends to be slow. Normally, hundreds of iterations are required which means enormous rounds of finite element analysis (FEA). Therefore, effective designs cannot be guaranteed of timely delivery. Additionally, TO mainly supports freeform design, while CAD tools are dominated by feature modeling. This gap hinders the integration of TO and CAD tools. To fill this gap, post-treatment is common by manually building the feature model according to the freeform topological design

© Springer International Publishing Switzerland 2015
A. Bikakis and X. Zheng (Eds.): MIWAI 2015, LNAI 9426, pp. 262–270, 2015.
DOI: 10.1007/978-3-319-26181-2_25

(a) design domain   (b) freeform topological   (c) feature-based topological
                                design   design after post-treatment

Fig. 1. TO plus post-treatment

(see Fig. 1). However, manual reconstruction is tedious and time-consuming, and it may uncontrollably relax the optimality and violate the constraints.

In this work, a design history retrieval based TO method (Fig. 2) is developed to overcome the afore-mentioned limitations. Normally, there are plenty of history design cases which have already gone through the engineering process. The problem setups and design results are well documented and could be referred by new design cases. Therefore, a new design case can start by retrieving design history for cases with similar design domain, boundary condition (BC), and problem setup. Because of these similarities, the design results are predictably similar, especially for topology structure. Therefore, design result of the best-match history case can be applied as input of the new case. In this way, the close-to-optimal topology structure is inherited, and the complex TO problem is simplified into a sizing optimization problem to find the optimal feature parameter set. For sizing optimization, surrogate modeling based method is applied for its simplicity and efficiency. Given benefits, feature-based topological design is generated without manual reconstruction. Additionally, the design efficiency is greatly improved. For instance, only a few times of FEA are required.

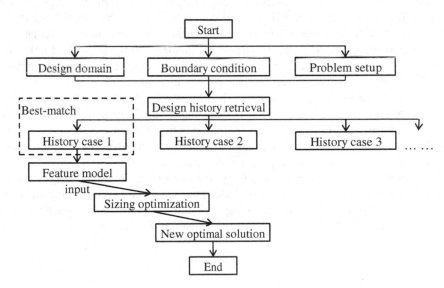

Fig. 2. Overall process of the design history retrieval based TO

2 Design History Retrieval

In mechanical design field, design history retrieval is mainly applied to solid model reuse, which is conducted by matching the shape descriptions. Both geometry similarity and topology similarity are evaluated in order to find out the best match. Several retrieval methods are available, and comprehensive reviews can be found in [6, 7].

This paper presents a novel scheme of reusing the past TO records to accelerate new design progress. For a TO case, it involves three main aspects: the design domain (solid model), the BC, and the problem setup. To ensure a history case reusable, all these aspects should be confidentially similar. To evaluate design domain similarity, the methods in [6, 7] could be applied. This work inherited the shape descriptor and matching method from [8] with certain adaptions to better serve our purpose. For BC similarity evaluation, it is rarely explored in literature. An innovative method has been developed to describe the BC by strain energy density (SED) distribution. For problem setup, there could be different combinations which generate very different optimization results. For the sake of simplicity, compliance minimization under material volume constraint is employed as the only problem setup.

2.1 Topology Simplification

Normally, the many topology details within a solid model complicate the topology similarity evaluation. As presented in [8], multi-resolution skeleton graph was developed to assess the topology similarity. For our interest, it is not necessary to make the topology similarity assessment so complicated, because TO is mainly applied in the conceptual design stage, and the initial design domain generally employs simple topology structure. For the only a few topology details, they have little impact on the optimization result and therefore can be omitted. Hence, a topology simplification procedure is developed and applied to all the design domains, with the following steps: first, recognize machining features from the initial design domain; second, calculate the size of each recognized machining feature. If it's smaller than the predefined threshold value (like 10 % of the total size), fill the machining feature area with materials; otherwise, keep the area void. See Fig. 3 for a few examples.

After simplification, topology structure is described by the hierarchical tree graph. Normally, tree graphs are compared by calculating a similarity index indicating the topology similarity. But in this work, isomorphism is required between design domains with Boolean result, because the topology structure has already been simplified and only one resolution of the tree graph is employed.

An example is shown in Fig. 4, topology structure of the simplified design domain 1 (see Fig. 3b) is demonstrated by the nodes and lines inside the frame; comparatively, for the simplified design domain 2, topology structure is defined with three additional nodes. Therefore, Boolean result of the isomorphism assessment is false.

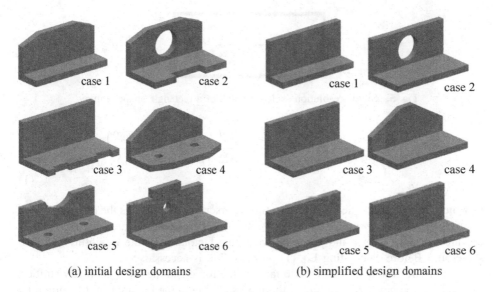

(a) initial design domains   (b) simplified design domains

Fig. 3. Topology simplification

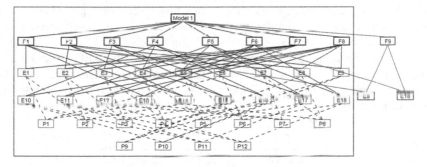

Fig. 4. Hierarchical tree graphs of the topology structures

2.2 Shape and BC Similarity

Through topology simplification and isomorphism assessment, a few history cases are removed from the candidate list. In Fig. 3, assume case 1 to be the query case and both case 2 and 4 are removed. For the rest, shape and BC are under similarity evaluation. About shape similarity, there are plenty of existing methods [6, 7]. We follow the method of [8], in which multi-resolution skeleton graph of the shape was developed. In this work, only single-resolution skeleton graph is required, because of the simplified topology. The skeletons are drawn based on the Vironoi diagram (see Fig. 5). Then, shape similarity is evaluated according to the skeleton-related data. To be specific, Eq. (1) is applied [8].

Fig. 5. Single-resolution skeleton graph of a rectangle shape feature

$$ShapeSI = \sum_{i=1}^{n} w_i \{1 - \alpha \frac{\left|A(S_i^{query}) - A(S_i^j)\right|}{\max\left(A(S_i^{query}), A(S_i^j)\right)} - \beta \frac{\left|AD(S_i^{query}) - AD(S_i^j)\right|}{\max\left(AD(S_i^{query}), AD(S_i^j)\right)}\}$$

(1)

in which, S_i represents the *ith* skeleton, the superscript *query* means the query case and *j* means the jth history case in the library. A and AD represent the area and the average distance, respectively. α and β are the weighting factors. w_i is the weighting factor of skeleton i. Before calculating Eq. (1), voxelization is necessary.

For the BC, there are mainly two factors: the force and the fixed geometry entities. To evaluate the similarity, an initial thought is to extend the hierarchical tree graph by adding nodes of BC elements. Normally, BC elements are clearly imposed on geometry entities which makes it trivial to extend the tree graph. Afterwards, the isomorphism assessment could be performed again for further history case filtering. However, this approach will make the similarity evaluation too conservative because non-isomorphism BCs could lead to similar load bearing effects.

On the other hand, the direct response of imposing BC is the SED distribution (see Fig. 6). Similar BCs would generate close SED distributions. Therefore, Eq. (2) is developed to compare the SED distributions, which in fact reflects the BC similarity. It is noted that, the development of Eq. (2) is inspired by Eq. (1).

$$BCSI = \sum_{i=1}^{n} w_i \{1 - \bar{\alpha} \frac{\left|E(S_i^{query}) - S(E_i^j)\right|}{\max\left(E(S_i^{query}), E(S_i^j)\right)} - \bar{\beta} \frac{\left|AE(S_i^{query}) - AE(S_i^j)\right|}{\max\left(AE(S_i^{query}), AE(S_i^j)\right)}\}$$

(2)

in which, E and AE represent the total strain energy and the weighted average strain energy of the skeleton, respectively. And AE is calculated by:

$$AE(S_i) = \frac{1}{m} \sum_{k=1}^{m} E(V_k) * D(V_k)$$

(3)

where, m is the number of voxels of skeleton S_i. V_k is the kth voxel of skeleton S_i. $S(V_k)$ and $D(V_k)$ represent the SED and the distance of V_k, respectively.

In summary, the total similarity index is calculated as: *TotalSI = ShapeSI * BCSI*.

To prove the effectiveness, we continue with the 6 cases presented in Fig. 3. For the 4 remaining cases: 1, 3, 5, 6, the BCs are demonstrated in Fig. 7 in which the arrows represents the forces and the dark blue colors represent the fixed boundary entities. Assume case 1 to be query case while the other three as the history cases. The shape and BC similarity evaluation results are listed in Table 1. It can be observed that, model 6 employs the highest similarity indexes regarding to both shape and BC.

Fig. 6. SED distribution of case 1 (darker color means smaller energy density) (Color figure online)

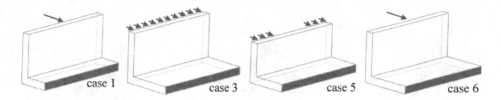

Fig. 7. Boundary conditions

Table 1. Data of shape and BC similarity evaluation

	Query case	History case	History case	History case
Case name	case 1	case 3	case 5	case 6
Shape SI	–	97.83 %	97.75 %	97.83 %
BC SI	–	93.78 %	90.79 %	97.21 %
Total SI	–	91.74 %	88.75 %	95.10 %

3 Surrogate Model Based Sizing Optimization

Among the TO methods, level set method supports both topology and sizing optimization. However, the parametric level set method [9] for sizing optimization is non-trivial in implementation and is not supported by any software tool. Therefore, an alternative surrogate model based sizing optimization method, inspired by [10], is applied, of which the procedures are well supported by commercial software tools.

The general procedures are illustrated below:

1. Apply design of experiments (DOE) and regression analysis to construct the surrogate models of the objective function (compliance) and the material volume ratio;
2. Configure the surrogate models into a new but simple optimization problem, as:

$$min.\boldsymbol{B}^T X s.t. \bar{\boldsymbol{B}}^T X \leq V_{max} \tag{4}$$

in which, $\boldsymbol{B}^T X$ is the surrogate model of compliance and $\bar{\boldsymbol{B}}^T X$ is the surrogate model of material volume ratio; $\boldsymbol{B}$ and $\bar{\boldsymbol{B}}$ are the coefficient vectors $(\beta_1, \beta_2, \ldots \ldots)^T$ and

$(\overline{\beta_1}, \overline{\beta_2}, \ldots \ldots)^T$; X is the factor vector $(1, x1, x2, \ldots, x1^2, x2^2, \ldots)$, which is determined by the regression model applied, e.g. linear mode, and pure quadratic mode.

3. Solve the optimization problem to derive the optimal parameter set $(x1, x2, \ldots)$.

After illustration of the methodology, here we continue with the example presented in Sect. 2. Case 6 has been identified as the best-match history case; therefore, the outcome of case 6 (Fig. 8a) is adopted as input of the query case (case 1). It is noted that, the feature model input needs to be adjusted in scale because the design domains are similar but not identical. As shown in Fig. 8, the feature model (Fig. 8a) is shrunk proportionally into Fig. 8b. Then, the sizing optimization is performed with the material volume constraint of 0.4, of which the result is demonstrated in Fig. 8c.

   (a) outcome of case 6   (b) model shrinkage   (c) sizing optimization result

Fig. 8. Surrogate model based sizing optimization

4 Case Study

In this section, a cantilever design problem is investigated. As shown in Fig. 9, there list the query and three history cases, for which topology simplification is not required and the topology structures are isomorphism.

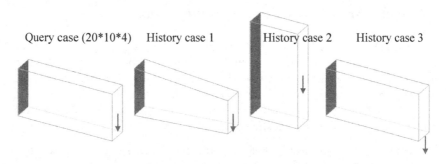

Query case (20*10*4)   History case 1   History case 2   History case 3

Fig. 9. Design domains and BCs

Following Eqs. (1) and (2), the similarity indexes are calculated and the results are presented in Table 2. It is observed that the BC similarity indexes play a significant role in determining the total similarity, because the shape similarity indexes are close.

Table 2. Data of shape and BC similarity evaluation

	Query case	History case 1	History case 2	History case 3
Shape SI	–	99.17 %	100 %	100 %
BC SI	–	96.81 %	88.80 %	95.85 %
Total SI	–	96.01 %	88.80 %	95.85 %

According to the results, history case 1 is determined as the best match. Its design result, as shown in Fig. 10a, is used as input of the query case. Then, the pure quadratic model is applied to build the surrogate models of the objective function and the material volume ratio. Detailed procedures of building the surrogate models will not be illustrated here for its simplicity. At the end, Eq. (4) is solved through multi-start gradient-based method with the material volume constraint of 0.55, and the optimal solution is derived as shown in Fig. 10b.

(a) feature model input from history case 1   (b) the final optimal solution

Fig. 10. Feature model input from history case 1 and the final optimal solution

5 Conclusion

This paper presents a design history retrieval based TO method. Any new design case starts by retrieving similar history cases through evaluating both the design domain geometry and BC similarities. As the retrieval result, the best-match history case offers the feature model input. Surrogate model based sizing optimization optimizes the feature parameters to derive the final optimal solution.

The advantage of this method is clear that it saves great computational expense by omitting the discrete TO process, as well as the post-treatment effort. Given the limitation, a well-documented collection of history design cases is indispensable.

References

1. Bendsoe, M.P., Sigmund, O.: Topology Optimization – Theory, Methods and Applications. Springer, Heidelberg (2003)
2. Xie, Y.M., Steven, G.P.: A simple evolutionary procedure for structural optimization. Comput. Struct. **49**(5), 885–896 (1993)
3. Wang, M.Y., Wang, X.M., Guo, D.M.: A level set method for structural topology optimization. Comput. Methods Appl. Mech. Eng. **192**(1–2), 227–246 (2003)

4. Rozvany, G.I.N.: A critical review of established methods of structural topology optimization. Struct. Multi. Optim. **37**(3), 217–237 (2009)
5. Sigmund, O., Maute, K.: Topology optimization approaches. Struct. Multi. Optim. **48**(6), 1031–1055 (2013)
6. Iyer, N., Jayanti, S., Lou, K., Kalyanaraman, Y., Ramani, K.: Three-dimensional shape searching: state-of-the-art review and future trends. Comput. Aided Des. **37**(5), 509–530 (2005)
7. Tangelder, J.W.H., Veltkamp, R.C.: A survey of content based 3D shape retrieval methods. Multimed. Tools Appl. **39**(3), 441–471 (2008)
8. Gao, W., Gao, S.M., Liu, Y.S., Bai, J., Hu, B.K.: Multiresolutional similarity assessment and retrieval of solid models based DBMS. Comput. Aided Des. **38**(9), 985–1001 (2006)
9. Liu, J.K., Ma, Y.S.: 3D level set topology optimization: a machining feature-based approach. Struct. Multi. Optim. (2015). doi:10.1007/s00158-015-1263-7
10. Cho, C.S., Choi, E.H., Cho, J.R., Lim, O.K.: Topology and parameter optimization of a foaming jig reinforcement structure by the response surface method. Comput. Aided Des. **43** (12), 1707–1716 (2011)

Online Detection of Moving Object in Video

Maryam Azimifar[1], Farhad Rad[1], and Hamid Parvin[1,2(✉)]

[1] Departement of Computer Science, Yasuj Branch,
Islamic Azad University, Yasuj, Iran
[2] Young Researchers and Elite Club, Nourabad Mamasani Branch,
Islamic Azad University, Nourabad Mamasani, Iran
parvin@iust.ac.ir

Abstract. Moving car discovery is one of the most essential problems in image processing. It is a very challenging problem that attracts many attentions recently. Major part of previous moving car discovery methods engages radar signals. Nevertheless, those face some troubles in special cases, for example they have difficulty in detection of moving cars in zigzag movements. Machine learning methods can be utilized to conquer these inefficiencies. For online moving car discovery, we propose to employ hierarchical partitioning over the features extracted from image. Each moving car is corresponds to a partition. Unlike the traditional partitioning algorithms, the threshold distance in the proposed method is not fixed. This threshold value is tuned by a Gaussian distribution. Harris features are applied to capture the corner features. Experimentations show the proposed method outperforms other competent methods.

Keywords: Object recognition · Video processing · Moving car discovery

1 Introduction

Traffic monitoring methods have been important in intelligent cars. Object recognition and tracking in an online manner on a street that is full of obstacles such as cars, pedestrians and many other things, are really challenging and hard problems. The performance of an online discovery method is a decisive matter for managers from a safety point of view. Therefore these algorithms have to be robust under different conditions. Moving car discovery has been a challenging topic for intelligent machines. Street environment will be very different if traffic situations, number of cars, lighting conditions, weather conditions, construction of streets, tunnels and many other involved things change. Then a fully adaptive and parametric method is required respect to the existing conditions. Near or mid-range moving car discovery is another effective topic in the structure of environment and the parameters of the camera. Different sensors are available for use in driver assistance methods such as lidar, radar, ultrasound and embedded camera.

There exist four situations according to the movement of camera and object. 1 – stationary camera, constant object 2 – stationary cameras, moving object 3 – moving camera, stationary object 4 – moving camera, moving object. The second situation is more practical and more common [1]. Fixed camera located on the highway for speed

© Springer International Publishing Switzerland 2015
A. Bikakis and X. Zheng (Eds.): MIWAI 2015, LNAI 9426, pp. 271–278, 2015.
DOI: 10.1007/978-3-319-26181-2_26

control is example of second kind. But camera and moving objects are in the navigation and driver assistance methods that are the most sophisticated state. Due to background changing and inapplicability of differential techniques the moving object discovery with moving camera are challenging task.

Most commercial moving car discovery methods are based on radar, a sensor that has a lot of limitation like angular constraint and temporal resolution. Usually, Radar-based methods are for detecting moving cars located directly in front of the observer car, and marginal cars with different angles not be detected. In this case, any change in car line may be dangerous. Also radar-based methods for high zigzag and steep roads may be problematic. On the other hand, the camera is inexpensive, has low power consumption and easy manageable to capture information from the environment. Visual data processing is Complex, but provides valuable information of environment.

In this paper we focus on obtaining low-level features and we use them to determine the location of moving cars. The shape and structure of moving cars are not intended. Since the moving car discovery method is in a dynamic environment and ego-moving car and other moving cars are moving, size and position of the devices can be varied over time. Therefore, an adaptive method based on the position can be helpful. This adaptability can be used in different component of method including setting thresholds described by the proposed method.

2 Backgrounds

Since most online moving car discovery methods detect moving cars in line with the observer, the moving cars can be seen from the rear by ego_car. There are many Horizontal lines at the back of the moving cars, including shadow lines, window line, top and bottom lines of the moving car. This feature is relatively stable against light changing and scaling. Matthews et al. [2] have used edge discovery and determination of vertical lines to indicate the left and right margins of moving car. Edge features are favorite feature for researchers in the field of moving car discovery [3, 4].

Most moving cars are quadrilateral forms that have corners. Corner is an important feature of video images. Corners maintain in their position on the moving cars and background, so they have useful information about the different components of image. Bertozzi et al. [5] have proposed a corner-based method for moving car discovery.

Texture is a surface with quantified and recognizable features, such as smoothness, regularity, and so on. Texture processing is required to generate co-occurrence metric and gray level descriptors [6]. Each moving car is represented by local variation of gray level with a certain texture. Each certain texture region can be processed further to make accurate discovery. Entropy calculation based on neighboring pixels is the texture determination criteria. Areas with high entropy are chosen for further processing [7]. Kalinke et al. [8] have been used texture to focus algorithm on the areas that have a lot of information. Shannon has been introduced local entropy to measure each image patch's information [9].

Symmetry is an invariant feature at the rear of moving cars that is stationary in different light conditions and scaling. Many researchers have used this feature to detect moving cars [10, 11]. A symmetry- and edge-based moving car discovery method using edge oriented histogram (EOH) and support vector machine (SVM) consequently for approximately moving car locating and post-processing for improved result, is proposed in [12]. The determination of symmetry is based on window with several sizes.

Optical Flow (OF) is an informative motion-based feature that provides information about the direction and speed of moving objects in video frames [13, 14]. Pixel-based and feature-based methods are two main approaches for optical flow computation. In pixel-based method, all of OF vectors for individual pixels have to be calculated. Time-consuming and sensitivity to noise are disadvantages of pixel-based method that made this method impractical in real time works. In contrast, in feature-based methods, OF vectors for the pieces of image with certain features, such as corners and colored bubbles, have to be calculated. In OF computation, (u, v) feature point mapping in (I_t) and (I_{t+1}) are done so that relation 1 minimized [15, 16].

$$e\,(dx, dy) = \sum_{x=u-w}^{u+w} \sum_{y=v-w}^{v+w} (I\,(x, y) - I\,(x + dx, y + dy)) \tag{1}$$

Shadow is used as a feature for moving car bounding. The lower part of moving cars in different lighting conditions has different shadows, and this feature can be used to determine the moving car's underside [17]. Due to different light and weather conditions the shadows have different gray levels. So determining shadow gray level threshold is challenging.

Another feature used to separate the moving cars and the background is color [18, 19]. The RGB color method [19] and The L*A*B color method [20] are more conventional. The LED lighting is used to detect and track the moving cars at night.

3   Proposed Method

Image features, beneficial for moving car discovery, have been introduced in previous section. In this paper, we use multiple low-level features and focus on the grouping and partitioning of these features to identify moving cars. The ROI (region of interest) is determined based on Gaussian probability distribution function.

As it is described later; there are many horizontal lines at the back of the moving cars, including shadow lines, window line, top and bottom lines of the moving car. This feature is relatively stable against light changing and scaling. Because in some frames, moving cars are not exactly in front of the camera and are in margins, horizontal or vertical lines are not clear and complete algorithm for edge discovery methods such as canny [21] would be more appropriate. In this paper, unlike other methods that use only horizontal or vertical lines, a fully edge discovery algorithm was utilized.

Corner is an important feature of video images that commonly were used in moving car discovery methods. The most common corner extraction method is based on harris features [22]. Each frame has a lot of details around the road. So the ROI determination for further process is important from two aspects. First, appropriate ROI placing has a direct effect on the accuracy of discovery, and second, more precise determination of the area, the less time for next level processing. According to The camera settings and its circumstance on ego_car, ROI will vary in size and details.

In this paper, some parameters such as ROI determination and marginal detail parameters as the camera parameters are specified. If the installed camera is angular then the captured image will be contained more details and the sky will be the greatest part of image. Conversely, the image will included more moving cars and less marginal details, if the camera position and angle relative to the horizon is less. λ and θ is considered as ego-car's height and the camera's angle with respect to the horizon respectively. The parameter values are inverted to interval [0, 1].

The probability that each image pixels be a moving car pixel computed according to its position X(x, y) and the bi-variant Gaussian distribution function mentioned in Eq. 2. In this distribution function for more adaptation, the parameters λ and θ are used to create covariance matrix (Eq. 3).

$$p\left(X; \mu, \Sigma\right) = \frac{1}{(2\pi)^{\frac{n}{2}} |\Sigma|^{\frac{1}{2}}} \exp\left(-\frac{1}{2}(X - \mu)^T \Sigma^{-1}(X - \mu)\right) \tag{2}$$

$$\Sigma = \begin{bmatrix} C*\lambda & 1 \\ 1 & R*\theta \end{bmatrix} \tag{3}$$

In Eq. 3, C and R are image's columns and rows size respectively. Since the Gaussian distribution is symmetrical relative to the variables mean, only the positive part has been used. As it has been described, the edge and corner features, are used to moving car discovery. Candidates for further processing are selected from the intersection of dilated edges and corners as Eq. 4.

$$C = corners \cap (edge \oplus se) \tag{4}$$

The edges and corners are features that extracted in previous stage and se is dilation mask like circle with 5 pixel radius. Further processes are done on the C defined in Eq. 5.

$$C = \left\{ c_i = \left(x_i, y_i\right) i = 1, 2, \ldots, K \right\} \tag{5}$$

As regard, the image size of nearby moving cars, from camera, are large and distant moving cars are small, then applying a uniform threshold for drawing bounding box and partitioning centers are not reasonable. Therefore, an adaptive threshold based on the candidate position (x, y) and the likelihood $p(x_i, y_i, \mu, \Sigma)$ is used.

$$AD_i = patch_{size} * p\left(x_i, y_i, \mu, \Sigma\right) \tag{6}$$

To draw a bounding box around the initial center points and then partitioning them, we use $Adaptive_{Distance}$ (AD) mentioned in Eq. 6 threshold. So the center that is deemed more likely being moving car, have an even greater margin.

To obtain more accurate results, Centers are clustered using the Euclidean distance measure and the AD threshold computed for each center.

$$D\left(c_i, c_j\right) = \sqrt{\left(x_i - x_j\right)^2 + \left(y_i - y_j\right)^2} \tag{7}$$

$$TD\,(i,j) = \min\left(AD_i, AD_j\right) \tag{8}$$

$$G\left(c_i, c_j\right) = \begin{cases} 1 & \text{if } D\left(c_i, c_j\right) \le TD\,(i,j) \\ 0 & o.w \end{cases} \tag{9}$$

In this grouping method, the number of clusters is variant and will be determined during the partitioning according the cluster merging. After partitioning, the cluster centers have to be determined for subsequent processing according to the following equation.

$$\underbrace{CL_k}_{k=1..N} = \left\{\left[\; mean(x_i)\quad mean(y_i)\;\right] | ViG\left(c_i, c_k\right) == 1\right\} \tag{10}$$

where N is the number of pre-generated clusters. After the initial partitioning, the final step to moving car determination is done by re-partitioning with the larger patch size and averaging based on members number. Neighbor points are selected by teammate selection based on certain correlation and average computation. In this stage, certain correlation between all central points is measured pairwisely so a multivariate data adjacency matrix is created. The certain correlations between points are computed based on gray levels difference and Euclidean distances of centers coordinate as bellow.

$$CR\left(c_i, c_j\right) = \frac{1}{\left(\left|v_i - v_j\right|\right)^\rho + \left(dist\left(c_i, c_j\right)\right)^\gamma + 1} \tag{11}$$

CR takes a value in the interval (0, 1]. The self adjacency value for each center point is 1. The CR value tends to zero by increasing Euclidian distance between the centers and difference between gray levels. γ and ρ parameters determine effectiveness of the Euclidean distance and gray level differences respectively. After multivariable adjacency matrix construction, the recruitment is done for same samples grouping. A question arises, how many centers with how much adjacency amounts can be a cluster? To do this, the CR values weighted averaging is used in such a way that the average is weighted based on the number of samples it takes. The weight grows while the number of samples increases.

In fact, the recruitment and weight increase goes a head while the monotony property of averaging not violated in recruiting. By imposing this constraint, the close points made a cluster. Even though the distant points selection increase averaging weight it is not enough to increase total amount of weighted average with respect to the previous value. So our goal is recruiting more such that the adjacency value of members not less than of a threshold. As previously described, after applying weighted averaging and recruitment, an adaptive partitioning step based on different thresholds is applied to obtain final results.

4   Experimental Study

The proposed method have evaluated by the LISA dataset available in "http://cvrr.ucsd.edu/LISA/index.html". There are some well-known performance measures like precision (P), recall (R) [23], average false positive per frames ($AFPPF$), average false positive per object ($AFPPO$) and average true positive per frame ($ATPPF$) as evaluation criteria.

Here is a problem, only the rears of the moving cars are tagged. If a moving car is seen except the rear part of it, then this discovery is not considered and the algorithms that detect these moving cars will be double losses in terms of method evaluation. Because the correctly detected moving car (TP) will be considered as the method error (FP) so the method performance will be failed. The emerged results are shown in Table 1. Parameters used in each dataset are chosen expertly and with a little trial and error.

Table 1. Experimental result in the three datasets of LISA dataset

Criteria	#1	#2	#3
ATPPF	1	2.966	2.67
AFPPF	2.133	1	2.22
Recal	1	0.989	0.621
False discovery rate	0.519	0.223	0.399
AFNPF	0	0.033	0.193
ACNPF	1	3	4.38
AFPPO	2.133	0.333	0.507
ATPPO	1	0.989	0.621
AFNPO	0	0.011	0.044
Precision	0.48	0.768	0.60

5   Conclusion

This paper focuses on an important task to detect moving cars with an in-car camera. Our approach is based on several general features that characterize the moving cars robustly such as "edges" and "corners". The discovery is done at several stages hierarchically. After feature extraction, masks are applied to determine the ROI based on the probability of the position of each pixel,. Then a feature partitioning step is performed according to the listed parameters. Next, a weighted average based sampling and teammate selection is done and the final partitioning is performed by the new parameters. As the results show, the high TPR (recall) is the strength of proposed method. Due to the complexity and difficulty of dataset#3, results are not so good. Parameters for complex environments can be improved by optimization algorithms such as genetic algorithms.

References

1. Nadimi, S., Bhanu, B.: Multistrategy fusion using mixture model for moving object discovery. In: International Conference on Multisensor Fusion and Integration for Intelligent Methods, 2001. MFI 2001, pp. 317–322 (2001)
2. Matthews, N., An, P., Charnley, D., Harris, C.: Vehicle discovery and recognition in greyscale imagery. Control Eng. Pract. **4**, 473–479 (1996)
3. Betke, M., Haritaoglu, E., Davis, L.S.: Real-time multiple vehicle discovery and tracking from a moving vehicle. Mach. Vis. Appl. **12**, 69–83 (2000)
4. Srinivasa, N.: Vision-based vehicle discovery and tracking method for forward collision warning in automobiles. In: Intelligent Vehicle Symposium, 2002. IEEE, pp. 626–631 (2002)
5. Bertozzi, M., Broggi, A., Castelluccio, S.: A real-time oriented method for vehicle discovery. J. Meth. Archit. **43**, 317–325 (1997)
6. Yilmaz, A., Javed, O., Shah, M.: Object tracking: A survey. ACM Comput. Surv. (CSUR) **38**, 13 (2006)
7. Ten Kate, T., Van Leewen, M., Moro-Ellenberger, S., Driessen, B., Versluis, A., Groen, F.: Mid-range and distant vehicle discovery with a mobile camera. In: Intelligent Vehicles Symposium, 2004 IEEE, pp. 72–77 (2004)
8. Kalinke, T., Tzomakas, C., Seelen, W.V.: A texture-based object detection and an adaptive model-based classification. In: Proceedings of the IEEE Intelligent Vehicles Symposium 1998, pp. 341–346 (1998)
9. Shannon, C.E.: A mathematical theory of communication. ACM SIGMOBILE Mob. Comput. Commun. Rev. **5**, 3–55 (2001)
10. Kuehnle, A.: Symmetry-based recognition of vehicle rears. Pattern Recogn. Lett. **12**, 249–258 (1991)
11. Zielke, T., Brauckmann, M., Vonseelen, W.: Intensity and edge-based symmetry discovery with an application to car-following. CVGIP: Image Underst. **58**, 177–190 (1993)
12. Teoh, S.S., Bräunl, T.: Symmetry-based monocular vehicle discovery method. Mach. Vis. Appl. **23**, 831–842 (2012)
13. J, Diaz.Alonso., Ros Vidal, E., Rotter, A., Muhlenberg, M.: Lane-change decision aid method based on motion-driven vehicle tracking. IEEE Trans. Veh. Technol. **57**, 2736–2746 (2008)
14. Perrollaz, M., Yoder, J.-D., Nègre, A., Spalanzani, A., Laugier, C.: A visibility-based approach for occupancy grid computation in disparity space. IEEE Trans. Int. Transp. Meth. **13**, 1383–1393 (2012)
15. Lucas, B.D., Kanade, T.: An iterative image registration technique with an application to stereo vision. In: IJCAI, pp. 674–679 (1981)
16. Choi, J.: Realtime online vehicle discovery with optical flows and haar-like feature detectors. Urbana **51**, 61801 (2007)
17. Mori, H., Charkari, N.M.: Shadow and rhythm as sign patterns of obstacle discovery. In: IEEE International Symposium on Industrial Electronics, Conference Proceedings, ISIE 1993, Budapest, pp. 271–277 (1993)
18. Crisman, J.D., Thorpe, C.E.: Color vision for road following. In: 1988 Robotics Conferences, pp. 175–185 (1989)
19. Buluswar, S.D., Draper, B.A.: Color machine vision for autonomous vehicles. Eng. Appl. Artif. Intell. **11**, 245–256 (1998)

20. Guo, D., Fraichard, T., Xie, M., Laugier, C.: Color modeling by spherical influence field in sensing driving environment. In: Proceedings of the IEEE Intelligent Vehicles Symposium, IV 2000, pp. 249–254 (2000)
21. Canny, J.: A computational approach to edge discovery. IEEE Trans. Pattern Anal. Mach. Intell. **8**, 679–698 (1986)
22. Harris, C., Stephens, M.: A combined corner and edge detector. In: Alvey Vision Conference, p. 50 (1988)
23. Olson, D.L., Delen, D.: Advanced data Mining Techniques (electronic resource). Springer, Heidelberg (2008)

Robotics

Trajectory Planning to Optimize Base Disturbance of 7-DOF Free-Floating Space Manipulator Based on QPSO

Tongtong Hu, Jianxia Zhang, and Qiang Zhang[✉]

Key Laboratory of Advanced Design and Intelligent Computing,
Dalian University, Ministry of Education, Dalian 116622, China
zhangq26@126.com

Abstract. Because of the movement of space manipulator is in accordance with the law of conservation of momentum. All the motions will pose interference on the position and posture of the base. So the interference on the base generated by the motions of manipulator must be reduced. In this paper, the nonholonomic redundancy features of the free-floating space manipulator system are used. The position and posture of 7-DOF space manipulator are planned at the same time. Firstly, the kinematics equations of manipulator are established. Secondly, the angles of joints are parameterized by the sine polynomial function, and then objective function is designed according to the precision index of the position and posture. Finally, quantum-behaved particle swarm optimization (QPSO) is applied to optimize the base disturbance. The model of space manipulator system is set up, which is composed of a free- floating base and a 7-DOF manipulator. It can be seen from the experimental results, the proposed method could find the global optimal value quickly. Moreover, this method has less correlation parameters. And the planned joint path is smooth and It meets the ranges of angle of joints, angular velocity and angular acceleration. The theoretical analysis and simulation results show that the proposed method is feasible and suitable to optimize the base disturbance.

Keywords: Free-floating space manipulator · 7-DOF · QPSO · Position and posture · Base disturbance

1 Introduction

Along with the rapid development of national economy and defense industry, the number of spacecraft and satellites is getting larger and larger. So the development of space manipulator has been paid more attention by people in the space technology. However, owing to the various reasons, such as satellite malfunctions, failures or task termination, useless satellites are left in the air and they will have a huge impact on the track resources and other spacecraft safety [1]. The primary task of on-orbit service of manipulator is to plan the right path.

At present, the problem of base stability is studied by a growing number of scholars [2]. A two-way planning method was presented as nonholonomic constrains of manipulator by S.R. Li et al. [3], and Lyapunov function was introduced to design

© Springer International Publishing Switzerland 2015
A. Bikakis and X. Zheng (Eds.): MIWAI 2015, LNAI 9426, pp. 281–293, 2015.
DOI: 10.1007/978-3-319-26181-2_27

feedback control in the joint space. It can coordinate the base and manipulator, but it still can't smooth the joint angular motion. In 1991, M.A. Torres et al. [4] put forward the method of enhancing the interference figure, although this control method can reduce the posture disturbance, it represents a significant storage space and low efficiency due to the low speed. According to the nonholonomic path planning of free-floating system, VAFA et al. [5] put forward the self-correcting motion method and perturbation method. But the correction method can't change the final state of the joint and it can only adjust the base posture. In 2006, P.H. Huang [6] proposed the method of optimal path planning about the minimum base reaction, and it was based on the genetic algorithm. But genetic algorithm is more complicated than the particle swarm optimization (PSO). The optimal trajectory planning was presented by X.S. Ge et al. [7] based on genetic algorithm, it can solve the path planning problem of manipulator system. Kun Wang et al. [8] proposed a method of global path planning about mobile robots and it was based on QPSO.

In this paper, QPSO is firstly applied to optimize base disturbance of 7-DOF free-floating space manipulator. The process of this paper is as follows. The kinematics equations of 7-DOF space manipulator are first established in the second part. Then, the angles of joints are parameterized and the objective function is given in the third part. Next, this paper gives specific description of QPSO. At last, the simulations and conclusions are presented.

2   Kinematics Equations of 7-DOF Space Manipulator

In this paper, the quality characteristics of space manipulator are as shown in Table 1.

Table 1. Quality characteristics of space manipulator

	l_1	l_2	l_3	l_4	l_5	l_6	l_7
$m(kg)$	3.691	60.416	3.678	35.653	3.460	2.624	0.113
$l(m)$	0.130	1.500	0.120	1.240	0.220	0.070	0.010
$I_{xx}(kg \cdot m^2)$	0.013	13.556	0.012	1.235	0.0619	0.006	0
$I_{yy}(kg \cdot m^2)$	0.005	2.274	0.010	7.309	0.028	0.004	0
$I_{zz}(kg \cdot m^2)$	0.015	15.756	0.005	6.104	0.035	0.004	0

2.1   Description of Rigid-Body Position and Posture

In the coordinate system, a 3*1 position vector can determine the position of space point. For rectangular coordinate system {A}, the position of any point in space can be represented by a 3*1 column vector.

$$A_p = \begin{bmatrix} p_x \, p_y \, p_z \end{bmatrix}^T \tag{1}$$

Where, p_x, p_y and p_z is the three coordinate component of the P in {A}.

There are many ways to represent the posture of a rigid body, such as rotation matrix, Euler angles or unit quaternion. Quaternion is used to describe the posture of the rigid body system by C.K.C. Jack [9]. Quaternion has the following advantages: (1) there will not be a singular condition. (2) Quaternion is a linear equation, which has a small amount of calculation and high efficiency [10]. A quaternion equation is defined as follows:

$$Q = \eta + q_1 \vec{i} + q_2 \vec{j} + q_3 \vec{k} = \eta + q \in R^4 \tag{2}$$

2.2 Kinematics Equations of 7-DOF Space Manipulator

The positional equation of 7-DOF end-effector is as follows:

$$p_e = r_0 + b_0 + \sum_{j=1}^{7} (p_{j+1} - p_j) \tag{3}$$

The linear velocity of end-effector can be obtained by differentiating (3) with respect to time as follows:

$$v_e = \dot{p}_e = v_0 + \omega_0 \times (p_e - r_0) + \sum_{j=1}^{7} [k_j \times (p_e - p_j)] \cdot \dot{\theta}_j \tag{4}$$

Similarly, the angular velocity of end-effector is written as follows:

$$\omega_e = \omega_0 + \sum_{j=1}^{7} k_j \dot{\theta}_j \tag{5}$$

The different form of them in matrix of kinematical equation is as follows:

$$\begin{bmatrix} v_e \\ \omega_e \end{bmatrix} = J_b \begin{bmatrix} v_0 \\ \omega_0 \end{bmatrix} + J_m \dot{\Theta} \tag{6}$$

Therefore, v_0 and ω_0 can be expressed as follows:

$$\begin{bmatrix} v_0 \\ \omega_0 \end{bmatrix} = -I_b^{-1} I_{bm} \dot{\Theta} = \begin{bmatrix} J_{vb} \\ J_{\omega b} \end{bmatrix} \dot{\Theta} \tag{7}$$

Finally, the final result can be described as follows:

$$\begin{bmatrix} v_e \\ \omega_e \end{bmatrix} = [J_m - J_b I_b^{-1} I_{bm}] \dot{\Theta} = J^*(\Psi_b, \Theta, m_i, I_i) \dot{\Theta} \tag{8}$$

Where, $\mathbf{J}^*(\mathbf{\Psi}_b, \mathbf{\Theta}, m_i, \mathbf{I}_i)$ is named Generalized Jacobian Matrix of space manipulator, and it is the function about base posture, joint angle of manipulator, the mass and inertia of rigid-body.

3   Description of Trajectory Planning Problem of Joint Space

3.1   Equation of System State

X_b, Q_b, P_b mean the base pose, the base posture and base position respectively. The equation of system state is defined as follows:

$$X_b = \begin{bmatrix} Q_b \\ P_b \end{bmatrix} \in R^{7 \times 1} \tag{9}$$

This equation of system state can be calculated by numerical integration as follows:

$$Q_b(t) = \int_0^t \frac{1}{2} \begin{bmatrix} -q_b^T \\ \eta_b I - \tilde{q}_b \end{bmatrix} J_{bm\_\omega} \dot{\theta} dt \tag{10}$$

$$P_b = \int_0^t J_{bm\_v} \dot{\theta} dt \tag{11}$$

The formula (10) is used to update the posture by using Generalized Jacobian Matrix and formula (11) is used to update the position similarly.

X_{b0} and X_{bf} are the initial and final pose of base respectively during the time $[0, t_f]$. The aim of motion planning is to make the final position and posture of base approach the expected values. In other words, that is closed to the initial values as follows:

$$\left\| X_{b0} - X_{bf} \right\| \to 0 \tag{12}$$

3.2   Parameterization of Joint Trajectory

In order to guarantee the smooth motion of manipulator, the initial and final state of joint should satisfy the following conditions:

$$\theta_i(0) = \theta_{i0} \quad \theta_i(t_f) = \theta_{id} \tag{13}$$

$$\dot{\theta}_i(0) = \ddot{\theta}_i(0) = 0 \quad \dot{\theta}_i(t_f) = \ddot{\theta}_i(t_f) = 0 \tag{14}$$

$$\theta_{i\_\min} \leq \theta_i(t) \leq \theta_{i\_\max}, i = 1, 2, \ldots, 7 \tag{15}$$

Because the sinusoidal function can directly restrain the range of joint angle, the joint angle could be parameterized by sinusoidal function. In this paper, each angle of joint is parameterized by the five-order polynomial of sinusoidal function [11] as follows:

$$\theta_i(t) = \Delta_{i1} \sin(a_{i5}t^5 + a_{i4}t^4 + a_{i3}t^3 + a_{i2}t^2 + a_{i1}t + a_{i0}) + \Delta_{i2} \tag{16}$$

$$\Delta_{i1} = \frac{\theta_{i\_max} - \theta_{i\_min}}{2}, \quad \Delta_{i2} = \frac{\theta_{i\_max} + \theta_{i\_min}}{2} \tag{17}$$

Here, $i = 1, 2, \ldots, 7$, $a_{i0} \sim a_{i7}$ are undetermined coefficients. Δ_{i1}, Δ_{i2} are defined according to the scope of the joint angle and $\theta_i \in [\theta_{i\_min}, \theta_{i\_max}]$.

The joint function, joint angle velocities and angular acceleration are accordingly expressed as follows:

$$\theta(t) = \Delta_{i1} \sin\left(a_{i5}t^5 - \frac{5}{2}a_{i5}t_f t^4 + \frac{5}{3}a_{i5}t_f^2 t^3 + \sin^{-1}\left(\frac{\theta_{i0} - \Delta_{i2}}{\Delta_{i1}} \right) \right) + \Delta_{i2} \tag{18}$$

$$\dot{\theta}(t) = \Delta_{i1} \cos\left(a_{i5}\left(t^5 - \frac{5}{2}t_f t^4 + \frac{5}{3}t_f^2 t^3 \right) + \sin^{-1}\left(\frac{\theta_{i0} - \Delta_{i2}}{\Delta_{i1}} \right) \right)\left(a_{i5}\left(5t^4 - 10t_f t^3 + 5t_f^2 t^2 \right) \right) \tag{19}$$

$$\ddot{\theta}_i(t) = -\Delta_{i1} \sin\left(a_{i5}\left(t^5 - \frac{5}{2}t_f t^4 + \frac{5}{3}t_f^2 t^3 \right) + \sin^{-1}\left(\frac{\theta_{i0} - \Delta_{i2}}{\Delta_{i1}} \right) \right)\left(a_{i5}\left(5t^4 - 10t_f t^3 + 5t_f^2 t^2 \right) \right)^2$$
$$+ \Delta_{i1} \cos\left(a_{i5}\left(t^5 - \frac{5}{2}t_f t^4 + \frac{5}{3}t_f^2 t^3 \right) + \sin^{-1}\left(\frac{\theta_{i0} - \Delta_{i2}}{\Delta_{i1}} \right) \right)\left(a_{i5}\left(20t^3 - 30t_f t^2 + 10t_f^2 t \right) \right) \tag{20}$$

The values of parameters can be obtained as follows:

$$a_{i0} = \sin^{-1}\left(\frac{\theta_{i0} - \Delta_{i2}}{\Delta_{i1}} \right), \quad a_{i1} = a_{i2} = 0, \quad a_{i3} = \frac{5}{3}a_{i5}t_f^2, \quad a_{i4} = -\frac{5}{2}a_{i5}t_f \tag{21}$$

Eventually, it is only one undetermined parameter a_{i5}:

$$a = [a_{15}, a_{25}, a_{35}, a_{45}, a_{55}, a_{65}, a_{75}] \tag{22}$$

So, the state of system X is the function about parameter 'a'. We can carry out the trajectory planning of joint space when parameter 'a' is determined.

3.3   Definition of Fitness Function

The 'a' is intended as the independent variable and the fitness function is defined as follows:

$$F(a) = \frac{\|\delta q_b\|}{J_q} + \frac{\|\delta p_b\|}{J_p} + \frac{L_{\dot{\Theta}}}{J_{\dot{\Theta}}} + \frac{L_{\ddot{\Theta}}}{J_{\ddot{\Theta}}} \tag{23}$$

Where, δq_b, δp_b are the quaternion error of base posture and error of base position respectively. $L_{\dot{\Theta}}$, $L_{\ddot{\Theta}}$ are the constraints of joint angular velocities and accelerations respectively. J_q, J_p, $J_{\dot{\Theta}}$, $J_{\ddot{\Theta}}$ are the weighting coefficient and they are determined by accuracy requirement. Where, $\|x\| = \sqrt{x^T \cdot x}$.

4   Description of Trajectory Planning Based on QPSO

Particle Swarm Optimization (PSO) is an evolutionary computation technique and an optimization tool based on iteration. It is introduced in 1995 [12], which derived from the research of migration and cluster behavior during the process of birds foraging. In the PSO, the movement of particle is decided by speed and position. The search space is limited result from the trajectory of the particles is determinate and the speed is restricted in some level over time. However, PSO has the feature of premature convergence as genetic algorithm (GA). VAN et al. [13] had already demonstrated that PSO can't guarantee global convergence. SUN et al. [14] presented QPSO on the basis of the fundamental convergence properties in 2004.

QPSO mainly uses superposition state properties and probability expression characteristics of the quantum theory. The diversity of population is increased because the superposition state properties make single particle can express more states. Besides, probability expression characteristics express the particle position by a certain probability. Therefore, QPSO could converge to the global optimum rather than local optimum, and its robustness is better and has a fast convergence. Moreover, the method has very few parameters than others and has high computational efficiency.

M is set to the particle number, and the steps of trajectory planning based on QPSO are as follows:

Step 1: Set t = 0, initializing the positions $X_i(0)$ of particles in problem space $(i = 1, 2, \ldots M)$, and set $P_i(0) = X_i(0)$. Where, P_i is the optimal position of individual.

Step 2: Calculating the average best position of particles according to the following formula.

$$C_j(t) = \frac{1}{M} \sum_{i=1}^{M} P_{i,j}(t) \tag{24}$$

Where, j $(j = 1, 2, \ldots 7)$ is referred to as dimension of space.

Step 3: Calculating the fitness values $F_i(a)$ of each particle according to the formula (23) and updating the positions of particles. In other words, if $F[X_i(t)] < F[P_i(t-1)]$, then set $P_i(t) = X_i(t)$, otherwise $P_i(t) = P_i(t-1)$.

Step 4: Updating global optimal position $G(t)$, that is to say, comparing the fitness value of $P_i(t)$ with the fitness value of global best position $G(t-1)$, if $F[P_i(t)] < F[G(t-1)]$, then set $G(t) = P_i(t)$, else $G(t) = G(t-1)$.

Step 5: Calculating and gaining the location of random point according to the following formula:

$$\begin{cases} p_{i,j}(t) = \varphi_j(t)P_{i,j}(t) + [1 - \varphi_j(t)]G_j(t) \\ \varphi_j(t) \sim U(0,1) \end{cases} \qquad (25)$$

Here, $\varphi_j(t)$ obey the uniform distribution on the $(0,1)$.

Step 6: The new position of each particle is updated as follows:

$$\begin{cases} X_{i,j}(t+1) = p_{i,j}(t) \pm \alpha \times \left| C_j(t) - X_{i,j}(t) \right| \times \ln\left[\frac{1}{u_{i,j}(t)}\right] \\ u_{i,j}(t) \sim U(0,1) \end{cases} \qquad (26)$$

Where, $u_{i,j}(t)$ obey the uniform distribution on the $(0,1)$. α is called contraction-expansion coefficient of QPSO. It is the only control parameter of the algorithm except group scale and the number of iterations. α keep a dynamic changes as follows:

$$\alpha = \frac{(m - n) * (N - t)}{2} + n \qquad (27)$$

Parameter α diminishes from m to n linearly as the iteration, usually $m = 1$, $n = 0.5$. And N denotes the maximal number of iterations.

Step 7: If the termination condition of algorithm is not satisfied, then $t = t + 1$ and turn Step 2, otherwise stop the algorithm.

The flow chart of QPSO is as Fig. 1.

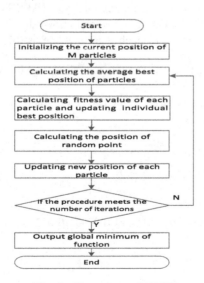

Fig. 1. Flow chart of QPSO

5 Simulations

To verify the accuracy and usefulness of the method in this paper, a model composed of a flight base and 7-DOF space manipulator is set up. The simulation environment is MATLAB R2013b.

In addition, the initial and desired joint angles are respectively defined as follows:

$$\Theta_{b0} = [0,0,0,0,0,0,0], \ \Theta_{bd} = [0,0,0,0,0,0,0] \tag{28}$$

The initial position and posture of base are as follows:

$$[P_{b0} \ Q_{b0}] = [0 \quad -0.0813 \quad -0.1880 \quad 1 \quad 0 \quad 0 \quad 0] \tag{29}$$

The desired position and posture of base are as follows:

$$[P_{bd} \ Q_{bd}] = [0 \quad -0.0813 \quad -0.1880 \quad 1 \quad 0 \quad 0 \quad 0] \tag{30}$$

As well as, the ranges of velocity and acceleration of joint angle are as follows:

$$|\dot{\theta}_i| = 60°/s = 1.0472 rad/s, |\ddot{\theta}_i| = 60°/s^2 = 1.0472 rad/s^2 \tag{31}$$

The weighting coefficients of fitness function are determined according to their precision as follows:

$$K_p = 2 \times 10^{-3}, \ K_q = \sin(\frac{\pi}{360}), \ K_{\dot{\theta}} = K_{\ddot{\theta}} = 0.01 \tag{32}$$

The fitness functions are optimized using the PSO, SAPSO and QPSO respectively. In the same environment of MATLAB, the minimum value can be obtained.

The experimental parameters of each algorithm are as follows:

$$M_p = 40, N = 100, \omega = 0.5, c1 = 2.05, c2 = 2.05, lamda = 0.5 \tag{33}$$

The optimal parameter and target value can be obtained by using PSO, SAPSO and QPSO respectively as follows:

$$\begin{cases} a_{PSO} = [0.05043, 0.00001, -0.08065, 0.00015, -0.18631, -0.18966, 0.08257] \times 10^{-3} \\ F_{PSO} = 2.93143 \times 10^{-6} \end{cases}$$
$$\tag{34}$$

$$\begin{cases} a_{SAPSO} = [-0.02697, -0.00000, 0.15230, 0.00002, 0.03323, 0.01223, 0.04255] \times 10^{-3} \\ F_{SAPSO} = 1.28968 \times 10^{-6} \end{cases}$$
$$\tag{35}$$

$$\begin{cases} a_{QPSO} = [-0.08423, 0.00010, 0.10952, -0.0008, 0.35371, 0.10001, -0.45127] \times 10^{-4} \\ F_{QPSO} = 5.86872 \times 10^{-8} \end{cases}$$

$$(36)$$

As can be seen from the above results, the minimum of QPSO is optimal. The results show that the QPSO algorithm is superior to the other two algorithms.

Figure 2 shows the base position of QPSO contrasting with the PSO and SAPSO. Figure 3 shows base attitude of QPSO contrasting with them. In addition, the red line shows PSO, the green line shows SAPSO, and the blue line shows QPSO.

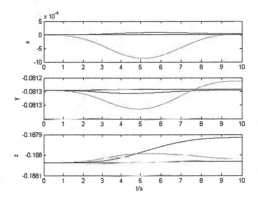

Fig. 2. Comparison chart of base position (Color figure online)

From Fig. 2, the curves of the position based on SAPSO have the largest disturbance, followed by PSO. The curves of SAPSO and PSO are not stable. The curves of QPSO are most stable. That is to say, the trajectories planning of QPSO have the minimum disturbance to base. The experimental results prove the validity and practicability of the algorithm in this paper.

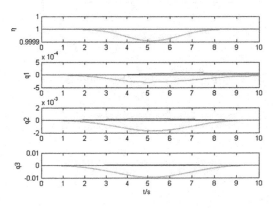

Fig. 3. Comparison chart of base attitude (Color figure online)

From Fig. 3, the curves of the attitude based on SAPSO are most instable. The curves of QPSO are a little better than them of PSO.

In the QPSO, Fig. 4 shows the curves of base position, Fig. 5 shows the curves of base attitude. And Figs. 6, 7 and 8 shows the joint angles, joint velocity and acceleration respectively.

From Figs. 4, 5, 6, 7 and 8, the curve of joint trajectory is smooth and it is easy to control.

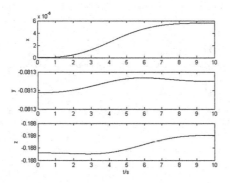

Fig. 4. Curves of base position

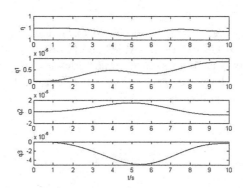

Fig. 5. Curves of base attitude

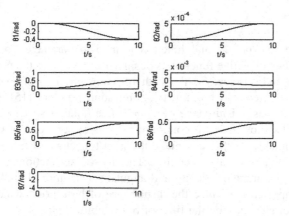

Fig. 6. Curves of joint angles

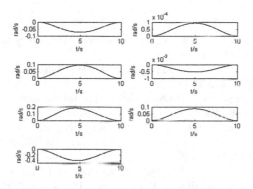

Fig. 7. Curves of joint velocity

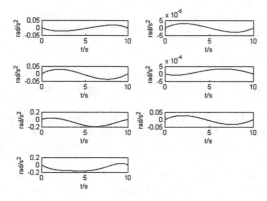

Fig. 8. Curves of joint acceleration

6   Conclusion

In this paper, the sine polynomial function is used to parameter joint trajectory of manipulator and then objective function is designed according to the precision index of the base position and attitude. So the path planning problem of free-floating space manipulator is converted into non-linear optimization problem [15]. QPSO is firstly applied to optimize base disturbance of 7-DOF free-floating space manipulator. The experimental results show that the continuous optimum solutions of QPSO approach global optimum quickly. QPSO can make the joint trajectory of manipulator smooth. So the space manipulator is easy to control. The sinusoidal functions can guarantee the ranges of joint angle, angular velocity and angular acceleration. Finally, the proposed method in this paper can optimize the disturbance of base pose, which can not only save resources, but also shorten the time of base stability.

References

1. Li, Y., Dang, C.P.: The development of orbital servicing technology in space. Ordnance Ind. Autom. **31**(5), 79–82 (2012)
2. Zhu, J.J., Huang, P.F., Liu, Z.X., Zhang, B.: Attitude stabilization method of space robot. Comput. Simul. **28**(7), 177–179 (2011)
3. Li, S.R., Ma, H.C., Jiao, D.Y.: Robust tracking control for mobile robots using backing stepping. Comput. Eng. Appl. **49**(8), 266–270 (2013)
4. Steven, D., Miguel, A.T.: Path planning for space manipulators to minimize spacecraft attitude disturbances. In: Proceeding of IEEE International Conference on Robotics and Automation, pp. 2522–2528 (1991)
5. Vafa, Z., Dubowsky, S.: On the dynamics of space manipulator using the virtual manipulator with application to path planning. J. Astronaut. Sci. **188**, 45–76 (1993)
6. Huang, P.F., Chen, K., Xu, Y.S.: Optimal path planning for minimizing disturbance of space robot. In: Proceeding of IEEE Ninth International Conference, pp. 139–144 (2006)
7. Ge, X.S., Chen, L.Q., Lv, J.: Nonholonomic motion planning of a space manipulator system using genetic algorithm. J. Astronaut. **26**(3), 262–267 (2005)
8. Wang, K., Zhang, H., Yang, L., Liu, Z.L.: Path planning for robots based on quantum-behaved particle swarm optimization. Microcomput. Inf. **26**(4), 156–165 (2010)
9. Chou, J.C.K.: Quaternion kinematic and dynamic differential equations. IEEE Trans. Robot. Autom. **8**(1), 53–64 (1992)
10. Xu, W.F., Liu, Y., Liang, B., Yang, Y.S., Qiang, W.Y.: Autonomous path planning and experiment study of free-floating space robot for target capturing. J. Intell. Rob. Syst. **51**, 303–331 (2008)
11. Xu, W.F., Liang, B., Li, C., Xu, Y.S., Qiang, W.Y.: Path planning of free-floating robot in cartesian space using direct kinematics. Int. J. Adv. Rob. Syst. **4**, 17–26 (2007)
12. Poli, R., Kennedy, J., Blackwell, T.: Particle swarm optimization. Swarm Intell. **1**, 33–57 (2007)
13. Bergh, F.D., Engelbrecht, A.P.: A new locally convergent particle swarm optimizer. In: IEEE Conference on Systems Man and Cybernetics, pp. 96–101 (2002)

14. Sun, J., Feng, B., Xu, W.B.: Particle swam optimization with particles having uqantum behavior. Evol. Comput. **1**(1), 325–331 (2004)
15. Shi, Y., Liang, B., Wang, X.Q., Xu, W.F.: Cartesian non-holonomic path planning of space robot based on quantum-behaved particle swarm optimization algorithm. J. Mech. Eng. **47** (23), 65–72 (2011)

Cognitive Robotics

Eric Chown[1] and Wai K. Yeap[2]([✉])

[1] Bowdoin College, Brunswick, ME, USA
echown@bowdoin.edu
[2] Auckland University of Technology, Auckland, New Zealand
wai.yeap@aut.ac.nz

Abstract. Cognitive robotics presents unique research challenges as it straddles the boundaries between two very different fields. Its name promises robots that exhibit cognitive behaviour and suggests that the research would be beneficial and interesting to researchers in both robotics and cognitive science. However, this is not generally the case. Many roboticists are disinterested in the developments of cognitive robots unless these robots are proven to be superior in terms of speed, efficiency and accuracy. Similarly, cognitive scientists are disinterested in the work of roboticists unless the robot's behavior is fit to some known empirical data and the models implemented are cognitively plausible. These different requirements, each from their own field, are often at odds, and has hampered the development of cognitive robotics, causing researchers to use robots only as a platform for simulating cognitive ideas or to use cognitive science as a weak source of ideas for robot mapping. In this article, we argue that a better synthesis of ideas from both fields must be encouraged and that cognitive robotics should move beyond its artificial limitations and in the process better serve robotics and cognitive science.

Keywords: Cognitive mapping · Robotics · Cognitive science · Cognitive robotics

1 Introduction

Academia is structured in such a way as to make interdisciplinary research difficult. Every field and subfield has unique methodology, terminology, and systems of evaluation. When stepping from one discipline to another one must quickly learn all of the differences between fields or success will be extremely difficult. In recent years computational scientists have bucked this general trend because they have brought new tools to nearly every field from the Humanities to the Sciences. Even so, within other disciplines there is always strong resistance (e.g. see Drucker 2012; McPherson 2012). Often these new methods from other fields are seen as a threat to existing methodologies and to old ways of doing things. By contrast, cognitive science has long since embraced computational methods. Computer models of psychological theories have a long history going back to Newell, Shaw and Simon's work on GPS (Newell et al. 1959). However, as with other fields, the intersection of computational methodologies with cognitive science is not always smooth. In the case of cognitive

© Springer International Publishing Switzerland 2015
A. Bikakis and X. Zheng (Eds.): MIWAI 2015, LNAI 9426, pp. 294–305, 2015.
DOI: 10.1007/978-3-319-26181-2_28

robotics it is not so much a resistance to computation, as the fact that cognitive science and robotics as fields have such different methods of evaluation. Here we will broadly characterize the two fields while at the same time recognizing that the truth is far more nuanced.

Broadly speaking the goal of Robotics is to build ever more capable robots. This means robots that can do more, are smarter, and are more efficient. As a subfield of computer science this is reflective of computer science as a whole. In its memo on tenure the Computer Research Association had this to say, "When one discovers a fact about nature, it is a contribution per se, no matter how small. Since anyone can create something new [in a synthetic field], that alone does not establish a contribution. Rather, one must show that the creation is better." (Patterson et al. 1999) The memo goes on to say that contributions will lead to "better results." This is perfectly sensible for computer science as a discipline. So in publishing in computer science, and by extension robotics, it is generally necessary to prove that one's work is "better" than what came before. In robotics, meanwhile, "better" can often be directly quantified. It might mean a more accurate map, a faster path, or more efficient computation. It should be no surprise that these are standard means of evaluating robot models.

The situation is quite different in cognitive science. The field has many goals, but most of them revolve around understanding human cognition. In many cases such understanding comes through building models. The goal of such models is not to be "better" in the same sense as computer science, but to be a more accurate reflection of human cognition and therefore be a potentially more useful tool in understanding how cognition works. When models are judged by characteristics of speed, accuracy and efficiency, it is not to be certain that such quantities are maximized, but rather that they closely reflect data on their human counterparts.

Mapping provides a simple and concrete example of how the two ways of working are at odds and throughout this article, we will use this example when discussing the problems cognitive robotics face. In robotics, the mapping literature is currently dominated by SLAM (Simultaneous Localization and Mapping) methods (e.g. see Thrun 2008). In SLAM, a robot's goal is to navigate an environment and simultaneously build an accurate map of the environment. SLAM models strive to build as precise a map of their environment as is possible. The better the map the better the robot will be able to use it later to navigate. Meanwhile human navigation is a completely different story. While humans are capable of amazing navigational feats, it has been repeatedly shown that their internal maps are actually quite distorted and sketchy. Roboticists interested in cognition might find themselves with a typical robot setup involving a small wheeled-robot with a laser rangefinder. They could then use such a robot to build human-like maps of its environment. From the perspective of mainstream robotics such maps are going to look poor compared to the best SLAM methods. From the perspective of cognitive science such maps may not be interesting simply because they were created using inputs from lasers; the latter is so unhuman.

Due to these differences and the problems inherent in evaluating such models, the development of cognitive robots has been hampered. Cognitive robotics researchers often end up with models that interest cognitive science or robotics but not both. Consequently, cognitive researchers have shown little interest in robotics beyond as a platform for implementing and testing their ideas and roboticists have shown little interest in cognitive science beyond borrowing some simple ideas to improve upon their models. For example, the idea that SLAM models might have anything to teach cognitive scientists is seen as absurd within much of cognitive science. Indeed we are unaware of any mainstream models of human navigation or cognitive mapping that have been significantly influenced by SLAM methods. Again, on the face of it this seems reasonable, after all robots are very different than humans and most SLAM researchers would not claim that their methods are reflective of human navigation. We believe, however, that such a view is shortsighted. To get a glimpse of why this is the case one need look no further than the work on animal navigation. Cognitive research has been greatly impacted by work on many different species, from rats whose brains share a significant amount of structure with humans to ants whose brains have little in common with human brains. The common element of all such species is that they all face challenges in navigation that they have evolved to overcome. By studying a wide range of species it is possible to start finding principles that are common to all of them. Such principles may be implemented very differently for a given animal, but if the principle is powerful enough, it is likely that many different systems will have "discovered" it through the process of evolution. The very fact that diverse animals are using the same principles is evidence of the power and utility of such principles.

We are proposing that by viewing robots and humans as different species solving the same problem (an idea first proposed in Yeap 2011b), there is much to be learned from each other. From this perspective, cognitive robotics is not restricted to the use of robots for testing cognitive ideas but also to find solutions to problems that baffled cognitive scientists. The latter leads to discovering new navigational principles and new algorithms that would benefit research in robotics and provide new insights into spatial cognition. Different models (i.e. species) can then be built to evaluate the applicability of the principles identified. Indeed, the different capabilities of robots – having wheels, sonar, lasers, and others – provides a test of the generality of the methods and principles that are being proposed.

In the rest of this article we discuss the nature of interactions between cognitive science and robotics through cognitive robotics, both in the successes and the challenges that have emerged over the years but also in suggesting that much more is possible, especially with regard to the contributions that both sides have to gain from each other.

2   What Do Cognitive Models Offer Robotics?

There are many reasons why robotics researchers might not be interested in cognitive models of navigation. For instance, it would be easy to conclude that the human brain is just too complex and little understood. While cognitive models offer significant insights, they often lack the detail necessary for an implementation, and in particular, the details most lacking are normally computational. Further, human navigation makes use of an unparalleled object recognition system that dominates even the most sophisticated machine vision system. Meanwhile, even when navigating humans are constantly doing other things and have many other cognitive processes that impact navigation performance. These range from emotional factors to things that take attention away from navigation. From the point of view of robotics it would appear that building systems that do not have such distractions would ultimately lead to better performance. It therefore stands to reason that doing things the way humans do them may not be the best path for robots. Further, the history of Artificial Intelligence (AI) suggests that eventually robots will be better than humans at navigation, just as with chess and with many other problems once thought to be "hard." And just as with chess and other problems, the solutions that AI finds may have little or nothing to do with the way that humans perform them. So what does cognition have to offer robotics? Why not simply pursue a path of ever more refined mathematical models?

In the early days, AI researchers were eager to learn from nature for two reasons. First, nature provides a rich source of ideas and second, nature's solutions are both interesting and tested by the need to survive. For mapping, roboticists have already borrowed many ideas from cognitive mapping albeit mainly at the structural level. For example, topological maps, inspired by human "route maps" have long been a staple of robotics (e.g. see Thrun and Bücken 1996). "Gateways", first developed as part of a theory of human cognitive mapping (Chown et al. 1995) found their way into numerous robot systems (e.g. Beeson et al. 2010). In both of these cases roboticists took an initial idea from cognitive models and used it as a starting point to end up with something new.

To the extent that cognitive ideas have had a positive impact on robotics, it is because these ideas can easily be translated onto a robot whatever its capabilities. For example, a topological map consists of a network of "landmarks." Such representations are powerful and useful regardless of whether the landmarks were learned by using vision, sonar or lasers. The navigational principles that these representations encapsulate are powerful enough that they can be implemented in an almost unlimited number of ways. The Gateway notion has a similar history. In humans Gateways occur where there is a visual occlusion followed by an opening. In other words Gateways are places where new information becomes available. A robot does not need the powerful human visual system to take advantage of the principle that locations in the environment where new information becomes available are important. One of the first uses of gateways (Kortenkamp and Weymouth 1992), for example, used sonar, a sensor that is notoriously noisy. For a robot moving down a corridor, however, sonar makes it extremely easy to identify gateways. As the robot is moving past walls it should get relatively constant, if noisy, reading. However, when a new corridor opens up, or there is an open door, the readings will jump dramatically, far beyond the magnitude of normal noise. Thus the principle that a sudden large environmental change should drive new representations is easily exploited by robots with a wide variety of sensory capabilities. This principle is powerful enough that it can be used to organize a map even for a perceptually weak robot.

The landmark example is instructive for the larger point that we are making. Topological maps have been a staple of cognitive theories of navigation going back to Piaget (Piaget and Inhelder 1967). These theories then made their way into robotics. Once robotics got ahold of them there were essentially two camps working on topological models. One consisted of cognitive roboticists who tended to slavishly try to mimic the exact details of human navigation. The second camp consisted of roboticists who wanted to make their robots navigate as effectively as possible. Both of these camps have narrow goals that naturally limited the scope of their work. The robot camp, for example rarely considers the full complexity of landmarks. Since their only goal is to build robots that navigate as effectively as possible they focus squarely on using perception to identify landmarks. A researcher, for example, using lasers for input, will work to engineer solutions optimized for lasers. A similar story will be true for robots with cameras or sonar. In each case landmarks are taken to be any object in view that has a unique perceptual signature, and consequently landmarks are perceived almost everywhere by robots because there is almost always something in any given view that stands out.

By contrast, humans remember a much smaller number of objects as landmarks, apparently using a much different, more global, criteria, and consequently each landmark is more important and more memorable. It is much more difficult to implement this on a robot because it is still not well understood exactly how humans select landmarks, and of course the human vision system is a crucial part of the process.

In this way the two points of view represent a kind of continuum of strategies that a cognitive being might use in navigation. Different points along this continuum might have different strengths and weaknesses with regard to different environments. For example, some environments, such as a dessert or prairie, are known to be landmark poor for humans. A robot might have better success in creating landmarks on the fly in such a place. On the other hand, with so many landmarks in their maps, robots may be more susceptible to becoming confused when changes in the environment occur as happens so often in so many environments. Robotic solutions along these lines are often accused of being brittle. This is one of the oldest criticisms of AI models. For example, consider the application of probabilistic solutions in robot soccer competitions such as RoboCup (Chown and Lagoudakis 2015). While these algorithms perform well in it, even small changes to the domain can cause them to completely break. This would appear to be in stark contrast to human intelligence. Humans appear to effortlessly adapt to even large changes in domain rules. For example, not only can a group of five year olds instantly adapt to the outdoor conditions of a new soccer field, they can even create a field on the fly using sticks, trees, and bushes and then localize beautifully on the ad hoc pitch. It would be more than a stretch to suggest that such children are creating highly precise internal maps to accomplish such feats. Indeed it may well be the case that the lack of such precise models is a key to such adaptability.

Computer science and AI have championed the idea that problems can be viewed as search. It is certainly possible to view "navigation" as such a problem. What we are seeing is that a very small part of that space is being explored right now. There are researchers who are exploring the space that most closely resembles human navigation and there are researchers who are looking to optimize what robots with specific sets of capabilities can do. In search terms both groups are essentially looking for local maxima using different criteria for defining their maxima. The "better" required in robotics is a kind of hill climbing as is the drive for ever more realistic models in cognitive science. Cognitive robotics affords the chance to pursue more global strategies. Strategies that might provide a deeper understanding of the space and could ultimately lead to solutions that combine the flexibility of human navigation with the precision of robot navigation.

3 What Does Robotics Offer Cognition?

Traditionally the major, if not the only, reason for a researcher interested in cognition to use robots has been to test ideas. The benefit of using robots is concreteness. If a model cannot be implemented on a robot, or if it simply does not work when implemented, then these are strong evidence of a model's deficiencies. In turn such deficiencies are indicative of places where existing theories need to be updated or discarded. Ironically then, success, in terms of lessons learned through robotics, typically has come through failures in implementation. However, when a model is successfully implemented, it is easy to dismiss that success. Some models are dismissed out of hand on the grounds of differences in hardware, with others it is because the model used is too general. In fact it is difficult to prove anything by building a model. Generally the best that can be done is to use the model to make novel predictions that can later be checked experimentally (e.g. by testing human subjects). Despite this apparently pessimistic view where knowledge only comes through failure, the act of making theories concrete via the use of robot is invaluable.

However, while the importance of making theories concrete is hard to overstate, an important lesson learned in developing early AI models of cognition is that the theory must be formulated and tested at the appropriate level. For example, in the early history of AI and cognitive science, many models of cognition were developed at a high level of thinking (Langley 2007). Conscious thinking at that time was deemed to be the most interesting and important part of cognition. Perhaps the greatest lesson of robotics and related fields such as machine vision, is that they provided stark and unequivocal evidence of just how difficult and important perception is. This is evidenced by the famous story of how in the early days of AI Marvin Minsky assigned some of his students to solve the problem of using a camera to identify objects as a good "summer project." More than 50 years later and it is still nowhere close to being solved. Robotic provides an excellent platform for developing concrete ideas about the perceptual process that cognitive theories often lack but require. It is surprising then that this aspect of the advancements from robotics has largely been ignored. Research on SLAM provides an instructive example.

Many cognitive theories of navigation have converged around the idea that people learn a series of views as one form of navigation (Yeap 1998; Chown et al. 1995; Franz et al. 1998). All of these theories, even the ones implemented on robots, have not fully tested the implications of this idea. Meanwhile SLAM models have been implemented and tested on enormous scale at levels far beyond what the cognitively inspired models have managed. What SLAM researchers have found is that the process of building a global map of an environment based on integrating successive views of the environment has a fundamental problem – the accumulation of error. As has long been known in robotics, when an agent moves through an environment little errors tend to accumulate. Over time and space these little errors turn into large errors. Imagine, for example, that you want to head north, but your heading is off by a small amount. The further you go on that heading the more you stray from true north. Over large distances you will end up a long way away from your goal. SLAM researchers have had to learn to cope with this problem and have come up with techniques that allow them to correct for the errors that cannot help but occur when building a global map.

If humans do indeed build maps out of successive views, than this process is necessarily going to run into the same problems of accumulating errors. It is also possible, and may even be likely, that the solutions being found by SLAM researchers have also been "found" by evolution. At the very least SLAM is leading to a more thorough understanding of the problems and issues inherent in such processes. The question is what if any of this knowledge has found its way back into psychological models and testing? The answer, as far as we can tell, is "little or none." Cognitive scientists by and large are not interested in SLAM because it isn't a cognitive model. Worse, from their point of view, it is often implemented on robots with very different capabilities than humans. A cognitive scientist might say that SLAM isn't relevant because humans can resolve their errors through the use of their superior vision systems or in some other way that is different than SLAM. It is possible that this is even true but what would such a model be? Cognitive science has not proposed any as of yet. It is also possible that the general principles used by SLAM systems may be similar, or even the same, as the principles used in human cognition. If this is the case then it is clearly a good idea to identify those principles. A different cognitive scientist might say that it doesn't matter anyway since human cognitive maps are known to be distorted and sketchy. Indeed one of the co-authors of this article has leveled both of these criticisms at SLAM in the past. Saying a cognitive map is distorted, however, does not mean that it can be distorted without limit. Indeed this suggests an interesting line of research on the nature of distortions in internal maps and just how distorted they can become while still being functional. More to the point these are things that are not known. There is still a great deal to be learned about the nature of cognitive maps, about the nature of dealing with errors in maps, and about how navigation works in general. Robotics provides a tool for exploring these questions and the answers that roboticists are finding may provide critical information for better understanding how humans navigate. Cognitive roboticists, and by extension cognitive scientists, would be wise to pay more attention to these explorations and to use them as springboards for their own work.

4 Cognitive Robotics – on the Edge of Discovering New Ideas

Early work on cognitive robotics, rightly so, has emphasized on the use of robots as test beds for evaluating cognitive theories. However, the fact that robots are so different from humans has meant that cognitive scientists are unlikely to view the work of cognitive roboticists as making a significant contribution to their understanding of how the mind works. Similarly, the fact that cognitive ideas are so semantically laden and so specific to humans has meant that roboticists are unlikely to view cognitive robotics as practically significant. What we are proposing in this article is that cognitive robotics should pay more attention to the questions cognitive scientists raise and attempt to find answers to those questions using robots often borrowing developments in robotics to do so. Initially, these questions should center on perceptual problems since this is where cognitive theories tend to be weakest and where robotics must necessarily find solutions. In this section, we provide two examples of exploring such questions.

4.1   Contributions to Cognitive Science

While cognitive scientists in general and psychologists in particular have made significant discoveries concerning cognitive mapping, their theorizing often lacks important computational details concerning the underlying process itself. Such omissions are not a matter of mere details since without those details the correctness of any such theory is impossible to verify. This is because any representation inferred from behavioral results could be challenged with an alternative representation. For example, even the most fundamental idea of cognitive mapping – that humans and animals compute a map of their environment, first proposed by Tolman (1948) and later given significant support from two prominent pieces of work: Lynch (1960) and O'Keefe and Nadel (1978) – is controversial precisely because proponents of the theory have failed to demonstrate exactly the kind of map computed and how it is learned. Consequently, by providing alternative explanations to account for the behavior observed, many have challenged the very idea that a map is computed at all. For example, for rats searching for food next in a radial arm maze, Brown (1992) argue that they could have considered only which alley not visited and thus would not need to use a spatial map. In the water maze problem, Benhamou (1996) argue that rats could use some orientation mechanisms and not a spatial map to locate the platform in the water maze (for further discussions, see Yeap 2011b). Even for humans, the idea that we compute a map in the head has been challenged recently. Wang and Spelke (2002) argue that humans maintain only a transient egocentric map that allows awareness of their immediate surroundings but not an enduring non-egocentric map. The latter is too empowering (Yeap 2014) and based on their observation that other lower animals do not compute such a map and their belief that all animals navigation abilities should build on a common set of mechanisms, they conclude that no such map is computed at the perceptual level.

This gap in cognitive science research is what cognitive roboticists could fill via experimentations using robots. To do so, cognitive roboticists need to address how the key ideas/representations identified in cognitive science are physically realized and in ways that match their characterization by cognitive scientists. For example, cognitive roboticists, unlike traditional roboticists, must not only show how a map of the environment is computed, but the map must also bear many of the characteristics of a cognitive map. For example, one distinctive characteristic of a cognitive map is that it is fragmented and inexact. Successfully implementing such a process on a mobile robot, even with sensors that differ from cognitive agents, would provide insights into the nature of the process. Recently, Yeap (2011a) describes one such process implemented on a mobile robot equipped with a laser sensor and an odometer (see also Yeap et al. 2011). Unlike SLAM, Yeap's process eschews error corrections, continuous updating, and continuous self-localization. Instead it takes a snapshot of the environment as a kind of map of a local environment that it is about to explore, and as it moves out of its current bounded space, it takes a new snapshot corresponding to the next local environment that it will explore and so on. What is computed as it explores is a trace of the individual local maps. When moving in each of the local areas, objects within the area are tracked in each subsequent view by the moving robot and these tracked objects enable the robot to recover its pose (its position and orientation in space) in these maps, thereby allowing the robot to generate a global metric map. This metric map ends up being incomplete and inexact, but it is not as distorted as a similar map generated via integrating successive views. The map produced is accurate enough to allow the robot to orient itself in the environment.

While Yeap's process is not an exact analog of how human and/or animal cognitive mapping works, it nonetheless bears many interesting commonalities with human cognitive mapping, especially when compared with the SLAM approach. For example, the central idea is that tracking objects as the robot moves can compensate for the normal errors that might build up during the mapping process. The objects provide a natural source of error correction as the robot moves. Global correctness is not necessarily important. This is as opposed to the SLAM-based approach whereby a robot has to continuously localize and correct its position in the map. In SLAM, robots are constantly trying to place themselves in the correct position in the global map, whereas in Yeap's approach the robot is merely trying to solve the simpler problem of determining where it is relative to nearby landmarks. As already noted, human navigation makes use of an unparalleled object recognition system and thus it would be natural that any process proposed for human mapping would take maximum advantage of such a system. Imagine our hypothetical traveller heading to the north. As they head north they will naturally accumulate error. But then if they see a known landmark along the way the landmark will naturally and automatically correct the errors that they have accumulated. Such a traveller need not have a precise map in their head, they need only have one good enough to get to the next landmark. As has been noted before in robotics, "the world is its own best model" (Brooks 1991). Given this, it is hard to imagine that the humans would compute an exact and complete map, and indeed it is well known that they do not.

4.2 Contributions to Robotics

Within its own space, roboticists are proud of finding a solution to the principle problem that they have identified – namely how to correct the sensor errors and produce a correct map while simultaneously exploring the environment – and rightly so. Their confidence in this approach has led some to predict that the future key challenges lie in developing ever larger, more persuasive demonstrations of the approach, such as mapping a city, or massive structures such as the Barrier Reef or the surface of Mars (Bailey and Durrant-Whyte 2006; Durrant-Whyte and Bailey 2006). Not surprisingly then, these algorithms have been extended for handling dynamic environments (e.g. Fox et al. 1999; Hahnel et al. 2003), for creating maps in large outdoor environments (Thrun and Montemerlo 2006; Folkesson and Christensen 2007), for creating 3D maps (Nüchter et al. 2007; Pathak et al. 2010), for creating sub-maps as places in a topological map (Konolige et al. 2011; Ranganathan and Dellaert 2011), and for use with vision (Ho and Newman 2007; Schleicher et al. 2010). More recently SLAM-based approaches have become a popular choice for use with drones.

Despite all of this work and all of these successes with SLAM, and despite the fact that the basic processes involved in SLAM are similar to what cognitive scientists are interested in with humans, cognitive science has not paid attention to the results of SLAM research, even though some of these results indicate problems that must be resolved by cognitive models going forward. Humans may not resolve the accumulation of error in the same way that SLAM does, for example, but SLAM research has very effectively shown that it must be resolved in some way, it cannot simply be ignored. Conversely, nature must have discovered an alternative approach to the one being studied and adopted by roboticists because there are so many examples of successful animal navigators. As discussed above, one such alternative, as illustrated in Yeap (2011a), could provide an alternative paradigm for robot mapping. As we have already witnessed in the development of SLAM-based approaches, if the robotics community take interests in such alternative approaches, they will help accelerate the development and understanding of the alternative paradigms and thus more effectively help to search the space for solving the larger problem of general purpose navigation, both for humans and robotics.

5   Concluding Remarks

When two different fields intersect in a new way it is natural for the early work at the intersection to run into the problems inherent in attempting to please two different masters with two distinct sets of needs. This has certainly been the case with cognitive robotics. Ultimately the different standards of evaluations stemming from the two fields can be restrictive, stifling and even self-defeating. In this article we have proposed that cognitive robotics should move in new directions aimed less at slavishly modeling human navigation and instead focus more on processes and principles. In the end we expect that all of the fields involved will gain. Cognitive science can benefit from the work done in robotics in exploring the problems of situating real agents in the real world. This work continues to uncover new problems and alternative solutions to such problems. Meanwhile, robotics can benefit from ideas stemming from systems that are far more general and flexible than any produced by man to date.

References

Bailey, T., Durrant-Whyte, H.: Simultaneous localization and mapping (SLAM): Part II. IEEE Robot. Autom. Mag. **13**(3), 108–117 (2006)

Beeson, P., Modayil, J., Kuipers, B.: Factoring the mapping problem: Mobile robot map-building in the hybrid spatial semantic hierarchy. Int. J. Robot. Res. **29**(4), 428–459 (2010)

Benhamou, S.: No evidence for cognitive mapping in rats. Anim. Behav. **52**, 201–212 (1996)

Brown, M.: Does a cognitive map guide choices in the radial-arm maze? J. Exp. Psychol. Anim. Behav. Process. **18**, 56–66 (1992)

Brooks, R.A.: Intelligence without representation. Artif. Intell. **47**, 139–159 (1991)

Chown, E., Kaplan, S., Kortenkamp, D.: Prototypes, location and associative networks (PLAN): towards a unified theory of cognitive mapping. Cog. Sci. **19**, 1–52 (1995)

Chown, E., Lagoudakis, M.: The standard platform league. In: Bianchi, R., Akin, H., Ramamoorthy, S., Sugiura, K. (eds.) RoboCup 2014. LNCS, vol. 8992, pp. 636–648. Springer, Heidelberg (2015)

Durrant-Whyte, H., Bailey, T.: Simultaneous localization and mapping (SLAM): Part I. IEEE Robot. Autom. Mag. **13**(2), 99–110 (2006)

Drucker, J.: Humanistic theory and digital scholarship. In: Gold, M.K. (ed.) Debates in the Digital Humanities, pp. 85–95. University of Minnesota Press, Minneapolis (2012)

Folkesson, J., Christensen, H.I.: Graphical SLAM for outdoor applications. J. Field Robot. **24**(1/2), 51–70 (2007)

Fox, D., Burgard, W., Thrun, S.: Markov localization for mobile robots in dynamic environments. J. Artif. Intell. Res. **11**, 391–427 (1999)

Franz, M.O., Schölkopf, B., Mallot, H.A., Bülthoff, H.H.: Learning view graphs for robot navigation. Auton. Robots **5**, 111–125 (1998)

Hahnel, D., Burgard, W., Fox, D., Thrun, S.: An efficient FastSLAM algorithm for generating cyclic maps of large-scale environments from raw laser range measurements. In: Proceedings of the IEEE/RSJ International Conference on Intelligent Robots and Systems (IROS 2003), pp. 206–211 (2003)

Ho, K.L., Newman, P.: Detecting loop closure with scene sequences. Int. J. Comput. Vision **74**(3), 261–286 (2007)

Konolige, K, Marder-Eppstein, E., Marthi, B.: Navigation in hybrid metric-topological maps. In: Proceedings of IEEE International Conference on Robotics and Automation (ICRA 2011), pp. 3041–3047 (2011)

Kortenkamp, D., Weymouth, T.: Using gateways to build a route map. In: Proceedings of the IEEE/RSJ Conference on Intelligent Robots and Systems, pp. 2209–2214 (1992)

Langley, P.: Artificial intelligence and cognitive systems. In: Cohen, P. (ed.) AI: The First Hundred Years. AAAI Press, Menlo Park (2007)

Lynch, K.: The Image of the City. MIT Press, Cambridge (1960)

McPherson, T.: Why are the Digital Humanities so white? Or Thinking the histories of race and computation. In: Gold, M.K. (ed.) Debates in the Digital Humanities, pp. 139–160. University of Minnesota Press, Minneapolis (2012)

Newell, A., Shaw, J.C., Simon, H.A.: Report on a general problem solving program. In: Proceedings of the International Conference on Information Processing, pp. 256–264 (1959)

Nüchter, A., Lingemann, K., Hertzberg, J., Surmann, H.: 6D SLAM-3D Mapping outdoor environments. J. Field Robot. **24**(8–9), 699–722 (2007)

Pathak, K., Birk, A., Vaskevicius, N., Poppinga, J.: Fast registration based on noisy planes with unknown correspondences for 3D mapping. IEEE Trans. Rob. **26**(3), 424–441 (2010)

Ranganathan, A., Dellaert, F.: Online probabilistic topological mapping. Int. J. Robot. Res. **30**(6), 755–771 (2011)

O'Keefe, J., Nadel, L.: The Hippocampus as a Cognitive Map. Oxford University Press, Oxford (1978)

Patterson, D., Snyder, L., Ullman, J.: Best practices memo: evaluating computer scientists and engineers for promotion and tenure. Computing Research Association, Special Insert (1999)

Piaget, J., Inhelder, B.: The Child's Conception of Space. W.W. Norton, New York (1967)

Schleicher, D., Bergasa, L.M., Ocana, M., Barea, R., Lopez, E.: Real-time hierarchical stereo visual SLAM in large-scale environments. Robot. Auton. Sys. **58**(8), 991–1002 (2010)

Tolman, E.C.: Cognitive maps in rats and men. Psychol. Rev. **55**, 189–208 (1948)

Thrun, S.: Simultaneous localization and mapping. In: Jefferies, M.E., Yeap, W.K. (eds.) Robotics and Cognitive Approaches to Spatial Mapping. STAR, vol. 38, pp. 13–41. Springer, Heidelberg (2008)

Thrun, S., Bücken, A.: Integrating grid-based and topological maps for mobile robot navigation. In: Proceedings of the Thirteenth National Conference on Artificial Intelligence, Portland, OR, pp. 944–950 (1996)

Thrun, S., Montemerlo, M.: The GraphSLAM algorithm with applications to large-scale mapping of urban structures. Int. J. Robot. Res. **25**(5–6), 403–429 (2006)

Wang, R.F., Spelke, E.S.: Human spatial representation: Insights from animals. Trends Cogn. Sci. **6**(9), 376–382 (2002)

Yeap, W.K.: Towards a computational theory of cognitive maps. Artif. Intell. **34**, 297–360 (1998)

Yeap, W.K.: A computational theory of human perceptual mapping. In: Proceedings of the Cognitive Science Conference, Boston, USA, pp. 429–434 (2011a)

Yeap, W.K.: How $Albot_0$ finds its way home: a novel approach to cognitive mapping using robots. Top. Cogn. Sci. **3**(4), 707–721 (2011b)

Yeap, W., Hossain, M., Brunner, T.: On the implementation of a theory of perceptual mapping. In: Wang, D., Reynolds, M. (eds.) AI 2011. LNCS, vol. 7106, pp. 739–748. Springer, Heidelberg (2011)

Yeap, W.: On egocentric and allocentric maps. In: Freksa, C., Nebel, B., Hegarty, M., Barkowsky, T. (eds.) Spatial Cognition 2014. LNCS, vol. 8684, pp. 62–75. Springer, Heidelberg (2014)

The Design and Experiment of the Leg Model Based on Galvanic Coupling Intra-Body Communication

Yueming Gao[1,2(✉)], Juan Cai[1,2], Zhumei Wu[1,2], Željka Lučev Vasić[3],
Min Du[1,2], and Mario Cifrek[3]

[1] Key Lab of Medical Instrumentation and Pharmaceutical Technology of Fujian Province,
Fuzhou, China
fzugym@163.com
[2] College of Physics and Information Engineering, Fuzhou University, Fuzhou, China
[3] Faculty of Electrical Engineering and Computing, University of Zagreb, Zagreb, Croatia

Abstract. The phantom model of the leg based on galvanic coupling IBC (intra-body communication) was established to simulate the channel characteristics of human body. The 3D reconstruction of human's leg was carried out by using the data of real man. It's outer contour was printed by the 3D printer, then filled it with the special solution to obtain the phantom model. To verify whether the phantom model can be conformity with reality, the in vivo experiment was done. The experimental results showed that the outcome of the phantom experiment was consistent with that of the in vivo experiment, while the channel length range from 4 cm to 30 cm and the frequency was at 10 kHz–500 kHz. The absolute value of the error of corresponding data was less than 7 dB. It turned out that the phantom model restructured by the real human image data could simulate the real galvanic coupling IBC channel better.

Keywords: Galvanic coupling intra-body communication · Channel modeling · 3D human body image · Phantom model

1 Introduction

IBC is a new technology for short distance communication, which is used human body as the transmission medium of the electric signal [1]. Compared with the traditional communication technology, such as cable communication, infrared, Bluetooth and ZigBee. IBC is characterized by no antenna design, low power consumption, low radiation, anti interference [2]. These characteristics can be applied to the field of remote health monitoring, and meet the need of the design of wearable medical devices [3]. IBC researches are mainly concentrated in the stage of modeling and application, the purpose of modeling is to study the nature of IBC signals, but also the basis of application.

The existing modeling of IBC was divided into two types, including the equivalent circuit model and the electric field model. The human tissue was abstracted as a two dimensional, even one-dimensional RC circuit network in the equivalent circuit model [4, 5]. Its main characteristics was simple, fast-setting, easy to achieve. But the communication performance of the circuit model was low, and could not describe the distribution of the electromagnetic field in the organism. Relatively, the electric field model can

© Springer International Publishing Switzerland 2015
A. Bikakis and X. Zheng (Eds.): MIWAI 2015, LNAI 9426, pp. 306–313, 2015.
DOI: 10.1007/978-3-319-26181-2_29

better reflect the characteristics of the human tissue. The EF model equivalent the human body to a multi-layer concentric cylinder. Simulation results were obtained by finite element method or finite difference method [6–8], and then compared with the actual measurement results of human body [9]. This approach had not fully considered the impact of real human conditions on signal transmission, and simulation environment of finite element model could not be completely consistent with the actual. Therefore, to design and make the phantom model based on the digital human is in demand, which can be test in real measurement environment for the research of IBC modeling.

This paper was based on the experimental results of a multi-layer concentric cylinder, using the data of real man to design and make the legs model in kind. The phantom model was placed in the same environment as in vivo, to observe the electrical characteristics of the model. The results were compared with the results of in vivo to verify the validity of the reconstructed model. The phantom experiment will provide an experimental reference for the further design and improvement of the finite element model.

2   The Manufacturing Method of Material Object

In the process of IBC, what influenced the signal transmission most was the skin, fat and muscle layer [10]. While the effect of periosteum and bone marrow was very small. When the frequency was near 40 kHz, the conductivity of the fat layer and the skin layer were almost the same [11], so the fat layer and the skin layer were equivalent to a layer as the outer layer. That is, the multi-layer phantom model was mainly divided into two layers, whose inner layer was muscle layer (Table 1).

Table 1.   Frequency-conductivity of each layer

Frequency (Hz) Conductivity(S/m)	10000	40000	70000	100000	150000
muscle	0.34083	0.34977	0.35579	0.36185	0.37265
fat	0.00293	0.02167	0.04475	0.06583	0.09399
skin-wet	0.02383	0.02419	0.02433	0.02441	0.02451

2.1   Reconstruction of Human Body Image

In this paper, the human body image data was used male lateral anatomy tomographic of the visible human project data set, which was established by the American NLM and Colorado university. Figure 1(a) shows that the image feature is: 1 mm spacing, 1 mm spacing, 2048 pixels × 1216 pixels, 24 bit color images, resolution of 0.33 mm × 0.33 mm, about 4 M bytes per image [12]. The step of reconstruction of the leg contour model was as follows:

(1)   The real human image data was imported into the mimics10.0 to transform the discrete 2D image into 3D sequence, then the 3D reconstruction was completed;

(2) The 3D reconstruction data was put into geomagic 2013 to realize surface fitting, because of the reconstructed data can not constitute the surface.
(3) The fitting curved surface data was substantiated in SolidWorks 2013, to obtain the outer contour of real human model.

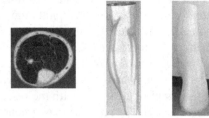

Fig. 1. (a) Cross-section slice of leg; (b) The effect figure of reconstruction; (c) The outer contour of phantom model

The leg model was selected from the reconstructed human body model, including the middle section of the knee joint, and the length was about 30 cm, as shown in Fig. 1(b). The outer contours of the inside layer and outside layer were printed by the 3D printer. The individual has differences in size. Considering the physical characteristics of the volunteer, who will participate in the experiment, the resolution of the leg model was adjusted to 0.2 mm * 0.2 mm.

2.2   The Method of Making Phantom Model

The agar, HEC (Hydroxyethyl cellulose) and distilled water were mixed as a certain ratio, in order to ensure that the material has a certain hardness, and not easy to deterioration [13]. Then a certain amount of industrial sugar and potassium chloride were added. The electrical conductivity of the model was changed by adjusting the quality of potassium chloride, and using industrial sugar to fine tune, in order to approximate the characteristics of human muscle layer and fat layer respectively. Then the biological stain called methylene blue was added to the outer layer, to distinguish between internal and external layers by the naked eye. The powder which mentioned above was dissolved in the distilled water. The mixture was heated until it boiled and formed a uniform solution.

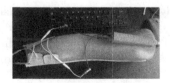

Fig. 2. Phantom model

The uniform solution was injected into the outer contour of muscular layer, standing for several hours until it become condensed. The obtained material was put into the outer contour in the corresponding position, similar to the method above to obtain the uniform solution with the stain, pumping it into the outer contour of outer layer, then a non separable solids was obtained, the phantom model was shown below (Fig. 2).

2.3   Measurement on Electrical Conductivity of Phantom Model

The method for measuring the electrical conductivity of the phantom model was indirect measurement. A length of phantom model both in the inner and outer layers were cut respectively, which were marked as sample 1 and sample 2, measured their resistance by partial pressure. Two pieces of copper sheets were placed in the ends of the sample. The function signal generator generated a signal, whose V-pp was 2 V, frequency was 40 kHz, The value of series resistance was adjusting until the voltage value of the resistance was close to 1 V, measured by the AC millivoltmeter. So that the resistance values of the outer layer and inner layer can be obtained. Then their electrical conductivity can be got by the following formula.

$$\sigma = \frac{1}{\rho} = \frac{l}{RS} \tag{1}$$

Where R the resistance of the sample, l the height of the sample, S the contact area of the sample and the copper sheet, σ the electrical conductivity (Fig. 3).

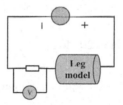

Fig. 3. Circuit model of electrical conductivity measurement

After multiple measurements, the electrical conductivity of the inner layer and the outer layer were obtained and shown in the following table (Table 2).

Table 2. Measurement on electrical conductivity of each layer.

Times Conductivity(S/m)	first	second	third	mean
Inner layer	0.2982	0.3244	0.3117	0.3114
Outer layer	0.0212	0.0225	0.0251	0.0229

3   Phantom Experiment and the Vivo Experiment

3.1   Phantom Experiment

The leg model was placed horizontall on the test bench, a pair of 4 cm × 4 cm transmitting electrodes were affixed to the surface, the distance to the bottom of the leg model was about 3 cm. At the same time, a pair of receiving electrodes were placed parallel with the transmitting electrode, then changed the channel length by moving the receiving electrode. The galvanic coupling IBC experiment block diagram was as follows (Fig. 4).

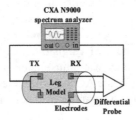

Fig. 4.   Phantom experiment block diagram

The CXA Agilent N9000 spectrum analyzer was used in the experiment. An 0dBm output signal from the extremity of TX was injected into the body through the transmitting electrode. To avoid the results being affected by the human body and the ground connecting to form a looping-in, the receiving signal was detected by the differential probe, then shown on the screen by the form of voltage gain.

$$Gain\,(\text{dB}) = 20 \cdot \log_{10}\left(\frac{V_r}{V_t}\right) \tag{2}$$

Where Vr the receiving voltage, Vt the sending voltage.

3.2   The In Vivo Experiment

In order to demonstrate the accuracy of the model, the in vivo experiment was carried out on the corresponding sites of the volunteers. The experimental procedure was the same as the phantom experiment, took 2 cm as the step size of the movement of the receiving electrode, the leg channel characteristics of 5 volunteers were measured. During the experiment, their leg was kept free force.

The results of the experiment showed that the cl-gain characteristic curves (the relationship between channel length and gain) of different individuality were very similar in shape, in the same experimental conditions. Therefore, the average value of the experimental data obtained through testing for many times, days and different volunteers was used as the final data to analyze, and the absolute value of the maximum difference was not more than 5 dB.

3.3   Comparison Results Between Phantom Experiment and Vivo Experiments

It can be found that both of the cl-gain curves and f-gain curves (the relationship between frequence and gain) of the phamtom experiment were in line with those of the in vivo experiment.

Once the signal frequency was constant, the voltage gain was gradually decreased as the increase of the channel length. The attenuation increased linearly when the channel length was relatively short. As shown in Fig. 5(a), it is obvious that there were little difference between the phantom model experiment and the in vivo experiment in the cl-gain of channel characteristics.

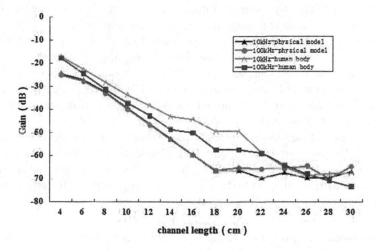

Fig. 5.   The characteristic curve of cl-gain (phantom model V·S human body)

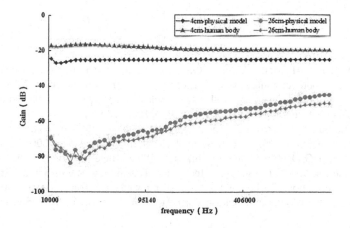

Fig. 6.   The characteristic curve of f-gain (phantom model V·S human body)

According to the Fig. 6, under different channel lengths, the f-gain curves of the phantom experiment and the in vivo experiment were consistent with each other, and the absolute value of the error was less than 7 dB. When the channel was short (such as at 4 cm), the peak of the voltage gain was reached at about 30 kHz, since that the value of voltage gain was almost unchanged. But the attenuation would decrease with the increase of frequency at long channel transmission.

4   Conclusion

In this paper, the phantom model of legs based on galvanic coupling IBC was designed which was reconstructed by using real human image data. The characteristics of the phantom model were definite through the experimental verification.

The whole regulation of phantom experiment were consistent with the in vivo experiment. In case that the frequency was fixed, the voltage gain almost decayed linearly with the increase of the channel length. Once the channel length was constant, the value of voltage gain would increase and then decrease with rising frequency. It reached the peak at 30 kHz, after that the variation trend was stationary. However, if the channel length was relatively large, the curve would have a upward trend.

It was proved that the phantom model can simulate the real human communication channel well. The next step is to expand the leg model to the entire human body, then the whole model can be used as the research object. To some degree, the study of the phantom model will give support to the application of implantable IBC and the improvement of the finite element model in the future.

Acknowledgement. This paper has been supported by the National Science Foundation of China, grant number 61201397, 51047001; International Cooperation Project of Chinese Ministry of S&T, grant number 2013DFG32530; The Project of Education Department of Fujian, grant number JA13027.

References

1. Gao, Y.M., Pun, S.H., Mai, B.Y., et al.: Quasi-static model and transceiver design for galvanic coupling intra-body communication. J. Electron. Meas. Instrument. **26**(8), 732–737 (2012)
2. Liu, Y.H., Zhang, S., Qin, Y.P., et al.: Development and prospect of implantable intra-body communication technology. Chin. Sci. Pap. **9**(1), 16–23 (2014)
3. Kibret, B., Lai, D.T.H.: Investigation of galvanic coupling intrabody communication using human body circuit model. IEEE J. Biomed. Health Inf. **18**(4), 1196–1206 (2014)
4. Zimmerman, T.G.: Personal area networks (PAN): near-field intra-body communication, in media art and science. Massachusetts Institute of Technology, Massachusetts (1995)
5. Cho, N., Yoo, J., Song, S.-J., et al.: The human body characteristics as a signal transmission medium for intrabody communication. IEEE Trans. Microw. Theory Tech. **55**(5), 1080–1086 (2007)
6. Gao, Y.M., Pan, S.H., Du, M., et al.: Quasi-static field modeling and validation for intra-body communication. In: 3rd IEEE International Conference on Bioinformatics and Biomedical Engineering, ICBBE 2009, pp. 1–4 (2009)

7. Gao, Y.M., Pun, S.H., Mak, P,U., et al.: A multilayer cylindrical volume conductor model for galvanic coupling intra-body communication. In: IEEE 7th International Conference on Information, Communications and Signal Processing, ICICS 2009, pp. 1–4 (2009)
8. Gao, Y.M., Pun, S.H., Mai, B.Y., et al.: Model establishment and analysis of feeble current coupling signal transmitting through human limb. Chin. J. Tissue Eng. Res. **15**(52), 9738–9741 (2011)
9. Song, Y., Hao, Q., Zhang, K., et al.: The simulation method of the galvanic intrabody communication with different signal transmission paths. IEEE Trans. Instr. Meas. **60**(4), 1257–1266 (2011)
10. Zeng, Z.X., Gao, Y.M., Pun, S.H., et al.: Effects of muscle conductivity on signal transmission of intra-body communications. J. Electron. Meas. Instrument. **27**(1), 21–25 (2013)
11. IFAC-CNR. http://niremf.ifac.cnr.it/tissprop
12. Wang, X., Gui, Y., Yang, P.: Visible human data set and it's application. J. Biomed. Eng. **16**(1), 109–111 (1999)
13. Gao, Y., Pun, S., Du, M., et al.: Modeling and analysis of the intra-body communication of implantable medical sensors. J. Electron. Meas. Instrument. **33**(12), 2661–2666 (2012)

AI in Bioinformatics

A Method of Motif Mining Based on Backtracking and Dynamic Programming

Xiaoli Song, Changjun Zhou, Bin Wang, and Qiang Zhang[(✉)]

Key Laboratory of Advanced Design and Intelligent Computing,
Dalian University, Ministry of Education, Dalian 116622, China
zhangq26@126.com

Abstract. Because of the complexity of biological networks, motif mining is a key problem in data analysis for such networks. Researchers have investigated many algorithms aimed at improving the efficiency of motif mining. Here we propose a new algorithm for motif mining that is based on dynamic programming and backtracking. In our method, firstly, we enumerate all of the 3-vertex sub graphs by the method ESU, and then we enumerate sub graphs of other sizes using dynamic programming for reducing the search time. In addition, we have also improved the backtracking application in searching sub graphs, and the improved backtracking can help us search sub graphs more roundly. Comparisons with other algorithms demonstrate that our algorithm yields faster and more accurate detection of motifs.

Keywords: Motif mining · Backtracking · Dynamic programming · Sub graph

1 Introduction

On completion of the human genome project, we entered the so-called post-genome era [1]. The main aim of bioinformatics in this post-genome era is to explore the functions and relationships of biological molecules in a living cell and the complex biological networks they constitute. There is increasing interest in researching complex biological networks [2]. Network motifs have been identified as the basic modules of biological function in such networks. Studies on network motifs are a major focus in functional genomics and bioinformatics research.

First described by Milo et al., a motif represents the smallest unit in a network [4–6]. Since then, many algorithms for motif mining have been proposed [7–15]. Kashtan et al. proposed an edge-sampling algorithm (ESA) [3], an improvement on the algorithm of Milo et al. for motif identification that focuses on sub graph mining. Wernicke proposed the high-efficiency ESU algorithm for motif mining, and then improved this with RAND-ESU [16], which overcomes the disadvantages of ESA. The RAND-ESU algorithm is efficient and has a software platform with a user-friendly interface for research. Hu et al. proposed an algorithm based on feature selection [17] that can accurately categorize sub graphs as isomorphic. It generates unique codes for sub graphs of any size whereby non-isomorphic sub graphs do not have the same code. Tian et al. proposed a partition method for identifying graph isomorphism [18]. Zhang et al. [19] proposed a motif mining algorithm based on compression of the network space in which the search and storage spaces for real and random graphs are

© Springer International Publishing Switzerland 2015
A. Bikakis and X. Zheng (Eds.): MIWAI 2015, LNAI 9426, pp. 317–328, 2015.
DOI: 10.1007/978-3-319-26181-2_30

compressed according to the characteristics of parity nodes. The Mnnsc [19] can detect motifs more quickly and is more stable in comparison to other algorithms.

Here we propose inclusion of dynamic programming and improved backtracking [20] in an algorithm for motif mining. In the first step of our method, all three-vertex sub graphs and their associated nodes and edges are enumerated by using the method ESU. Using a dynamic programming approach, we store these sub graphs and their associated nodes and edges as intermediate variables, and then search for (n + 1)-vertex sub graphs via reference to previously searched n-vertex sub graphs and their associated nodes and edges via improved backtracking and dynamic programming. This yields a highly efficient and accurate search among all sub graphs. Finally, we use both biological and non-biological networks to verify the performance of our algorithm.

2 Materials and Methods

2.1 Related Definitions

Sub Graph: Two graphs: G = (v, e) and $G_s = (v_s, e_s)$, if and only if $v_s \subseteq v$, $e_s \subseteq e$, and then we can call G_s is a sub graph of G.

Motif [4, 5]: There is no consensus definition of a motif. Although many recognize it as the smallest unit of a network, some researchers have other definitions of a motif. Here we use the more frequent definition of a motif. The following conditions determine whether a sub graph is a motif:

(1) The probability that the sub graph frequency in a random graph (N_{rand}) is greater than its frequency in the real graph (N_{real}) is very small, normally less than 0.01 ($P \leq 0.01$).
(2) The frequency of the sub graph in the real network is not less than a lower limit U (e.g., U = 4).
(3) The sub graph occurs more frequently in the real network than in a random network, usually expressed as $N_{real} > 1.1 \, N_{rand}$.

Z_{score}[1]: if a sub graph meets all the conditions for motif, we need to judge the importance of the sub graph in the real network. Z_{socre} is used to judge the importance of a sub graph, the greater value of the Z_{score}, the more important of the sub graph in the real network. Its expressions such as:

$$Z_{score} = \frac{N_{real} - N_{rand}^-}{std(<Nrand>)}).$$

(1)

N_{real} is the number of occurrences in the real network of the sub graph, N_{rand}^- is the average number of occurrences in the random networks of the sub graph, $std(<Nrand>)$ is the standard variance of the number of occurrences in n random networks of the sub graph.

Node In-degree and Out-degree [1]: In the graph $G(v, e)$, $v = (v_1, v_2, v_3, \ldots)$, $e = (e_1, e_2, e_3, \ldots)$, the in-degree is the number of edges pointing to a node, and the out-degree

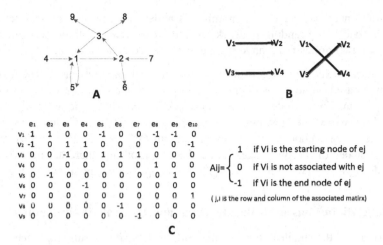

Fig. 1. A: the graph take for example B: an example of edge switching operator C: associated matrix of the graph shown in A.

is the number of edges of the node pointing to other nodes. According to this definition, for node 1 in Fig. 1A, the in-degree is 3 and the out-degree is 2.

Associated Nodes and Edges: We call a node the associated node if it is an adjacent node to a node in the sub graph, and the relevant edge is called the associated edge. To search for $(n + 1)$-vertex sub graphs, we refer to the associated nodes and edges of each n-vertex sub graph.

Associated Matrix [1]: In the first step of our algorithm, we need to store the real graph. An associated matrix is suitable for storing real graphs and is easy to use, so we chose an associated matrix for storage of real and random graphs. Consider Fig. 1A as an example. The graph has nine nodes and 10 edges, including a looped edge. In Fig. 1C, a row represents a node and a column represents an edge. The sum of all the associated matrix elements is zero; the sum of "–1" and "1" elements is equal to the number of edges, and there are the same number of "–1" and "1" elements.

Information Matrix: An information matrix has two rows. The first row stores the starting point of an edge, and the second row the end point of the edge. The number of columns is equal to the number of edges in the graph. The information matrix of the graph in Fig. 1A is shown in Table 1.

Table 1. Information matrix of the graph shown in Fig. 1A

Start-node	1	1	2	2	3	3	3	4	5	7
End-node	2	5	3	6	1	8	9	1	1	2

2.2   Random Network Generation

Random networks are an integral part of motif mining. We generate a series of random networks based on the features of a real network. These random networks meet the following requirements [22]:

(1) Random networks have same number of nodes and edges as the real network.
(2) Each node in the random network has a corresponding node in the real network, and the two nodes must have the same in-degree and out-degree.

We use an exchange algorithm to generate random networks. The exchange algorithm is based on degree sequences, and generates a random network according to a Markov chain. To generate a random network, we consider a network with directed edges and no two-way edges, and then make a series of Monte Carlo exchanges. For example, if we first choose two edges at random (V1 → V2, V3 → V4), then we change the nodes of the two edges under the condition that there are no two-way or looped edges (V1 → V4, V3 → V2) [1]. These operations are shown in Fig. 1B.

2.3   Motif Mining Based on Backtracking and Dynamic Programming

A network motif, the smallest unit of a network, is defined as a sub graph that occurs at higher frequency in the real network than in a random network. Research on network motifs can provide insight into the structural composition of biological networks and the mechanisms of action in organisms, and is highly important in bioinformatics.

To balance accuracy and complexity, we propose a new algorithm for motif mining that is based on backtracking and dynamic programming. Our approach is more efficient than other algorithms because we avoid a great deal of double counting using dynamic programming. We enumerate sub graphs to search for three-vertex sub graphs using the method ESU. We store the sub graphs identified and their associated nodes and edges as an intermediate variable, and then use dynamic programming and backtracking to search for sub graphs of other sizes. We have also improved the application of backtracking for sub graph searches. Our method identifies more sub graphs than the Mnnsc [19] does, therefore motif detection is more accurate. Dynamic invocation of pseudo code as follows:

```
Algorithm: application of dynamic programming in motif
mining
Input: all size-n (n ≥ 3) sub graph
Output: all size-n+1 (n ≥ 3) sub graph
  01 call a SubGraph_n(v,e) and its Associated_n(v',e') until call
out entirely
  02 for each v' ∈ Associated_n(v',e') do
  03 if v' ≠ v in SubGraph_n = (v,e)
  04 SubGraph_{n+1}(v,e) ←put v' and its e' into SubGraph_n(v,e)
  05 Call
Backtracksearch_sbgraph (SubGraph_{n+1}(v,e), Associated_n(v',e'))
  06 return

Backtracksearch_sbgraph (SubGraph_{n+1}(v,e), Associated_n(v',e'))
  E1 for each v' ∈ Associated_n(v',e')
  E2   if v' = v in SubGraph_{n+1}(v,e) && e' ≠ e in SubGraph_{n+1}(v,e) do
  E3     new SubGraph_{n+1}(v,e) ←put e' into SubGraph_{n+1}(v,e)
  E4 return
```

2.4   Backtracking

In traditional backtracking used in searching sub graphs, they compared each associated node with the nodes of a sub graph for only one time. Based on this, we made some improvements on backtracking. In our method, when we compare an associated node with the nodes of a sub graph (for example: an n-vertex sub graph), if the associated node is different from the nodes of the sub graph, we put the associated node in the sub graph, and we get a new sub graph (it's an n + 1-vertex sub graph) based on the n-vertex sub graph. What is different from method ESU, we will compare all of the associated nodes of the n-vertex sub graph with the nodes of the n + 1-vertex sub graph, if the associated node is same to one of the nodes of the n + 1-vertex sub graph and the associated edges is different from any edge of the n + 1-vertex sub graph, we put the associated edge in the n + 1-vertex sub graph. In this way, we can get another n + 1-vertex sub graph, and then compare the next associated node. This is what we have improved on the backtracking used in searching sub graph, and the improved backtracking can search sub graphs more roundly.

To describe the backtracking applied in our approach, we use the graph in Fig. 1A as an example and the corresponding information matrix in Table 1. In column 1 of Table 1, the start node represents the starting point of an edge and the end node represents the end of an edge. The number of columns denotes the number of edges in the graph shown in Fig. 1A. For example, in the final column, 7 is the starting node and 2 is the ending node.

We can identify 25 three-vertex sub graphs in Fig. 1A. We chose one of these sub graphs, along with its associated nodes and edges (show in Table 2), to explain how backtracking is applied. The first row in Table 2 lists the data name and the second row contains the relevant data. {V1, V2, V5} are the nodes, {E1, E2} are the edges, {V3, V4, V5, V3, V6, V7} are the associated nodes, and {E5, E8, E9, E3, E4, E10} are the associated edges of the three-vertex sub graph.

Table 2. A 3-vertex sub-graph and its associated nodes and associated edges of the graph

Nodes of the 3-vertex sub-graph	Edges of this sub-graph	Associated nodes	Associated edges
V1, V2, V5	E1, E2	V3, V4, V5, V3, V6, V7	E5, E8, E9, E3, E4, E10

Figure 2 shows how to search for four-vertex sub graphs from this three-vertex sub graph. We compare the first associated node selected from {V3, V4, V5, V3, V6, V7} with all of the nodes in the three-vertex sub graph ({V1, V2, V5}, {E1, E2}). If the associated node is not in this three-vertex sub graph, the associated node and its corresponding edge are put into the three-vertex sub graph to obtain the four-vertex sub graph ({V1, V2, V5, V3}, {E1, E2, E5}) with associated nodes {V4, V5, V3, V6, V7, V8, V9} and associated edges {E8, E9, E3, E4, E10, E6, E7}. Then we compare the associated nodes {V3, V4, V5, V3, V6, V7} and associated edges {E5, E8, E9, E3, E4, E10} of the three-vertex sub graph with the new four-vertex sub graph ({V1, V2, V5, V3}, {E1, E2, E5}). If the associated node is in the four-vertex sub graph but the corresponding associated edge is not, we put the associated edge in the four-vertex sub graph to yield a new four-vertex sub graph, ({V1, V2, V5, V3}, {E1, E2, E5, E9}). Backtracking, repetition of the second step yields two more four-vertex sub graphs, ({V1, V2, V5, V3}, {E1, E2, E5, E3}) and ({V1, V2, V5, V3}, {E1, E2, E5, E9, E3}). Backtracking to the three-vertex sub graph ({V1, V2, V5}, {E1, E2}) and repeating the above steps yields all of the four-vertex sub graphs.

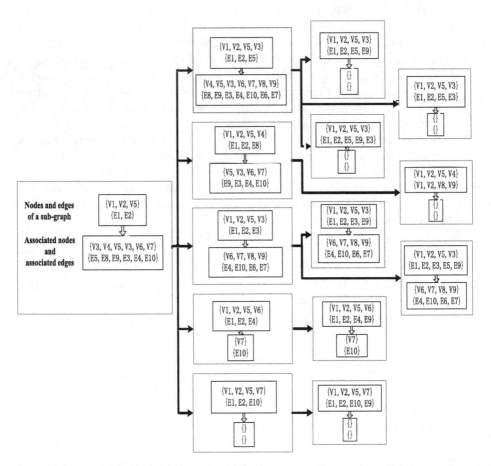

Fig. 2. An example of searching sub graphs by backtracking

3 Results and Discussion

Motif mining comprises three main subtasks: (1) identify all sub graphs of the input graph; (2) group the sub graphs into classes according to their isomorphism; and (3) distinguish sub graph classes that occur more frequently in the real graph than in random graphs [16]. In this paper, our main work is on the first subtask, searching for sub graphs in a network is the most important part of the motif mining process [19] because if sub graphs are not effectively and accurately identified, motifs cannot be correctly recognized.

We take use of the data of some real networks to test our algorithm, and you can get these data from: (http://wws.weizmann.ac.il/mcb/UriAlon/index.php?q=download/downloadable-data). The data include both biological and non-biological networks. Biological networks include: E.coli (V423, E519), sea urchin (V45, E83), protein2 (V53, E123), and protein (V95, E213), these data are gene regulatory networks, V means the number of gene, E means the relationship between the genes. Non-biological networks include: S838 (V512, E819), S420 (V252, E399), and S208 (V122, E189), these data are sequence logic electronic electric networks of ISCAS89benchmark, V means logic gates and triggers, E means the direction of the current.

3.1 Verify the Validity of Our Algorithm

In order to verify the correctness of our algorithm, we list the relevant data of motif obtained by our method in Table 3, and the Z-score data obtained by ESA are also listed in the table. In Table 3, the first column is the real networks. The second column is the number of nodes and edges of the corresponding networks. The third column is the size of motifs. The fourth column is the unique Code values represent the iso-morphic class, and the method of getting the Code values can reference paper 1. The fifth column is the number of the corresponding motif occurrences in real network. The sixth column is the topological structures of motifs, and the seventh and eighth column is the Z-score of the motifs.

According to the data shown in Table 3, we can know that most motifs are similar in the different networks, while small parts are different. If the topological structures of the motifs are the same, their Code values must be the same, for example all of the networks (E.coli, S208, S420, S838) contain the Code-2359520, the topological structures of the motif corresponds to the Code are same, and this proves the accuracy of our algorithm. Compared with the method ESA [4], the Z-score of different motifs obtained by our method are similar to that in Milo et al. [4], and this proves the correctness of our method. In addition, our method obtains the motifs in different kinds of networks, and it proves the correctness and widely feasibility of our method.

3.2 Comparison of the Number of Sub Graphs

To verify its effectiveness, we using the same hardware environment: an Intel(R) Core (TM) 2 Quad CPU Q6600 at 2.40 GHz and 3.25 GB of RAM, with a Windows 7, 32-bit operating system and VC++ 6.0 compiler. Motif is a basic function module, and it should be smaller than the module, it's not necessary to make an algorithm to search sub-graphs with too many vertexes. In our algorithm, we have only enumerate n-size sub-graphs ($3 \leq n \leq 5$), and the numbers of sub graphs identified by our method and Mnnsc [19] are shown in Table 4, respectively. Each row represents a real network, where V is the number of nodes, E is the number of edges, and n is sub graph size. Comparison the relevant data in Table 4 shows that the data for size 3 are the same

Table 3. The summary of Code values and Z-score in different networks

Networks	(Nodes,Edges)	Size-n	Code	Nreal	Motif	Z-score Our method	ESA
E.coli	(423,519)	3	-3068	42		10.747	10
		4	-2359520	520		20.460	13
S208	(122,189)	3	5668	10		8.456	9
		4	-2359520	8		10.189	3.8
			13698880	10		10.047	5
S420	(252,399)	3	5668	2		18.054	18
		4	-2359520	20		17.887	10
			13698880	22		17.792	11
S838	(512,819)	3	5668	40		38.345	38
		4	-2359520	44		25.291	20
			13698880	46		27.362	25

because our algorithm uses the same method as Mnnsc [19] to search for sub graphs of size 3. Take the protein network for example, our method finds 886 more size-4 sub graphs and 6787 more size-5 sub graphs than Mnnsc, demonstrating that our algorithm is more comprehensive in searching for sub graphs. This advantage becomes more obvious as the sub graph scale increases.

Figure 3 compares the number of sub graphs identified for the *E. coli* network. Our algorithm identified more sub graphs than the Wernicke and Mnnsc. It recognized 126 more three-vertex sub graphs than Mnnsc, and 1141 and 7328 more four-vertex sub graphs, and 62,958 and 270,751 more 5-vertex sub graphs than Mnnsc and Wernicke algorithms, respectively. Thus, our algorithm is more accurate in finding sub graphs.

Table 4. The number of sub graphs searched by our method and Mnnsc [19]

Networks	V	E	Size-3	Our method		Mnnsc [19]	
				Size-4	Size-5	Size-4	Size-5
Sea Urchin	45	83	608	7240	88849	6280	77626
Protein2	53	123	606	4504	34667	3956	27973
Protein	95	213	947	6125	39463	5239	32676
S838	512	819	2521	11282	63841	11119	59083
S420	252	399	1181	4854	24653	4775	22527
S208	122	189	533	2037	9461	2000	8611

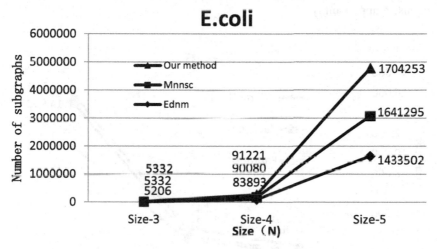

Fig. 3. Compare the number of sub graphs we have searched on E.coli network with Mnnsc [19] and Ednm [16]

Since sub graph identification is the most important step in motif mining, we can conclude that our method is more accurate for motif recognition.

3.3 Search Time Comparison

We compared search times for our method and other algorithms in searching for sub graphs in *E. coli* network. The results are listed in Table 5, time in seconds. It is clear that our method requires a shorter search time than Kavosh and Fanmod algorithms.

Figure 4 compares the search time for different algorithms. Compared to Kavosh and Fanmod, our algorithm has good time efficiency. For example, to search for three-vertex sub graphs of the *E. coli* network, our algorithm requires just 0.063 s, compared to 0.300 s for Kavosh and 0.810 s for Fanmod. Our search time is slightly longer than that of the Mnnsc, which compresses the search space using parity nodes, resulting in lower accuracy. Taking the *E. coli* network as an example, the original network has 519 nodes and 423 edges, which are reduced to 320 nodes and 256 edges

Table 5. Compare the search time of different sizes sub graphs obtained by different algorithms (data of E.coli, time in seconds)

	Size-3	Size-4	Size-5
Fanmod [21]	0.810	2.530	15.710
Kavosh [13]	0.300	1.840	14.910
Mnnsc [19]	0.016	0.266	4.141
Our method	0.063	1.435	10.561

after compression. Thus, the sub graphs are not accurately identified. This explains why the Mnnsc search time is shorter. By contrast, our algorithm achieves a better balance between time efficiency and accuracy. We can conclude that our algorithm is valid and has extensive applicability.

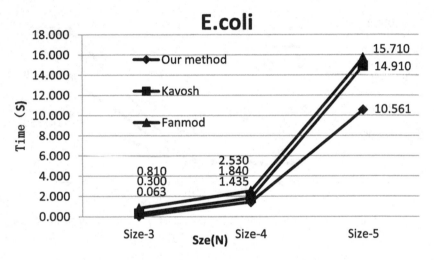

Fig. 4. Compare the search time of different algorithms

4 Conclusion

We proposed a new method for motif mining in biological networks, which is based on backtracking and dynamic programming. The method has two innovation points, one is that we put dynamic programing in the progress of searching sub graphs, this innovation point can help reduce the time of searching sub graphs; The other innovation point is that we have improved the backtracking used in the progress of searching sub graphs, and the innovation point can help us search sub graphs more roundly! Detailed steps of our method: In the first step, we enumerate all 3-vertex sub graphs through the method-ESU. Then we enumerate n + 1-vertex (n >= 3) sub graphs by calling n-vertex sub graphs we have searched before. In this way, we can enumerate all sub graphs efficiently. Experimental results confirmed the accuracy and effectiveness of our method, and demonstrated that the sub graph searches are more comprehensive. The number of sub graphs identified is greater and the search time is shorter in comparison to other algorithms. Thus, our method has wider applicability and is more efficient for sub graph searches than other algorithms.

Acknowledgments. This work is supported by the National Natural Science Foundation of China (Nos. 61425002, 61402066, 61402067, 31370778, 61370005, 31170797), the Basic Research Program of the Key Lab in Liaoning Province Educational Department (Nos. LZ2014049, LZ2015004), the Project Supported by Natural Science Foundation of Liaoning Province (No. 2014020132), the Project Supported by Scientific Research Fund of Liaoning Provincial Education (No. L2014499), and by the Program for Liaoning Key Lab of Intelligent Information Processing and Network Technology in University.

References

1. Xu, Y., Zhang, Q., Zhou, C.J.: A new method for motif mining in biological networks. Evol. Bionform. **10**, 155–163 (2014)
2. Kanehisa, M.: Post-genome Informatics, vol. 3, pp. 104–131. Oxford University Press, Oxford (2001)
3. Kashtan, N., Itzkovitz, S., Milo, S., Alon, U.: Efficient sampling algorithm for estimating sub graph concentrations and detecting network motifs. Bioinformatics **20**, 1746–1758 (2004)
4. Milo, R., Shen-Orr, S., Itzkovitz, S., Kashtan, N., Chklovskii, D., Alon, U.: Network motifs: simple building blocks of complex networks. Science **298**, 824–827 (2002)
5. Koyutürk, M., Subramaniam, S., Grama, A.: Introduction to network biology. Bioinformatics **5**, 1–13 (2011)
6. Alon, U.: Network motifs: theory and experimental approaches. Nat. Rev. Genet. **8**, 450–461 (2007)
7. Hu, H.Y., Yan, X.F.: Mining coherent dense sub graphs across massive biological networks for functional discovery. BMC Bioinformat. **21**, i213–i221 (2005)
8. Tanay, A.: Revealing modularity and organization in the yeast molecular network by integrated analysis of highly heterogeneous genome wide data. Proc. Natl. Acad. Sci. U.S.A. **101**, 2981–2986 (2004)
9. Pereira, J.B., Enright, A.J., Quzounis, C.A.: Detection of functional modules from protein interaction networks. Proteins **54**, 49–57 (2004)
10. Jiang, R., Tu, Z.D., Chen, T., Sun, F.Z.: Network motif identification in stochastic networks. PNAS **103**, 9404–9409 (2006)
11. Grochow, J.A., Kellis, M.: Network motif discovery using subgraph enumeration and symmetry-breaking. In: Speed, T., Huang, H. (eds.) RECOMB 2007. LNCS (LNBI), vol. 4453, pp. 92–106. Springer, Heidelberg (2007)
12. Alon, N., Dao, P., Hormozdiari, F.: Biomolecular network motif counting and discovery by color coding. Bioinformatics **24**, 241–249 (2008)
13. Kshani, Z., Ahrabian, H., Elahi, E., Nowzari-Dalini, A.: a new algorithm for finding network motifs. BMC Bioinform. **10**, 318 (2009)
14. Huafeng, D., Huang, Z.: Isomorphism identification of graphs: Especially for the graphs of kinematic chains. Mech. Mach. Theory **44**, 122–139 (2009)
15. Ribeiro, P., Silva, F.: G-tries: An efficient data structure for discovering network motifs. In: 25th Proceedings of the 2010 ACM Symposium on Applied Computing, pp. 1559–1565. ACM Press, Sierre (2010)
16. Wernicke, S.: Efficient detection of network motifs. Comput. Biol. **3**, 347–359 (2006)

17. Hu, J.L., Sun, L., Yu, L., Gao, L.: A novel graph isomorphism algorithm based on feature selection in network motif discovery (2011). http://www.paper.edu.cn/html/releasepaper/2011/09/56/
18. Tian, L.J., Liu, C.Q., Xie, J.Q.: A partition method for graph isomorphism. Phys. Procedia **25**, 1761–1768 (2012)
19. Qiang, Z., Xu, Y.: Motif mining based on network space compression. Biodata Min. **7**, 1–13 (2014)
20. Xie, P.: A dynamic model for processive transcription elongation and backtracking long pauses by multi subunit RNA polymerases. Proteins **80**, 2020–2024 (2012)
21. Wernicke, S., Rasche, F.: FFANMOD: A tool for fast network motif detection. Bionformatics **22**, 1152–1153 (2006)
22. Milo, R., Kastan, N., Itzkovitz, S., Newman, M., Alon, U.: Uniform generation of random graphs with arbitrary degree sequences. arXiv:cond-mat/0312028. 106, 1–4 (2003)

RNA Sequences Similarities Analysis
by Inner Products

Shanshan Xing, Bin Wang, Changjun Zhou, and Qiang Zhang[(⊠)]

Key Laboratory of Advanced Design and Intelligent Computing,
Dalian University, Ministry of Education, Dalian 116622, China
zhangq26@126.com

Abstract. According to 2-D chaos game representation of RNA secondary structures, we propose a new method of 3-D graphical representation which does not lose any biological information. Then we extract a new numerical feature which called inner products from 3-D graphical representation. It is regarded as a value to analyze the relationship between nine kinds of viruses. We find that our conclusion is almost consistent with the reported data and the realities of life. Finally, we use the method of inter-class analysis to analyze our results. The works show that our data improve 7.52 % on interclass distance and about 7 % in different class. It is easier to distinguish the different classes than previous results.

Keywords: RNA sequences · Graphical representation · Similarities degree

1 Introduction

Sequences similarities analysis is a hot bioinformatics issue, and it has great significance for the study of the evolutionary origin of species. Sequences comparison includes the following two methods: sequence alignment and non-sequence alignment.

Sequence alignment is to align the nucleotide of the sequences and make the score higher through a certain mechanism [1]. The score reflects the degree of similarity between sequences. Currently there are many sequence alignment methods mostly using dynamic programming. Needleman and Wunsch firstly proposed a sequence global alignment, namely Needleman-Wunsch algorithm which was a dynamic programming algorithm in 1970 [2], then Smith and Waterman proposed Smith-Waterman local alignment algorithm of double sequences in 1981 [3]. But the efficiency of dynamic programming algorithm increased exponentially with the number of sequences. Non-sequence alignment methods have been rapidly developed in recent year, because they could improve the computational efficiency greatly. This approach viewed the sequence as a whole and converted it to a mathematical object to be analyzed. One kind of method was converting sequences into algebraic objects, such as numerical sequences, vectors, matrices, etc., and using algebra to study the mathematical theories such as probability and statistics [11]. Another kind of method was the curve of the sequence, converting the sequence similarity analysis into curve similarity analysis comparison, such as two-dimensional representations and the three-dimensional representation of sequences [4–8]. Matrix was extracted from the graph by calculating the difference between the curves, and used to characterize differences in

© Springer International Publishing Switzerland 2015
A. Bikakis and X. Zheng (Eds.): MIWAI 2015, LNAI 9426, pp. 329–339, 2015.
DOI: 10.1007/978-3-319-26181-2_31

sequence similarity [23]. There were also using information theory methods, such as the complexity of Kolmogrov [9], Kullback-Leibler deviation method [10], the probability method [11].

DNA and RNA both have elementary structures, secondary structures and tertiary structures. Elementary structures refer to the order of the four bases. The DNA secondary structures refer to double helix structures formed by the antiparallel coiled of double helix. RNA secondary structures refer to stem-loop structures formed by RNA single-stranded folding back forms part of the base pairs and single-stranded itself. Thus RNA secondary structures are much more complex than DNA secondary structures. The DNA tertiary structures were formed by the DNA single chains and double chains folding [12]. RNA tertiary structures are formed by secondary structural motifs forming stable positioning and orientation through the interactions with each other. Ribosomal RNA, transfer RNA and messenger RNA are RNA species and their main functions are to participate in protein synthesis. Therefore, the study of RNA secondary structures can help to understand the relationship between them and proteins. There is only one function that encoded information of DNA, while RNA contains many species, for example ribozyme [13], snRNA [14] and snoRNA [15], and their functions are more than that of DNA.

There are many algorithms to analyze similarity degree of RNA secondary structures [16–19]. One class of method is based on RNA structures alignment. We usually use a distance function or a score function to insert, delete and substitute letters in the compared structures [19]. Another class of method is based on the graphical representations of sequences. This method uses graphs to represent RNA secondary structures and then derives some numerical invariants from graphs to compare RNA secondary structures [22, 23].

In this paper, we study the similarity of RNA sequence using the second class of method. RNA secondary structures can be transformed into linear sequences like DNA sequences by certain method [16–19]. Therefore we can use the graphical representation method, which deal with DNA sequences, to analyze some character of RNA secondary structures. Firstly, the RNA sequence is converted to graphics, and a new three-dimensional graphical representation is proposed on the bases of original graphical representation. Then the numerical characteristic of what we need is extracted from graphics, and we analyze the similarity of nine kinds of virus through the cross-correlation coefficient formula. Finally, we use the method of inter-class analysis to prove our results. The results show that our approach can easily extract data. It is easier to distinguish the different class than previous results. Furthermore we get rid of complicated square and radical operation in calculation. Thus our method increases efficiency of calculation.

2   3-D Representations of RNA Secondary Structure

RNA secondary structures are usually made up of free bases A, C, G, U and base pairs A–U, C–G, and A, U, G, C in pairs were represented by A', U', G', C' [20]. With this method, RNA secondary structures are transformed into elementary sequences, which called characteristic sequences [20]. For example, the substructure of AIMV-3 (see Fig. 1) corresponds to the characteristic sequence AUGCU'C'A'U'G'C'A'A AACU 'G'C'A'U'G'A'AUGCC'C'C'UAAG'G'G'AUGC (from 5' to 3') [20].

Let

$G' \rightarrow (1,0)$, $A' \rightarrow (\sqrt{2}/2, \sqrt{2}/2)$, $C' \rightarrow (0,1)$, $U' \rightarrow (-\sqrt{2}/2, \sqrt{2}/2)$

$A \rightarrow (-1,0)$, $U \rightarrow (-\sqrt{2}/2, -\sqrt{2}/2)$, $G \rightarrow (0,-1)$, $C \rightarrow (\sqrt{2}/2, -\sqrt{2}/2)$

We have eight uniformly placed on the base of a regular octagon of eight vertices.

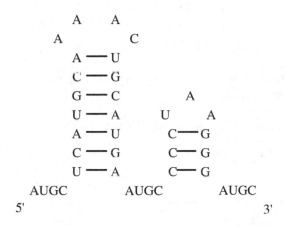

Fig. 1. The Substructure of AIMV-3

For any characteristic sequences of the RNA secondary structure $S = S_1 S_2 \ldots S_i \ldots S_N$, where N is the length of the signature sequence of RNA secondary structure. According to the CGR (chaos game representation), the recurrence formula of two-dimensional graphical representation of RNA is as follows [22]:

$$(x_i, y_i) = ((x_{i-1} + x_{s_i})/2, (y_{i-1} + y_{s_i})/2) \tag{1}$$

$i = 1,2,\ldots,n$, where s_i denotes the i th element in S, $(x_0, y_0) = (0,0)$, and x_{s_i}, y_{s_i} are calculated by the formula (2).

3-D graphical representation method subjoins one-dimensional based on the original 2-D representation approach in order to eliminate degradation of 2-D representation. For example, when nucleotides are the same in a sequence, 2-D graphical representation reciprocate in a straight line. This leads to the degradation. It degenerates to a straight line from a polyline. This doesn't lead to any kind of two-dimensional representation corresponds with the only base sequence. Our subjoined third dimensional is $1 - 1/i$ [8]. In this way, we can ensure that every base is in a column body, and not in the same plane. The third dimension values are not the same. It avoids the degradation. Each sequence is corresponding with the unique graphics. Eight RNA bases are initialized as follows:

$G' \rightarrow (1,0,0)$, $A' \rightarrow (\sqrt{2}/2, \sqrt{2}/2,0)$, $C' \rightarrow (0,1,0)$, $U' \rightarrow (-\sqrt{2}/2, \sqrt{2}/2,0)$

$A \rightarrow (-1,0,0)$, $U \rightarrow (-\sqrt{2}/2, -\sqrt{2}/2,0)$, $G \rightarrow (0,-1,0)$, $C \rightarrow (\sqrt{2}/2, -\sqrt{2}/2,0)$

$$(x_{s_i}, y_{s_i}) \begin{cases} = (1,0), & \text{if } s_i = G', \\ = (\sqrt{2}/2, \sqrt{2}/2), & \text{if } s_i = A', \\ = (0,1), & \text{if } s_i = C', \\ = (-\sqrt{2}/2, \sqrt{2}/2), & \text{if } s_i = U', \\ = (-1,0), & \text{if } s_i = A, \\ = (-\sqrt{2}/2, -\sqrt{2}/2), & \text{if } s_i = U, \\ = (0,-1), & \text{if } s_i = G, \\ = (\sqrt{2}/2, -\sqrt{2}/2), & \text{if } s_i = C. \end{cases} \tag{2}$$

Then the structural features of RNA sequences are mapped to a series of points, and their coordinates are (x_i, y_i, z_i).

$$(x_i, y_i, z_i) = ((x_{i-1} + x_{s_i})/2, (y_{i-1} + y_{s_i})/2, 1 - 1/i) \tag{3}$$

x_{s_i}, y_{s_i} are calculated by formula (2).

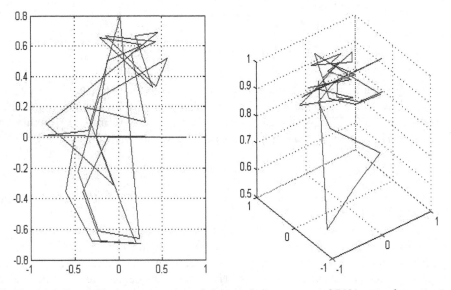

Fig. 2. (a) 2-D graphical representation of characteristic sequence of RNA secondary structure AIMV-3 (b) 3-D graphical representation of characteristic sequence of RNA secondary structure AIMV-3.

Thus, each RNA sequence is converted to a curve in three-dimensional space and each sequence corresponds to a unique curve. It can eliminate the degradation of 2-D representation. No matter how long the sequence is, its coordinates are in a positive internal octagonal column. The Fig. 2 is 2-D representations and 3-D representations of AIMV-3.

3 The Numerical Characteristics of RNA Secondary Structures

3.1 Numerical Characteristics of Correlation Coefficient

Bai et al. raised another three-dimensional graphical representation based on two-dimensional graphics of CGR ideas [20]. Let:

$A' \rightarrow (-1,-1,-1)$, $U' \rightarrow (1,-1,1)$, $G' \rightarrow (1,1,-1)$, $C' \rightarrow (-1,1,1)$
$A \rightarrow (-1,-1,1)$, $U \rightarrow (1,-1,-1)$, $G \rightarrow (1,1,1)$, $C \rightarrow (-1,1,-1)$

The 3-D representations are got as formula (4) [20].

$$R(x_{i+1}, y_{i+1}, z_{i+1}) = \frac{R(x_i, y_i, z_i) \mid S(x_{s_{i+1}}, y_{s_{i+1}}, z_{s_{i+1}})}{d} \tag{4}$$

Where, d is a non-negative real number, and $R(x_0, y_0, z_0) = 0$. Bai et al. mapped RNA secondary structures to 3-D space using the method of 3-D mapping. They mapped the bases to points in 3-D space. They described RNA secondary structures using the first 20 eigenvalues of L/L matrix as carrier. And they characterized the similarities of sequences using correlation coefficients of vectors. For an original sequence, the element in L/L matrix is the ratio of the distance between two points on the curve of the Euclidean distance and the sum of the distance. The correlation coefficients between $x_\alpha(n)$ and $x_\beta(n)$ were as follows [20]:

$$r_{\alpha\beta} = \frac{\sum\limits_{n=0}^{N-1} x_\alpha(n) x_\beta(n)}{\left[\sum\limits_{n=0}^{N-1} |x_\alpha(n)|^2 \sum\limits_{n=0}^{N-1} |x_\beta(n)|^2 \right]^{\frac{1}{4}}} \tag{5}$$

3.2 New Numerical Characteristics

The above 3-D mapping method has defect. When sequence is constituted by the same nucleotide, the graphic will be a straight line in three-dimensional space. We cannot get the unique nucleotide sequence from a graphical representation. Each graphical representation and the nucleotide sequence are not one to one. Thus we extract the numerical characteristics from the graphics of the second chapter. Using the method of the second chapter, we can get a three-dimensional numerical sequence of RNA sequences (x_i, y_i, z_i). The formula of correlation coefficient is complex. It is also worse than length-normalized inner products of vectors in terms of reflecting the relationship between two groups of data. So we prepare a new method to characterize the degree of similarity.

At first, we have all RNA sequences linked to a whole sequence, and it is seen as one long sequence. All other comparison sequences are compared with the one sequence. In this way, we can ensure that the final result is zero. If we measure distances between pairs of structures directly, the final result is not normalized.

We have graphical representation of each sequence of bases represented by 3-D coordinates. Then we can get length-normalized inner products of vectors between long sequence and each RNA sequence. Finally, Euclidean distance of each inner product is used as the final figure features to characterize the degree of similarity. Thus we extract the length-normalized inner products of vectors of numerical sequence r, which is defined using the following formula:

$$
r = \frac{\sum\limits_{n=0}^{N_Y-N_X} \sum\limits_{i=1}^{N_X} h_i^X h_{i+n}^Y}{N_X(N_Y-N_X)}, N_X < N_Y \tag{6}
$$

Where r is the number of length-normalized inner products, and N_X represents the length of the two sequences, N_Y represents all sequences to be compared to the overall connection, h_i^X is the coordinate of the i th base on x sequence.

Each RNA sequence is regarded as a small signal, and all connected into a long sequence is regarded as a large signal. We obtain the length-normalized inner products between every small signal with the large signal. This is the idea of the wavelet transforming, which can effectively reflect the location and frequency of large-signal information. It can extract useful information from large signal by translation and transformation of small signal. It is the length-normalized inner products between small signal and large signal that we extract. In the formula of the length-normalized inner products, we translate the small signal and compare with large signal. The resulting data can comprehensively reflect the correlation between small signal and large signal.

4   The Result of 9 Kinds of RNA Viruses

To demonstrate the feasibility of the method we proposed, we use nine different kinds of RNA viruses to analyze. The RNA secondary structures are converted into a series of coordinates using the formula (3). Then the coordinates are taken into the formula (6), and we get a series of length-normalized inner products. Finally, we calculate the Euclidean distance between the length-normalized inner products. The Euclidean distance shows that the more similar RNA secondary structures are, the smaller their similarity coefficients are. We list length-normalized inner products in Table 1 and similarities coefficient in Table 2.

Table 1. The data of RNA length-normalized inner products.

Species	AIMV-3	CiLRV-3	TSV-3	CVV-3	APMV-3	LRMV-3	PDV-3	EMV-3	AVII
Data	0.9244	0.9547	0.9459	0.9473	0.9145	0.9486	0.9235	0.9515	0.9562

Obviously RNA secondary structure is the most similar to itself, and its value is 0 which is the lowest value in similarity matrix. Therefore 0 can be used as a judgment of RNA secondary structures numerical boundary value of similarity degree [24]. The smaller Euclidean distance is, the more similar two kinds of RNA viruses are.

Table 2. Data of RNA similarity matrix

Species	AIMV-3	CiLRV-3	TSV-3	CVV-3	APMV-3	LRMV-3	PDV-3	EMV-3	AVII
AIMV-3	0	0.0303	0.0215	0.0229	0.0099	0.0242	0.0009	0.0271	0.0318
CiLRV-3		0	0.0088	0.0074	0.0402	0.0061	0.0312	0.0032	0.0015
TSV-3			0	0.0014	0.0314	0.0027	0.0224	0.0056	0.0103
CVV-3				0	0.0328	0.0013	0.0238	0.0042	0.0089
APMV-3					0	0.0341	0.0090	0.0370	0.0417
LRMV-3						0	0.0251	0.0029	0.0076
PDV-3							0	0.0280	0.0327
EMV-3								0	0.0047
AVII									0

We can find that AVII, CiLARV-3, LRMV-3, EMV-3 and CVV-3 are similar with each other from Table 2, and they are similar to TSV-3 in some certain. We can see them as a class. The others are regarded as difference class. Further AIMV-3, APMV-3 and PDV-3 also have a high degree of similarity. These conclusions are not only consistent with the realities of life, but also consistent with that reported in the literature [17–19, 21]. In reference [20], the result demonstrates that EMV-3 and CVV-3 are similar pairs, and then PDV-3 and APMV-3, EMV-3 and AVII, LRMV-3 and AVII, LRMV-3 and EMV-3, and CVV-3 and AVII are similar. While the result in other references is CiLRV-3 and TSV-3 are similar with them. The data in Table 3 come from Reference [17], while the data in Table 4 come from Reference [21]. The larger differences of similarity matrix elements contribute to the following cluster analysis, which show the advantage of our method.

Table 3. RNA similarity matrix reported in [17]

Species	AIMV-3	CiLRV-3	TSV-3	CVV-3	APMV-3	LRMV-3	PDV-3	EMV-3	AVII
AIMV-3	0	0.5439	0.3790	0.4862	0.2901	0.5227	0.2042	0.5766	0.6607
CiLRV-3		0	0.2275	0.0699	0.5620	0.2465	0.6339	0.1002	0.1665
TSV-3			0	0.2166	0.3790	0.3083	0.4293	0.3144	0.3767
CVV-3				0	0.5156	0.1994	0.5937	0.1222	0.1806
APMV-3					0	0.4594	0.2042	0.6366	0.6607
LRMV-3						0	0.6002	0.2917	0.2359
PDV-3							0	0.7001	0.7638
EMV-3								0	0.1492
AVII									0

For convenience, we use AVII and eight other viruses as the objects to analyze our data. From our data, AVII, CiLARV-3, CVV-3, LRMV-3, EMV-3 and TSV-3 are regarded as a class because they are quite similar, and other things are regarded as a class in that they do not have a similarity. Our data is easier to be extracted than the data reported in [17, 21]. Then we analyze the data by the inter-class analysis method.

The minimum in the Table 5 refers to minimum in their class in the similarity degree matrix, and the maximum refers to the maximum. For example, in Table 1, we

Table 4. RNA similarity matrix reported in [21]

Species	AIMV-3	CiLRV-3	TSV-3	CVV-3	APMV-3	LRMV-3	PDV-3	EMV-3	AVII
AIMV-3	0	0.3294	0.3467	0.4789	0.0296	0.5067	0.1160	0.5177	0.5320
CiLRV-3		0	0.0185	0.1523	0.3007	0.1782	0.2138	0.1890	0.2032
TSV-3			0	0.1341	0.3180	0.1603	0.2309	0.1712	0.1854
CVV-3				0	0.4504	0.0318	0.3630	0.0422	0.0560
APMV-3					0	0.4780	0.0874	0.4890	0.5033
LRMV-3						0	0.3907	0.0119	0.0258
PDV-3							0	0.4018	0.4161
EMV-3								0	0.0143
AVII									0

can see AVII, CiLARV-3, CVV-3, LRMV-3, EMV-3 and TSV-3 as a class. We can find that their corresponding values in the matrix are 0.0015, 0.0103, 0.0089, 0.0076, 0.0047, and the minimum is 0.0015 while maximum is 0.0103. Percentage is calculated as follows:

$$percentage = \frac{max - min}{max} \times 100\% \tag{7}$$

It can be seen that the smaller the percentage from the formula is, the smaller the gap between values are. In the Table 5 the percentage can reflect the difference of the same class. The closer genetic relationship of species is, the smaller the percentage is. In the same class, the smaller the species difference is, the more easily we distinguish the same type of species. Thus the smaller the percentage is, the better our proposed algorithm is. The percentage in our data improves 7.52 % compared with the data reported in [21] in the sáme class. But the percentage in normalized data is worse than that the data reported in [17] in same class.

Table 5. Data comparison in same class

Data	Minimum	Maximum	Percentage
Our data	0.0015	0.0103	85.44 %
Reference [17]	0.1492	0.3767	60.39 %
Reference [21]	0.0143	0.2032	92.96 %

In Table 6 the data in second column refer to the average in same class, while the data in the third column refer to the average out of same class. For example, in Table 2, AVII, CiLARV-3, CVV-3, LRMV-3, EMV-3 and TSV-3 are in same class, and their corresponding value in the matrix are 0.0015, 0.0103, 0.0089, 0.0076 and 0.0047. Their average is 0.0066. While 0.0318, 0.0417, 0.0327 are out of the same class and their average is 0.0354. Percentage is calculated as follows:

$$percentage = \frac{out\ of\ same\ class - same\ class}{out\ of\ same\ class} \times 100\% \tag{8}$$

Table 6. Data comparison in different class

Data	Same class	Out of same class	Percentage
Our data	0.0066	0.0354	81.36 %
Reference [17]	0.2218	0.6951	68.09 %
Reference [21]	0.0969	0.4832	79.97 %

It can be seen that the greater the percentage difference from the formula is, the greater the gap between values are. Between the same class and other classes, the greater the difference is, the easier it will be distinguished. The father genetic relationship of species is, the greater the percentage is. The greater percentage difference is, the better our proposed algorithm is. Compared with the data reported in [17], the percentage in our data increases by 13.27 % in different class. The percentage in our data improves 1.39 % than the data in [21] in different class. Our data is improved compared with that reported in [17, 21] in different class.

From the above analysis, we clearly conclude that our data have an advantage. It is relatively quick to identify our data in the same class. On the one hand, we use the idea of the wavelet transforming, which can effectively reflect the location and frequency of large-signal information. In our approach, it is applied to obtaining the length-normalized inner products of vectors between small signal and large signal. It can comprehensively reflect the correlation between small signal and large signal. On the other hand, in the process of extracting numerical feature, addition and multiplication operations are only used in formula (6). In the reference [20] it used time-consuming square and radical operations. The efficiency of addition and multiplication operations is higher than that of square and radical operations. In the calculation of computer, space and time of a radical operation are more than multiplication and addition operations. In the process of execution, it saves space and time. In reference [20], they first calculate the L/L matrix, and later re-calculating the similarity coefficient. But we calculate length-normalized inner products of vectors directly, which also saves time and space. Our method is relatively simple in the process of calculation.

While it also has shortcomings, for example, the data in [17] is better than that using our method in same class. It is greater than that in [17] in same class. We can see that our approach is not perfect.

5   Conclusions

Firstly, according to CGR idea we propose a new 3-D graphical representation which can eliminate the degradation and make RNA secondary structure correspond with the graphical representation. Secondly, we propose a new value feature - the length-normalized inner products of vectors to characterize RNA sequence similarity. Finally, we apply it to analyze the example of 9 kinds of virus and achieve relatively valuable results. It can verify the reasonableness of the similarity metric. In the graphical representation, the three-dimensional representation can eliminate degradation. In the process of extracting characteristic features, the calculation of the length-normalized inner products of vectors formula increases efficiency. We also apply the idea of the wavelet transforming to it. It can be more fully reflect the relevant information. There is a disadvantage that the percentage in same class is worse than that reported in [17]. Our approach is not perfect. Thus study on value characteristics is needed in the future.

Acknowledgement. This work is supported by the National Natural Science Foundation of China (Nos. 61425002, 61402066, 61402067, 31370778, 61370005, 31170797), the Basic Research Program of the Key Lab in Liaoning Province Educational Department (Nos. LZ2014049, LZ2015004), the Project Supported by Natural Science Foundation of Liaoning Province (No. 2014020132), the Project Supported by Scientific Research Fund of Liaoning Provincial Education (No. L2014499), and by the Program for Liaoning Key Lab of Intelligent Information Processing and Network Technology in University.

References

1. Sun, X., Lu, Z.H., Xie, J.M.: Bioinformatics Basis. Qinghua University Press, Beijing (2005)
2. Needleman, S.B., Wunsch, C.D.: A general method applicable to the search for similarities in the amino acid sequence of two proteins. J. Mol. Biol. **48**, 443–453 (1970)
3. Smith, T.F., Waterman, M.S.: Identification of common molecular subsequences. J. Mol. Biol. **147**, 195–197 (1981)
4. Yao, Y.H., Dai, Q., Ling, L., Nan, X.Y., He, P.A., Zhang, Y.Z.: Similarity/dissimilarity studies of protein sequences based on a new 2D graphical representation. J. Com. Chem. **31**, 1045–1052 (2010)
5. Liu, Z., Liao, B., Zhu, W., Huang, G.: A 2D graphical representation of DNA sequence based on dual nucleotides and its application. Int. J. Quantum Chem. **109**, 948–958 (2009)
6. Yao, Y.H., Dai, Q., Nan, X.Y., He, P.A., Zhang, Y.Z.: Analysis of similarity/dissimilarity of DNA sequences base on a class of 2D graphical representation. J. Com. Chem. **29**, 1632–1639 (2008)
7. Randic, M., Vracko, M., Lers, N.: Analysis of similarity/dissimilarity of DNA sequence base on novel 2D graphical representation. Chem. Phys. Lett. **371**, 202–207 (2003)
8. Tang, X.C., Zhou, P.P., Qiu, W.Y.: On the similarity/dissimilarity of DNA sequences based on novel 4D graphical representation. Chinese. Sci. Bull. **55**, 701–704 (2010)
9. Li, M., Badger, J.H., Chen, X., Kwong, S., Kearney, P., Zhang, H.Y.: An information-based sequence distance and its application to whole mitochondrial genome phylogeny. Bioinformatics **17**, 149–154 (2001)
10. Wu, T.J., Hsiech, Y.C., Li, L.A.: Statistical measures of DNA sequence dissimilarity under Markov chain models of based composition. Biometrics. **57**, 441–448 (2001)
11. Pham, T.D., Zuegg, J.: A probabilistic measure for alignment free sequence comparison. Bioinformatics **20**, 3455–3461 (2004)
12. Jeong, B.S., Bari, A.G., Reaz, M.R., Jeon, S., Lim, C.G., Choi, H.J.: Codon-based encoding for DNA sequence analysis. Methods **67**, 373–379 (2014)
13. Sczepansiky, J.T., Joyce, G.F.: A cross-chiral RNA polymerase ribozyme. Nature **515**, 440–442 (2014)
14. Karunatilaka, K.S., Rueda, D.: Post-transcriptional modifications modulate conformational dynamics in human U2-U6 snRNA complex. RNA **20**, 16–23 (2014)
15. Branddis, K.A., Gale, S., Jinn, S., et al.: Box C/D small nucleolar RNA (snoRNA) U60 regulates intracellular cholesterol trafficking. J. Mol. Biol. **288**, 35703–35713 (2013)

16. Yao, Y.H., Wang, T.M.: A 2D graphical representation of RNA secondary structure and the analysis of similarity/dissimilarity based on it. J. Chem. **26**, 1339–1346 (2005)
17. Zhang, Y., Qiu, J.: Comparing RNA secondary structures based on 2D graphical representation. Chem. Phys. Lett. **458**, 180–185 (2008)
18. Liu, L.W., Wang, T.M.: On 3D graphical representation of RNA secondary structure and their applications. J. Mol. Biol. **42**, 595–602 (2007)
19. Feng, J., Wang, T.M.: A 3D graphical representation of RNA secondary structure based on chaos games representation. Chem. Phys. Lett. **454**, 355–361 (2008)
20. Bai, F.L., Li, D.C., Wang, T.M.: A new mapping rules for RNA secondary structures with its applications. J. Math. Chem. **43**, 932–942 (2008)
21. Tian, F.C., Wang, S.Y., Wang, J., Liu, X.: Similarity analysis of RNA secondary structures with symbolic dynamics. J. Com. Res. Develop. **50**, 445–452 (2013)
22. Guo, Y., Wang, T.M.: A new method to analyze the similarity of the DNA sequences. J. Mol. Struc-THEOCHEM **853**, 62–67 (2008)
23. Wang, S.Y., Tian, F.C., Qiu, Y., Liu, X.: Bilateral similarity function: a novel and universal method for similarity analysis of biological sequences. J. Theor. Biol. **265**, 194–201 (2010)
24. Liu, X., Tian, F.C., Wang, S.Y.: Analysis of similarity/dissimilarity of DNA sequences based on convolutional code model. Nucleos. Nucleot. Nucl. **29**, 123–131 (2010)

A Combining Dimensionality Reduction Approach for Cancer Classification

Lijun Han, Changjun Zhou, Bin Wang, and Qiang Zhang[✉]

Key Laboratory of Advanced Design and Intelligent Computing, Dalian University,
Ministry of Education, Dalian 116622, China
zhangq26@126.com

Abstract. Because the original gene microarray data has many characteristics such as high dimension and big redundant, which is not good at classification and diagnosis of cancer. So it is very important to reduce the dimensionality and identify genes which contribute most to the classification of cancer. A method of dimensionality reduction based on the combination of mutual information and PCA is proposed in this paper. We adopted the SVM as the classifier in the experiment to evaluate the effectiveness of our method. The experimental results prove that the proposed method is an effective method for dimensionality reduction which can get very small subset of features and lead to a better classification performance.

Keywords: Gene expression data · Feature selection · Mutual information · PCA

1 Introduction

In recent years, microarray gene expression data analysis has become an effective auxiliary measure for disease diagnosis and treatment, particularly for cancer [1]. However, these microarray data are usually high dimensional data and most of these original genes may be irrelevant for cancer classification problem, which reduce the classification performance. So it is necessary to reduce the dimensionality of original data and select the useful genes for cancer classification before analyzing the gene expression data. It mainly dimensionality reduction includes two kinds of methods, feature selection and feature extraction [2].

Feature selection can quickly screen out the important genes that contribute most to the cancer classification. However it does not consider interaction within genes. For example: t-statistic [3], S2N [4], Mutual information [5, 6] etc. Among of them, Mutual information is to measure the correlation between features and categories and remove redundant information. It has achieved broadly attention in recent years, for example, Novovičová [5] presented two algorithms for feature (word) selection based on the improved mutual information for the purpose of text classification. Marohnic [6] presented an attribute filter based on the mutual information relevance measure. Feature extraction can not only reduce the dimensionality but also check useful variables implied in genes datasets, but it is high complexity calculation. The most classical method includes: ICA [7], LDA [8], PCA and

© Springer International Publishing Switzerland 2015
A. Bikakis and X. Zheng (Eds.): MIWAI 2015, LNAI 9426, pp. 340–347, 2015.
DOI: 10.1007/978-3-319-26181-2_32

KPCA [9, 10] etc. Because of the effectiveness of PCA in dimensionality reduction, it has been widely used in the fields of face recognition and image compression. Clausen [9] proposed a new method for encoding color images based upon principal component analysis. Gottumukkal [10] proposed an improved face recognition technique based on modular PCA approach. In addition, there was a lot of other combinatorial optimization methods has been developed, Lu [11] proposed a filter-based gene selection algorithm by combining information gain and genetic algorithms; Luo [12] proposed a mixed two-step feature selection method by a modified t-test method and Principal component analysis; Han [13] proposed a novel two-stage cancer classification method by combining the BSS/WSS ratio mechanism and SLPP.

Considering the advantages and disadvantages of feature selection and feature extraction, we proposed a combined approach of dimensionality reduction based on mutual information and PCA. This method first measure the size of the correlation between feature genes and classes by using mutual information, and then screen out some strong correlation feature genes. Secondly, in order to remove the redundancy and find out the characteristic variable implicit in datasets, we use the PCA to do further processing with the feature gene subsets. At last, we use the SVM as the classifier to measure the performance of the proposed method, the experiment is taken on the three publically available microarray dataset demonstrate that the proposed method can effectively improve the recognition accuracy.

2   Algorithm Description

2.1   Mutual Information

Mutual information (MI) [14] shows the relationship of the two variables. Suppose the two random variables X and Y, if their marginal probability distributions are $p(x)$ and $p(y)$, and their joint probability is $p(x, y)$, the $I(X;Y)$ is defined as:

$$I(X;Y) = \sum_{x \in X} \sum_{y \in Y} P(x, y) \log \left(\frac{p(x, y)}{p(x)p(y)} \right) \tag{1}$$

Moreover, mutual information can be expressed as the form of information entropy:

$$I(X;Y) = H(X) + H(Y) - H(X, Y) \tag{2}$$

$H(X)$, $H(Y)$ are respectively represented information entropy of variable X, Y. The $H(X, Y)$ represents the joint entropy of variable X and Y. Inside,

$$H(X) = - \sum_{x \in X} p(x) \log p(x) \tag{3}$$

$$H(X, Y) = - \sum_{x} \sum_{y} p(x, y) \log p(x, y) \tag{4}$$

By the above formulas, when the variable X and Y are mutual independence or entirely irrelevant, the mutual information is zero. On the contrary, the stronger their dependency is, the mutual information will be higher.

2.2 PCA

PCA maps the sample data in high-dimension to low-dimensional space through linear transformation. There is an original sample matrix X, and PCA reduces the dimension by constructing a new $p \times m\,(p < n)$ characteristic matrix Y instead of the original sample matrix. PCA algorithm steps as below:

(1) Data centralization processing.

$$x'_{ij} = x_{ij} - \frac{1}{m}\sum_{j=1}^{m} x_{ij} \tag{5}$$

(2) Calculate the covariance matrix by new sample matrix X'.

$$S = \frac{X'^T X'}{m} \tag{6}$$

(3) Calculate all the eigenvalues λ_1, λ_2, ..., λ_n and corresponding eigenvectors $e_1, e_2, ..., e_n$ of covariance matrix S.
(4) Rank the eigenvalues in descending order, and the eigenvectors changes relevant.
(5) Pick up p pcs of former eigenvectors $e_1, e_2, ...,e_p$ to constitute of projection matrix E, and then do projection on original sample matrix to get the reduction dimensional sample matrix.

3 Methods

In order to make microarray data better for cancer classification, the processed data should have two features: the dimension is small and contains as much information as possible. The feature selection based on mutual information can effectively remove the noise genes and screen out useful genes. The feature extraction based on PCA can reduce the data dimension as far as possible on the basis of guaranteeing basic invariable of the primitive variable information. The proposed method is divided into three stages, the main steps as shown in Fig. 1.

The first stage is feature selection based on mutual information. Take G_1 as an example, the following introduces the solving process of mutual information of genes G_1 and classes C. Firstly, estimate marginal probability distribution of gene G_1 and joint probability distribution of gene G_1 and classes C. $X_1 = (x_{11}, x_{12}, ... , x_{1m})$ is the expression values of gene G_1 in samples. To divide the value rang of X_1 into 15 disjoint intervals of the same distance. $d_i(x)$ is the number of X_1 in the i th small interval. $p_i(x)$ equals $d_i(x)$ divided by m. So we can get the marginal probability distribution of gene G_1 and joint probability distribution of gene G_1 and classes C. Secondly, based on the formula (3) and (4), get the $H(G_1)$, $H(C)$ and $H(G_1, C)$. Thirdly, according to the formula (2), compute the mutual information $I(G_1;C)$.

In the gene expression data, select the genes that have large correlation with disease classification by use the mutual information, that is to say select the former k genes as the primary subset. Its dimension is k.

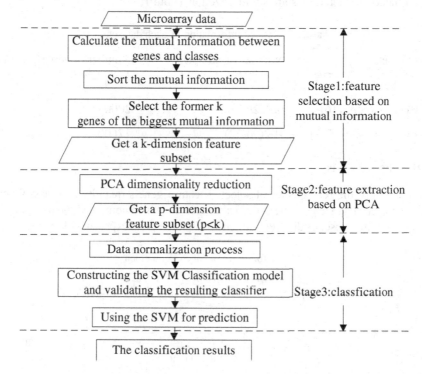

Fig. 1. The flowchart of combining dimensionality reduction for cancer classification

The second stage is the feature extraction based on PCA, PCA is used to further reduce the dimensionality for primary subset. In the gene expression data, we make gene expression matrix as the original sample matrix X, then each row is a feature data and each column is a sample. Finally get the new p-dimensional feature subset after projecting sample matrix.

The third stage is the data classification. We have already done data centralization processing when using PCA. In order to better analyze the data, the feature subset is normalized by using [0, 1] range before the data classification, and then using the SVM as the classifier for the feature subset's classification recognition, gets the corresponding classification accuracy.

4   Experiments and Results Analysis

In the paper, LS-SVM tool [15] as the classify tool to verify feature gene's performance of recognize sample classification, we choose linear kernel function as kernel function, set the parameters $gam = 0.5$ and $sig2 = 1$. The leave-one-out cross-validation is used in the classification performance evaluation process.

4.1   Data Sets

In this paper, three publicly available microarray data sets are applied to our experiment and the related characteristics are summarized in Table 1.

Table 1. Related information of each dataset.

Dataset	Number of Samples(+/−)	Number of genes
Leukemia [4]	72(47/25)	7129
Colon cancer [16]	62(22/40)	2000
Prostate cancer [17]	102(50/52)	12600

4.2   Results Analysis

In order to study the relationship of selected feature gene's quantity and classification recognition accuracy, we give the classification results which combined the primary gene subset and feature subset of different dimension in three gene expression datasets.

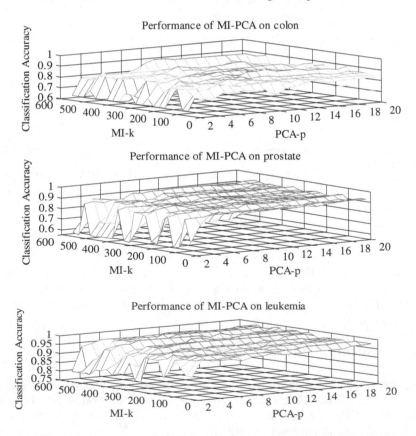

Fig. 2. Performance of MI-PCA on the three gene expression datasets.

In Fig. 2, the MI-k (MI-k = 25,50,75,100,...,550) is the dimension of primary gene subset screened by mutual information, PCA-p (PCA-p = 2,3,4,...,20) is the dimension of feature subset by use of PCA. MI-PCA is the method that we proposed. From blue to red indicates the accuracy rate steeled. We can see the accuracy changes according to the number of screened genes. When PCA-p less than 5, the accuracy rate has a larger fluctuation rang, but all revealed a trend of fast increase and then tend to steady and the change trend of accuracy on the three data sets are basically identical. When MI-k is equal to 500 and PCA-p is equal to 10, the proposed method can get the highest accuracy value on leukemia, colon, prostate, it is 98.61 %, 93.55 %, 94.12 %. The results verifies the rationality and certain universality of the MI-PCA algorithm, it can replace the original genetic data with small feature subset, and get higher classification accuracy.

In order to research the advantage of MI-PCA, we compared with MI and PCA on the three cancer datasets. The result is shown in Fig. 3, when the dimension of feature gene is greater than 9, MI-PCA algorithm is superior of MI and PCA, the accuracy reached the maximum and showed the steady state. This shows MI-PCA algorithm can take the advantage of these two kinds of algorithm, to extract more effective the classification of information to improve the efficiency of classification.

In additional, in order to study the performance of the proposed method, we compared the results with previously published method. Table 2 shows the results of the contrast. On the leukemia dataset, the classification accuracy obtained by the

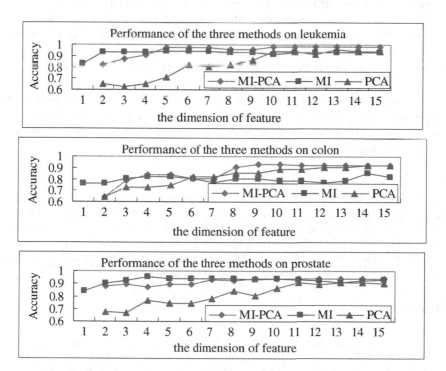

Fig. 3. Performances of the three methods on the three gene expression datasets.

Table 2. Reference comparison results.

Dimensionality reduction method	Leukemia	Colon	Prostate
IG+GA [11]	98.61	91.94	–
ERGS [18]	97.22	83.87	93.14
BSS/WSS ratio mechanism+SLPP [13]	96.8	82.8	–
Proposed method	98.61	93.55	94.12

proposed method same as reference [11], but besides that, the classification accuracy obtained by the proposed method is higher than others. The results prove that the proposed method is better than the existing method, which can get a higher classification accuracy rate (Table 2).

5   Conclusion

In this paper, by considering the advantages and disadvantage of those two methods of the feature selection and extraction, a novel combined approach of dimensionality reduction based on mutual information and PCA was proposed. The results show the proposed method can replace the origin gene data by the smaller gene subset and lead to a better classification performance after LISSVM classification. In addition, the proposed method does not rely on any of the classifiers, and it can be widely used in the high-dimensional data classification problem. Compared with other previously published methods, the MI-PCA algorithm has advantage in classification accuracy.

Although the proposed method can get good classification performance, but the specific genes associated with the cancer category could not be clearly found, this is because the new feature extracted by PCA algorithm is a linear combination of the original features. After that, we strive to work from the point of biological significance, to obtain more valuable features.

Acknowledgements. This work is supported by the National Natural Science Foundation of China (Nos. 61425002, 61402066, 61402067, 31370778, 61370005, 31170797), the Basic Research Program of the Key Lab in Liaoning Province Educational Department (Nos. LZ2014049, LZ2015004), the Project Supported by Natural Science Foundation of Liaoning Province (No. 2014020132), the Project Supported by Scientific Research Fund of Liaoning Provincial Education (No. L2014499), and by the Program for Liaoning Key Lab of Intelligent Information Processing and Network Technology in University.

References

1. Yuvaraj, N., Vivekanandan, P.: An efficient SVM based cancer classification with symmetry non-negative matrix factorization using gene expression data. In: International Conference on Information Communication and Embedded Systems, pp. 761–768. IEEE Press, Chennai (2013)

2. Su, Y., Wang, R., Li, C., Chen, P.: A dynamic subspace learning method for tumor classification using microarray gene expression data. In: 7th International Conference on Natural Computation, pp. 396–400. IEEE Press, Shanghai (2011)
3. Jafari, P., Azuaje, F.: An assessment of recently published gene expression data analyses: reporting experimental design and statistical factors. BMC Med. Inform. Decis. **6**(1), 27 (2006)
4. Golub, T.R., Slonim, D.K., Tamayo, P., et al.: Molecular classification of cancer: class discovery and class prediction by gene expression monitoring. Science **286**(5439), 531–537 (1999)
5. Novovičová, J., Malík, A., Pudil, P.: Feature selection using improved mutual information for text classification. In: Fred, A., Caelli, T.M., Duin, R.P.W., Campilho, A.C., Ridder, D. (eds.) SSPR&SPR 2004. LNCS, vol. 3138, pp. 1010–1017. Springer, Heidelberg (2004)
6. Marohnic, V., Debeljak, Z., Bogunovic, N.: Mutual information based reduction of data mining dimensionality in gene expression analysis. In: 26th International Conference on Information Technology Interfaces, pp. 249–254. IEEE Press, Cavtat (2004)
7. Huang, D.S., Zheng, C.H.: Independent component analysis-based penalized discriminant method for cancer classification using gene expression data. Bioinformatics **22**(15), 1855–1862 (2006)
8. Wang, S.-L., You, H.-Z., Lei, Y.-K., Li, X.-L.: Performance comparison of tumor classification based on linear and non-linear dimensionality reduction methods. In: Huang, D.-S., Zhao, Z., Bevilacqua, V., Figueroa, J.C. (eds.) ICIC 2010. LNCS, vol. 6215, pp. 291–300. Springer, Heidelberg (2010)
9. Clausen, C., Wechsler, H.: Color image compression using PCA and back propagation learning. Pattern Recogn. **33**(9), 1555–1560 (2000)
10. Gottumukkal, R., Asari, V.K.: An improved face recognition technique based on modular PCA approach. Pattern Recogn. Lett. **25**(4), 429–436 (2004)
11. Lu, H.J.: A Study of Cancer Classification Algorithms Using Gene Expression Data. Xuzhou, China (2012)
12. Luo, W., Wang, L., Sun, J.: Feature selection for cancer classification based on support vector machine. In: WRI Global Congress on Intelligent Systems, pp. 422–426. IEEE Press, Xiamen (2009)
13. Zhu, L., Han, B., Li, L., et al.: A novel two-stage cancer classification method for microarray data based on supervised manifold learning. In: 2nd IEEE International Conference on Bioinformatics and Biomedical Engineering, pp. 1908–1911. IEEE Press, Shanghai (2008)
14. Shannon, C.E.: A mathematical theory of communication. In: 5th ACM SIGMOBILE Mobile Computing and Communications Review, New York, pp. 3–55 (2001)
15. LS-SVM toolbox Download. http://www.esat.kuleuven.be/sista/lssvmlab
16. Alon, U., Barkai, N., Notterman, D.A., et al.: Broad patterns of gene expression revealed by clustering analysis of cancer and normal colon tissues probed by oligonucleotide arrays. P. Natl. Acad. Sci. **96**(12), 6745–6750 (1999)
17. Singh, D., Febbo, P.G., Ross, K., et al.: Gene expression correlates of clinical prostate cancer behavior. Cancer Cell **1**(2), 203–209 (2002)
18. Chandra, B., Gupta, M.: An efficient statistical feature selection approach for classification of gene expression data. J. Biomed. Inform. **44**(4), 529–535 (2011)

AI in Security and Networks

Parallel Multi-label Propagation
for Overlapping Community Detection
in Large-Scale Networks

Rongrong Li[1], Wenzhong Guo[1], Kun Guo[1(✉)], and Qirong Qiu[2]

[1] College of Mathematics and Computer Science,
Fuzhou University, Fuzhou 350108, China
{ditaps, fzugwz, guknl23}@163.com
[2] School of Economics & Management,
Fuzhou University, Fuzhou 350108, China
qqrkyc@fzu.edu.cn

Abstract. In recent years, with the rapid growth of network scale, it becomes difficult to detect communities in large-scale networks for many existing algorithms. In this paper, a novel Parallel Multi-Label Propagation Algorithm (PMLPA) is proposed to detect the overlapping communities in networks. PMLPA employs a new label updating strategy using ankle-value in the label propagation procedure during each iteration. The new algorithm is implemented in the Spark framework for its power in distributed parallel computation. Experiments on artificial and real networks show that PMLPA is effective and efficient in community detection in large-scale networks.

Keywords: Community detection · Label propagation · Large-scale network · Overlapping community

1 Introduction

Community structure is found to be a common characteristic in many real networks, such as social networks, biological networks, etc. [1]. Community structure refers to the occurrence of groups of nodes in a network that are more densely connected internally than the rest of the network. Community can represent a social group of person that has common interest or an organization. Community detection is helpful to reveal the topology structure and function of the real network. Consequently, it has become one of the hot topics in complex network research.

However, communities usually overlap each other. A vertex in the network may belong to more than one community, which is reasonable in real networks. For instance, if we divide students into many communities according to the student organizations they join, a person can join more than one student organizations. Thus, overlapping community detection seems to have higher research value.

With the development of network communication technology, the scale of social networks has seen explosive growth. As a result, community detection in social networks and many other networks has become an active research field.

© Springer International Publishing Switzerland 2015
A. Bikakis and X. Zheng (Eds.): MIWAI 2015, LNAI 9426, pp. 351–362, 2015.
DOI: 10.1007/978-3-319-26181-2_33

To date, community detection algorithms can be generally divided into some categories, including graph partition based methods [1–3], modularity optimization based methods [4–6], label propagation algorithms [7–9] etc.

Girvan and Newman proposed a community detection algorithm in Ref. [1]. Girvan-Newman (GN) algorithm, which is an algorithm based on graph partition, considers more about the edges connecting communities with high betweenness than those edges inside communities. The algorithm has to recalculate betweenness for all edges affected by the removal each time. Therefore, its running time is $O(n^3)$ on sparse graph, which makes it impractical for large graphs. However, this algorithm inspired many researchers to develop more community detection algorithms. Many later researchers proposed algorithms [2, 3] to lower the time cost of GN algorithm. However, some of those algorithms speed up the algorithm at the expense of the loss of clustering accuracy. Some of those algorithms can only applied to particular networks.

The optimization methods are based on maximization of the modularity proposed by Newman et al. [4]. Modularity can measure the quality of community. It has been widely used to measure the quality of the results of community detection methods. However, modularity optimization was found to have the resolution limit [10] and the modularity function Q exhibits extreme degeneracies [11].

Label propagation algorithm (LPA) is one of the fastest algorithms in the community detection algorithms with near-linear time complexity [7]. Each node in LPA is initialized with a unique label and its label is updated during each iteration by adopting the label of most of its neighbors. LPA simply uses the network structure and requires neither optimization of a predefined objective function nor prior information about the communities. LPA has attracted attention of many researchers for the advantages.

However, many proposed algorithms [1–7] are unable to discover overlapping communities. Gregory first applied label propagation algorithm to find overlapping community and proposed Community Overlap PRopagation Algorithm (COPRA) [8]. COPRA allows each node belong to v communities at most, where v is a parameter of algorithm. The parameter v is a global parameter, which means it is difficult to find a suitable value of v for all the nodes. Some methods were proposed to solve the problem. Wu et al. [9] proposed a method named BMLPA to avoid the limit of the number of communities a node can belong to. BMLPA uses a threshold parameter p to limit the belonging coefficient of labels. However, the threshold is not easy to determine.

With the rapid growth of the scale of networks, it is difficult for the stand-alone community detection algorithms to find communities in large-scale networks. Parallel community detection algorithm seems to be an urgent need.

Leung et al. [10] shows that label propagation is suitable to be parallelized, as it does not need much information about the network structures. Zhao et al. [13] proposed Parallel Structural Clustering Algorithm (PSCAN) for community detection in big networks. PSCAN implemented label propagation and structural clustering algorithm in MapReduce of Hadoop. However, PSCAN can only find non-overlapping communities.

In this paper, a novel Parallel Multi-Label Propagation Algorithm (PMLPA) is proposed to detect overlapping communities in large-scale networks. The proposed algorithm uses the ankle-value proposed in [16] to filter labels after label propagation. Therefore, no parameter is required.

The paper is structured as follows. Section 2 describes some label propagation algorithms. The proposed PMLPA is presented in Sect. 3. In Sect. 4, we show some experimental results. Finally, conclusions are presented in Sect. 5.

2   Label Propagation

The label propagation algorithm is proposed by Raghavan et al. [7]. The main procedures can be described as follows.

1. Initialization, each node is initialized with a label, usually expressed as an integer.
2. Iteration, the label of nodes is updated by the label most of its neighbor nodes have. If there is more than one label has the same number of neighbor nodes, one of them is chosen randomly.
3. If every node has a label that the maximum number of their neighbors have, then stop the algorithm. Else, go to 2, repeat iteration.
4. Split the nodes into communities according to node's label.

The time complexity of the algorithm is near-linear to the network size. The time required for the initialization is $O(n)$, where n is the number of nodes. It takes $O(m)$ time during each iteration, where m is the number of edges. Moreover, Raghavan claims that more than 95 % of the nodes can be classified correctly after 5 iterations in this algorithm.

However, asynchronous updating and label choosing randomly hamper its stability and robustness. Moreover, the LPA can only detect non-overlapping communities. In view of this, some researchers find other ways to improve LPA [8, 9]. Leung et al. found that synchronous updating takes more iterations than asynchronous updating to reach the maximum modularity. However, synchronous updating achieves more stable performance [12].

Gregory proposed COPRA [8], using the label propagation method, to find overlapping communities. Each node can belong to more than one community in COPRA. COPRA employs belonging coefficient to describe the strength of a node's membership with a community. The algorithm can be described as follows.

1. Initialization, each node is initialized with a pair of label and an initial belonging coefficient, values 1.
2. Iteration, the nodes update their labels and the belonging coefficient by summing and normalizing the belonging coefficient from the neighbors. Then delete the pairs whose belonging coefficient less than a threshold $1/v$, where v is a parameter.
3. Stop the iteration when the terminal conditions are satisfied.
4. Split the nodes into communities according to node's label.

By introducing the belonging coefficient, COPRA can limit the number of labels of nodes to be no more than v, where v represents the maximum number of communities a node can belong to. However, since v is a global parameter, it is hard to set a suitable value for v when there is some nodes in the network that are belong to more communities than the other nodes.

Wu et al. proposed BMLPA [9] to avoid limiting the number of communities a node can belong to. BMLPA filters the labels by Eq. (1) after label propagation.

$$\frac{b}{b_{max}} \geq p \tag{1}$$

where b is the belonging coefficient of label, b_{max} is the maximum belonging coefficient of all labels in the node, p is a parameter of algorithm using to filter labels. If a label with belonging coefficient b does not satisfied Eq. (1), the label will be removed from the node.

BMLPA includes a method called RC (Rough Core) to produce the initial labels instead of COPRA's putting unique label to each node. They show that RC can improve the quality and stability of community detection.

COPRA and BMLPA both show good performance in some networks. However, they both need a parameter in label update phase, and are hard to extend to large-scale networks. In view of this, a parallel label propagation algorithm was proposed in this paper to find communities in large-scale networks with high speed and accuracy.

3   Parallel Multi-label Propagation for Overlapping Community Detection

3.1   Parallel Multi-label Propagation Algorithm

Label propagation is very suitable to be parallelized because it is an iterative method and does not need much information about the network. In recent years, Spark [14] is a fast and general framework for large-scale data processing. Spark use Resilient Distributed Dataset (RDD) [15] for sharing data in cluster applications. So the proposed algorithm can be parallelized by transforming data into RDDs. Spark is particularly suited for iterative tasks. Consequently, we implement the proposed algorithm in Spark. The proposed algorithm can be parallelized by transforming data into RDDs.

Like COPRA, we label a node x with a set of pairs (l, bc), where l is a label identifier and bc is the belonging coefficient representing the strength of x's membership with the community with label identifier l. All the belonging coefficients of a node sums to 1.

The Rough Core algorithm [9] is used to initial the label in the initialization in our algorithm. RC can find the cliques in a network. Those cliques play the role of the core of communities. Therefore, it can greatly reduce the number of labels before propagation step.

The pseudo codes of the Parallel Multi-Label Propagation Algorithm (PMLPA) are shown in Algorithm 1.

Algorithm 1. Parallel Multi-Label Propagation Algorithm

Input: network G=(V,E), where V is the vertex set and
E is the edge set

Output: foundCommunities:Iterable[Set[Vertex]]
1: transform G into a RDD[(vertex,adjList)] graph,
   where adjList is list of adjacency vertex
2: RDD[(vertex,Map[label,bc])] nodeLabels,where key is
   the vertex,value is a Map of labels and label's belonging coefficient.
3: Initialize nodeLabels with the cores generated by
   RC(), and label's bc is 1
3: calculate idOld = distinct labels currently
4: **while** true
5:     propagating labels save to newLabels
6:     normalize labels' bc in each vertex
7:     use Ankle-value Filter() in each verte
8:     normalize labels' bc in each vertex
9:     calculate idNew = distinct labels currently
10:    **if** idOld == idNew
11:        break
12:    **else**
13:        id_old = id_new
           nodeLabels = newLabels
14:    **end if**
15:**end while**
16:remove those communities contained by other communities
17:foundCommunities=divide communities by vertexes'
   labels
18:**return** foundCommunities

The RC algorithm used in PMLPA is similar to Ref. [9]. The details of the Ankle-value Filter algorithm are described in the next subsection.

Since synchronous updating achieves better performance [10] and is suitable for parallel label propagation, the PMLPA uses synchronous updating strategy to update the labels after propagation. However, there are still some processes, like the RC algorithm, that cannot be parallelized.

3.2 Label Updating Strategy

The label updating strategy uses "ankle" value to update each node's label after label propagation. Huang uses "ankle" value to define a threshold point of sharp changing in k-dist values [16].

If we get a list of value $(v_1, ..., v_i, ..., v_n)$ with monotonically decreasing order, we can get two lines l_1, l_2 that connects v_1 to v_i and v_i to v_n, separately. The ankle value is the point that achieves minimal angle between l_1 and l_2. The calculation of the ankle value is given in Eq. (2).

$$\tan\theta = \frac{|m_1 - m_2|}{1 + m_1 \cdot m_2} . \tag{2}$$

where m_1 is the slop of l_1 and m_2 is the slope of l_2.

$$m_1 = (v_i - v_1)/i . \tag{3}$$

$$m_2 = (v_i - v_2)/i . \tag{4}$$

So (2) can be transformed to (5).

$$\theta_i = \arctan\frac{(n-i)(v_i - v_1)(v_n - v_i)}{i(n-i) + (v_i - v_1)(v_n - v_i)} . \tag{5}$$

So when θ_i get maximum value, we can get our ankle-value v_{ankle}.

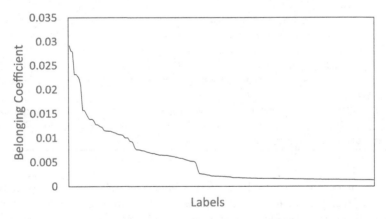

Fig. 1. Node x's labels and its belonging coefficients after the neighbors propagation their labels to node x.

The labels of node x was collected after label propagation during each iteration, as shown in Fig. 1. The labels are sorted by the normalized belonging coefficients. We can see that many labels' belonging coefficients are much smaller than the maximum belonging coefficient.

Ankle-value is used to find a threshold to filter those labels with small belonging coefficient. The ankle-value filter procedure is employed for all nodes after label propagation in each iteration. The details of the ankle-value filter algorithm are given in Algorithm 2.

Algorithm 2. Ankle-value Filter Algorithm

Input: List((l,bc)) a list of label and its belonging
        coefficient

Output: List((l,bc) a list of label and its belonging
        coefficient after filter
1: orderedLabelWeights is the input list, maxWgt is
   the maximum belonging coefficient, minWgt is the
   minimum belonging coefficient
2: n = orderedLabelWeights.size
3: **if** (n <= 2)
4:     **return** orderedLabelWeights
5: **end if**
6: ankleIndex = -1, maxAnkleValue = NegativeInfinity
7: **for** i = 1 to n
8:     v = orderedLabelWeights(i).bc
9:     ankleValue = Math.atan((n - i) * (v - minWgt) *
        (maxWgt - v) / (i * (n - i) + (v - minWgt) *
        (maxWgt - v)))
10:     **if** (ankleValue > maxAnkleValue):
11:         ankleIndex = i, maxAnkleValue = ankleValue
12:     **end if**
13:**end for**
12:**return** orderedLabelWeights.dropRight(n - an kleIn-
        dex

---

4 Experiments

4.1 Methodology

We evaluate the performance of the proposed algorithm on both artificial and real networks. The artificial networks are generated by benchmark proposed by Lancichinetti [17]. The real networks are collected from the SNAP websites [18].

All the experiments are conducted on a cluster of 5 computers. Each computer is equipped with 2 cores 2.0 GHz processors, 4 GB of memory, and Spark version 1.4.0 and Hadoop 2.4.0. One computer is used as the master, the others are used as the workers.

Normalized mutual information (NMI) [19] index is used to measure the performance of algorithm on the artificial networks. The NMI index evaluates algorithms' performance by comparing the found communities with the true communities. The value of NMI is in [0, 1]. Higher NMI value means better performance.

However, it is hard to know the true communities in real networks. In this case, it is commonly to use modularity Q [20] to evaluate the quality of the found communities. Since Q cannot handle overlapping communities, Q_{ov} [21] is proposed for overlapping communities.

4.2   Experimental Results

The proposed algorithm is compared with COPRA. The performance of the proposed algorithm is evaluated in benchmark networks with different *mu*, *N*. We also conducted experiments to measure the speedup of PMLPA. Table 1 shows some information of the artificial networks.

Table 1. Information of Artificial Networks

Networks	Parameters
Artificial networks	$N = 100$ k ~ 500 k, $mu = 0.3$, $k = 10 \sim 30$, $maxk = 50$, $minc = 50$, $maxc = 100$, $om = 5$, $on = 0.1*N$

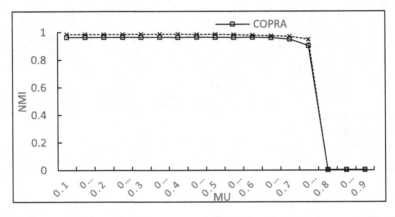

Fig. 2. NMI on networks with different *mu* of PMLPA and COPRA. $n = 100$ k, $k = 10$, $maxk = 50$, $minc = 50$, $maxc = 100$, $on = 1000$, $om = 5$, $mu = 0.1 \sim 0.9$

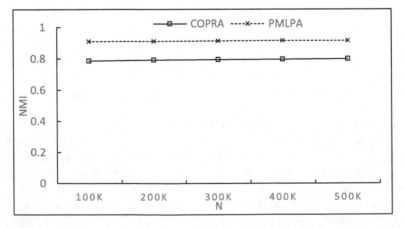

Fig. 3. NMI on networks with different scale of PMLPA and COPRA. $mu = 0.3$, $k = 30$, $maxk = 50$, $minc = 50$, $maxc = 100$, $om = 5$, $on = 0.1*N$, $N = 100$ k ~ 500 k

Figures 2 and 3 show the NMI values of PMLPA and COPRA on artificial networks with different *mu* and *N*. The results of Fig. 2 shows that PMLPA perform better than COPRA when *mu* smaller than 0.8, and when *mu* larger than 0.75, the NMI values of PMLPA and COPRA both get nearly zero. *mu* is the mixing parameters; High value of *mu* indicates that the communities are difficult to be discovered. Figure 3 also shows that PMLPA can get better results than COPRA on those large-scale networks.

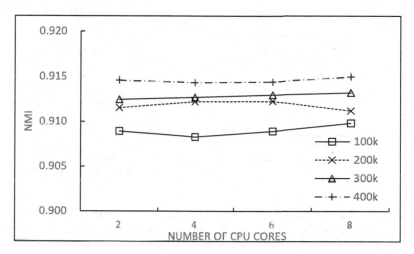

Fig. 4. NMI on different networks and different CPU cores of PMLPA. *mu* = 0.3, *k* = 30, *maxk* = 50, *minc* = 50, *maxc* = 100, *om* = 5, *on* = 0.1*N, *N* = 100 k ∼ 400 k, *cores* = 2 ∼ 8

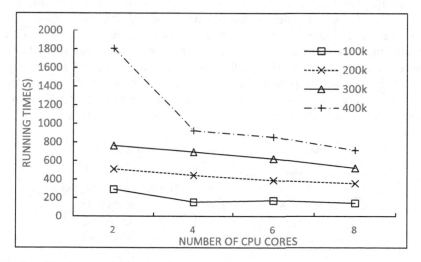

Fig. 5. Running time on different networks and different cores of PMLPA. *mu* = 0.3, *k* = 30, *maxk* = 50, *minc* = 50, *maxc* = 100, *om* = 5, *on* = 0.1*N, *N* = 100 k ∼ 400 k, *cores* = 2 ∼ 8

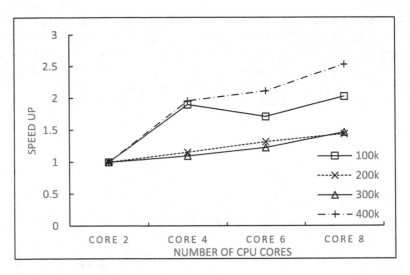

Fig. 6. Speed up on different networks and different cores of PMLPA. $mu = 0.3$, $k = 30$, $maxk = 50$, $minc = 50$, $maxc = 100$, $om = 5$, $on = 0.1*N$, $N = 100$ k $\sim$ 400 k, $cores = 2 \sim 8$

The results of Figs. 2 and 3 show that the proposed label updating strategy is more reasonable than COPRA's. Without a global limit of maximum number of communities vertex can belong to, PMLPA can detect communities more accurately.

Since PMLPA is a parallel algorithm, we also do experiments on different number of CPU cores. As Fig. 4 shows, the number of cores does not affect the accuracy of the found communities for PMLPA. It is because that the results of community detection depend on the algorithm itself but not the parallel computation framework and the number of CPU cores. Figures 5 and 6 shows that PMLPA has a good speedup performance. As the number of CPU cores increases, the running time of PMLPA decreases. Furthermore, larger scale of the dataset shows more obvious speedup when the number of cores increases. More CPU cores can do more tasks and the running time can be saved. However, the parallel framework would spent most of the running time to exchange data among machines on small networks. Consequently, PMLPA's advantage of great reduction in running time is more obvious on large networks.

Table 2. Accuracy and time cost on real networks

Networks	Nodes	Edges	COPRA		PMLPA	
			Q_{ov}	Time(s)	Q_{ov}	Time(s)
Amazon	334863	925872	0.604	4834	**0.698**	**142**
DBLP	317080	1049866	0.591	1918	**0.626**	**161**
Youtube	1134890	2987624	0.269	3793	**0.481**	**657**

Table 2 shows the results of the comparison of COPRA with PMLPA on real networks Amazon [22], DBLP [22] and Youtube [22]. The experimental results show that PMLPA achieves higher Q_{ov} than COPRA on Amazon, DBLP and Youtube networks. Moreover, the running time of PMLPA is much less than COPRA. It means that PMLPA is more efficient and scalable than COPRA. As PMLPA is implemented in Spark framework, it can deal with large-scale real networks, such as YouTube.

Generally, communities in the real networks are more difficult to detect than those in the artificial networks. Moreover, it is even harder to determine the parameter of the algorithms like COPRA. The proposed algorithm has the benefit of requiring no parameter. Since real networks are always large in scale, the PMLPA algorithm can detect communities efficiently.

5 Conclusion

In this paper, a new Parallel Multi-Label Propagation Algorithm (PMLPA) is proposed to detect overlapping communities in social networks. PMLPA use RC algorithm to generate the cores using to initialize the label of vertex. The algorithm takes a new label updating strategy using ankle-value after label probation in each iteration. The label updating strategy requires no parameter, which improve the applicability of the algorithm. The experiments on artificial and real networks show that PMLPA achieves high accuracy in overlapping community detection with reasonable running speed. In the future, we plan to study more parallel community detection algorithms and conduct more experiments on large-scale networks with larger computation clusters.

Acknowledgments. This work is partly supported by the National Natural Science Foundation of China under Grants No. 61103175 and No. 61300104, the Key Project of Chinese Ministry of Education under Grant No. 212086, the Fujian Province High School Science Fund for Distinguished Young Scholars under Grand No. JA12016, the Program for New Century Excellent Talents in Fujian Province University under Grant No. JA13021, the Fujian Natural Science Funds for Distinguished Young Scholar under Grant No. 2014J06017, and the Natural Science Foundation of Fujian Province under Grant No. 2013J01230.

References

1. Girvan, M., Newman, M.E.J.: Community structure in social and biological networks. Proc. Natl. Acad. Sci. **99**(12), 7821–7826 (2002)
2. Tyler, J.R., Wilkinson, D.M., Huberman, B.A.: E-mail as spectroscopy: automated discovery of community structure within organizations. Inf. Soc. **21**(2), 143–153 (2005)
3. Radicchi, F., Castellano, C., Cecconi, F., et al.: Defining and identifying communities in networks. Proc. Natl. Acad. Sci. U.S.A. **101**(9), 2658–2663 (2004)
4. Newman, M.E.J., Girvan, M.: Finding and evaluating community structure in networks. Phys. Rev. E **69**(2), 026113 (2004)
5. Newman, M.E.J.: Fast algorithm for detecting community structure in networks. Phys. Rev. E **69**(6), 066133 (2004)
6. Blondel, V.D., Guillaume, J.L., Lambiotte, R., et al.: Fast unfolding of communities in large networks. J. Stat. Mech: Theory Exp. **2008**(10), P10008 (2008)
7. Raghavan, U.N., Albert, R., Kumara, S.: Near linear time algorithm to detect community structures in large-scale networks. Phys. Rev. E **76**(3), 036106 (2007)

8. Gregory, S.: Finding overlapping communities in networks by label propagation. New J. Phys. **12**(10), 103018 (2010)
9. Wu, Z.H., Lin, Y.F., Gregory, S., et al.: Balanced multi-label propagation for overlapping community detection in social networks. J. Comput. Sci. Technol. **27**(3), 468–479 (2012)
10. Fortunato, S., Barthélemy, M.: Resolution limit in community detection. Proc. Natl. Acad. Sci. **104**(1), 36–41 (2007)
11. Good, B.H., de Montjoye, Y.A., Clauset, A.: Performance of modularity maximization in practical contexts. Phys. Rev. E **81**(4), 046106 (2010)
12. Leung, I.X.Y., Hui, P., Lio, P., et al.: Towards real-time community detection in large networks. Phys. Rev. E **79**(6), 066107 (2009)
13. Zhao, W., Martha, V., Xu, X.: PSCAN: a parallel structural clustering algorithm for big networks in MapReduce. In: 2013 IEEE 27th International Conference on Advanced Information Networking and Applications (AINA), pp. 862–869. IEEE (2013)
14. Apache Spark. http://spark.apache.org
15. Zaharia, M., Chowdhury, M., Das, T., et al.: Resilient distributed datasets: a fault-tolerant abstraction for in-memory cluster computing. In: Proceedings of the 9th USENIX Conference on Networked Systems Design and Implementation, p. 2. USENIX Association (2012)
16. Huang, J.B., Sun, H.L., Bortner, D., et al.: Mining hierarchical community structure within networks from density-connected traveling orders. J. Softw. (Chinese) **22**(5), 951–961 (2011)
17. Lancichinetti, A., Fortunato, S.: Benchmarks for testing community detection algorithms on directed and weighted graphs with overlapping communities. Phys. Rev. E **80**(1), 016118 (2009)
18. Leskovec, J., Krevl, A.: SNAP Datasets: Large Network Dataset Collection (2014)
19. Lancichinetti, A., Fortunato, S., Kertész, J.: Detecting the overlapping and hierarchical community structure in complex networks. New J. Phys. **11**(3), 033015 (2009)
20. Newman, M.E.J., Girvan, M.: Finding and evaluating community structure in networks. Phys. Rev. E **69**(2), 026113 (2004)
21. Nicosia, V., Mangioni, G., Carchiolo, V., et al.: Extending the definition of modularity to directed graphs with overlapping communities. J. Stat. Mech: Theory Exp. **2009**(03), P03024 (2009)
22. Yang, J., Leskovec, J.: Defining and evaluating network communities based on ground-truth. Knowl. Inf. Syst. **42**(1), 181–213 (2015)

ARP-Miner: Mining Risk Patterns of Android Malware

Yang Wang[1], Bryan Watson[1], Jun Zheng[1(✉)], and Srinivas Mukkamala[1,2]

[1] Department of Computer Science and Engineering,
New Mexico Institute of Mining and Technology, Socorro, NM 87801, USA
{yangwang,bwatson,zheng}@nmt.edu, srinivas@cs.nmt.edu
[2] The Institute for Complex Additive Systems Analysis,
New Mexico Institute of Mining and Technology, Socorro, NM 87801, USA

Abstract. Android applications need to request permissions to access sensitive personal data and system resources. Certain permissions may be requested by Android malware to facilitate their malicious activities. In this paper, we present ARP-Miner, an algorithm based on association rule mining that can automatically extract **A**ndroid **R**isk **P**atterns indicating possible malicious activities of apps. The experimental results show that ARP-Miner can efficiently discover risk rules associating permission request patterns with malicious activities. Examples to relate the extracted risk patterns with behaviors of typical malware families are presented. It is also shown that the extracted risk patterns can be used for malware detection.

1 Introduction

Android is the most popular mobile platform which makes it the primary target for malware attacks. It was reported by F-secure [1] that 97 % of mobile malware targeted Android and 99.9 % of them came from the third party Android app markets. Android uses a permission-based access control model to prevent malicious applications (apps) to access sensitive personal data like contact list, SMS messages, emails etc. and system resources like WiFi, GPS etc. The user has to approve all permissions requested by an app to install it on an Android device. However, it has been shown that Android permission system is ineffective and may confuse users because the large number of permissions are difficult for normal users to understand [5]. In this paper, we aim to extract risk rules that can associate permission request patterns with activities of malicious Android apps. It is expected that the extracted risk patterns can be used to help users better understand the risks of permissions requested by an app and efficiently flag potential malicious apps.

The rest of this paper is organized as follows. Related work is introduced in Sect. 2. In Sect. 3, we provide the overview of the Android risk pattern mining system, basic concepts for risk pattern mining, and the details of ARP-Miner algorithm. The performance of ARP-Miner is evaluated in Sect. 4 and we conclude the paper in Sect. 5.

© Springer International Publishing Switzerland 2015
A. Bikakis and X. Zheng (Eds.): MIWAI 2015, LNAI 9426, pp. 363–375, 2015.
DOI: 10.1007/978-3-319-26181-2_34

2   Related Work

Permission system is very important for protecting Android devices. The permissions requested by an app may indicate its potential malicious behavior. Therefore, many research focus on evaluating, improving or extending this system. Enck et al. proposed Kirin [4], a certification service that uses predefined rules to identify suspicious apps. However, all the rules are manually defined and some permissions in the rules were removed in recent Android versions. Zhou et al. [13] proposed DroidRanger, a system to detect malicious apps. Permission-based filtering is applied in this system. Sarma et al. [9] designed a set of signals triggered by potential malicious apps. They preselected some critical permissions, and found out the rare requested critical permissions or permission pairs. An app requesting certain number of critical permissions or permission pairs will trigger the risk signals. DroidRisk was proposed in [10] to quantitatively evaluate the risk of an Android app based on its requested permissions. Permlyzer was proposed in [11] as a general purpose framework to automatically analyze the permission use of Android apps. These studies prove that permissions could be used for detecting malicious behaviors of apps.

Permission request patterns of Android and Facebook apps were extracted in [6] using a probabilistic method for Boolean matrix factorization. They found that the permissions requested by low reputation apps are different with those request by high reputation apps. They concluded that the permission request patterns could be used as one of the indicators for app quality. There were a few works that extract permission request patterns of Android apps to detect malicious activities. Moonsamy et al. [8] proposed a contrast permission pattern mining algorithm to identify interesting permission request patterns. In their method, pattern mining has two steps: candidate permission generation and contrast permission pattern selection. The former step is derived from Apriori algorithm while the latter step depends on the difference of support in the benign and malware datasets. In [7], Droid Detective was proposed for permission combination mining and malware detection. The permission combination mining is based on the difference of permission request frequencies in malware family and corresponding benign apps. However, these methods could only find a few permission patterns or some simple patterns because of the efficiency of the method or the interesting metrics used in the method. ARP-Miner tries to extract more complex risk patterns that can better indicate malicious activities.

3   Methods

It is shown that the risk of an Android app can be represented by the risks of permissions requested by the app [10]. However, the assumption used in [10] that the requested permissions are independent may underestimate the effect of a combination of multiple permissions for certain malicious activities. For example, a malware needs to request permissions for both changing the network settings and sending SMS to upload the privacy information or send premium

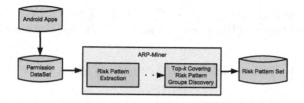

Fig. 1. The workflow of the Android risk pattern mining system

rate SMS without the awareness of the user. Therefore, mining the permission request patterns involving multiple permissions could improve the effectiveness in identifying the risk of an app. To this end, we propose the ARP-Miner algorithm which is derived from the well-known Apriori algorithm.

3.1 Overview of Android Risk Pattern Mining System

The workflow of the Android risk pattern mining system using the proposed ARP-Miner algorithm is shown in Fig. 1. It includes the following steps:

- Android apps are distributed as compressed APK files, which contain compiled source code and all necessary resources including configuration files, certificates, etc. Every Android app has an AndroidManifest.xml file which claimes all the requested permissions. In the system, permissions requested by an app are extracted from the Manifest file first by using the Android Asset Packaging Tool (aapt) included in Android SDK. The permission dataset is formed by extracting the permissions of apps in both the benign app dataset and the malware dataset.
- The ARP-Miner algorithm has two steps. The risk pattern extraction step derived from the popular association rule mining algorithm Apriori [2], is used to extract patterns with high values of adopted interesting measures. The size of the risk pattern set extracted by the first step may be too large. The top-k covering risk pattern groups discovery step based on the algorithm of [3] is used to reduce the pattern set size and find representative patterns.

3.2 Basic Concepts

Let $\mathbf{P} = \{p_1, p_2, ..., p_n\}$ is the set of all permissions that can be requested by an app. A permission request pattern Pa is defined as the set of some permissions requested by a group of apps and $Pa \subset \mathbf{P}$. The size of Pa is defined as $|Pa|$. A Pa containing k ($k \le n$) permissions ($|Pa| = k$) is called k-$pattern$.

Definition 1. *Sub-pattern and super-pattern:*
    Given two patterns Pa_1 and Pa_2, Pa_1 is defined as the sub-pattern of Pa_2 if $|Pa_1| < |Pa_2|$ and $Pa_1 \subset Pa_2$. Pa_2 is then called as the super-pattern of Pa_1. If $|Pa_2| - |Pa_1| = 1$, we call Pa_1 as Pa_2's direct child.

We define the class of an app as $C = \{M, B\}$ which indicates the app is malicious (M) or benign (B). Then we can define an association rule as follows.

Definition 2. *An association rule is the mapping of Pa to the class of an app, i.e. $Pa \rightarrow C$.*

To extract association rules, the well-known Apriori algorithm uses two interesting measures: support and confidence. Support is the ratio of samples in a dataset containing the specified pattern, which measures the frequency of a pattern occurring in the dataset. Confidence measures the strength of a rule, which is the ratio of samples containing the specified pattern that could result in certain consequence. Since the goal of ARP-Miner is to extract rules that can map Pa to M, we adopt two interesting measures which focus on the malware dataset: local support and likelihood.

Instead of using support, the local support of a pattern Pa in the malware dataset, $S(Pa)$, is used which is defined as:

$$S(Pa) = N_{M,Pa}/N_M \tag{1}$$

where N_M is the number of samples in the malware dataset, and $N_{M,Pa}$ is the number of malware samples requesting Pa. $S(Pa)$ measures the frequency of Pa in the malware dataset.

A pattern indicating malicious activities should have the property that it occurs much more frequently in the malware dataset than in the benign app dataset. Thus, likelihood is used as another interesting measure which is defined as the probability that an app is malicious if a pattern is requested by the app. The likelihood of Pa, $L(Pa)$, can be calculated by using Bayes' rule:

$$L(Pa) = P(M \mid Pa) = \frac{P(Pa \mid M) \times P(M)}{P(Pa)} \tag{2}$$

where $P(M \mid Pa)$ is the posteriori conditional probability that an app is malware when it requests the pattern Pa, $P(Pa \mid M)$ is the priori probability that Pa is requested by a malware, $P(M)$ indicates the priori probability of an app from the dataset being a malware, and $P(Pa)$ is the priori probability that Pa is requested by any app. A larger likelihood value implies that the pattern is more likely requested by a malware than a benign app. Note that the calculation of $L(P_a)$ needs both the malware dataset and the benign app dataset.

Pa in a risk rule must meet the two requirements: $S(Pa) > S_T$ and $L(Pa) > L_T$ where S_T and L_T are the thresholds for local support and likelihood, respectively. Thus, we can define risk rule and risk pattern as follows.

Definition 3. *A risk rule is an association rule that maps Pa to a malware ($Pa \rightarrow M$), and $S(Pa) > S_T$, $L(Pa) > L_T$. Pa in a risk rule is called as a risk pattern.*

Unlike confidence, likelihood is not monotonous. This means that the likelihood of a pattern may or may not be larger than that of its sub-pattern. Therefore, a pattern needs to be compared with every sub-pattern. The pattern will be kept if it has a larger likelihood than its sub-patterns. In this case, the pattern is called as an *advanced pattern*.

3.3 ARP-Miner Algorithm

The ARP-Miner algorithm involves two steps: risk pattern extraction and top-k covering pattern groups discovery. The notations used in the ARP-Miner algorithm and their meanings are shown in Table 1.

Risk Pattern Extraction. The first step of ARP-Miner algorithm is derived from the well-known Apriori algorithm as shown in Algorithms 1 and 2. Apriori is considered as an efficient algorithm for association rule mining, which uses two interesting measures, support and confidence, to extract rules. ARP-Miner replaces the two measures with local support and likelihood which can better relate a pattern with malicious activities. To extract risk patterns, the new pattern set $\mathbf{PA}_k$ is generated from $\mathbf{PA}_{k-1}$. A newly generated pattern will be immediately removed from the set if its local support is less than the threshold

Table 1. Notations used in the ARP-Miner algorithm

Notation	Description
$\mathbf{R}$	The set of extracted risk patterns
$\mathbf{PA_k}$	The set of $k-$patterns
Pa	A permission request pattern
Pa_i	The ith pattern
Pa_{new}	The new generated pattern
$Pa._L$	The likelihood attribute of a pattern
$Pa._{IL}$	The inherited likelihood attribute of a pattern
$Pa._{adv}$	The advance attribute of a pattern
$Pa[k]$	The kth requested permission of a pattern
$Pa[1...k]$	The first k requested permissions of a pattern
$\mathbf{P}$	The set of all permissions
p	A permission
$\mathbf{C}$	The set of child patterns
c	A child pattern
$\mathbf{M}$	The malware dataset
m	A malware
$\mathbf{R}_m$	The candidate risk pattern set requested by a malware m
$|\mathbf{PA}|$	The number of patterns in a pattern set

Algorithm 1. Risk Pattern Extraction

```
 1: procedure EXTRACT(S_T, L_T, P)
 2:     R ← ∅
 3:     PA₁ ← ∅
 4:     for all p ∈ P do
 5:         if S(p) > S_T then
 6:             PA₁ ← PA₁ ∪ p
 7:         end if
 8:     end for
 9:     R ← Sort(PA₁)
10:     k ← 2
11:     while PA_{k−1} ≠ ∅ do
12:         PA_k ← Gen(PA_{k−1})
                                        ▷ Generate new candidate patterns
13:         R ← R ∪_k PA_k
14:         k ← k + 1
15:     end while
16:     for all Pa ∈ R do
17:         if Pa.adv ≠ true then
18:             Remove(Pa, R)                       ▷ Remove Pa from R
19:         end if
20:     end for
21:     return R
22: end procedure
```

S_T. The patterns left in $\mathbf{PA}_k$ will then be used to generate $\mathbf{PA}_{k+1}$. However, at the end of Algorithm 1, only advanced patterns will be kept to form the risk pattern set $\mathbf{R}$. An advanced pattern has a larger likelihood than its sub-patterns and super-patterns. To reduce the comparison time, the inherited likelihood is introduced as an attribute of a pattern, which stores the maximum value of the likelihoods of all its sub-patterns and its own likelihood. Thus, we only need to compare a pattern's likelihood with its direct children's inherited likelihoods to determine if it is an advanced pattern. The $MaxIL(Pa)$ function in Algorithm 2 is used to find the inherited likelihood of a pattern Pa. The adv attribute is introduced to a pattern which is marked as true if the pattern is an advanced pattern.

Top-k Covering Risk Pattern Groups Discovery. After the first step, the extract risk pattern set may still contain a large number of patterns. Therefore, further processing needs to be performed to find the representative patterns. Gao et al. [3] proposed a novel algorithm to discover the top-k covering rule groups for gene expressions. The principle of the algorithm is that for each row of gene expression profiles, only the most significant k rules are extracted. This algorithm can significantly reduce the number of extracted rules. Based on their work, we proposed an algorithm to discover the top-k covering risk pattern groups as shown in Algorithm 3.

Algorithm 2. Generate New Candidate Patterns

1: **procedure** GEN($\mathbf{PA}_k, k$)
2:     $\mathbf{PA}_{k+1} \leftarrow \emptyset$
3:     **for** $i \leftarrow 1, |\mathbf{PA}_k|$ **do**
4:         **for** $j \leftarrow i+1, |\mathbf{PA}_k|$ **do**
5:             $Pa_i' \leftarrow Pa_i[1...k-1]$
                    ▷ When $k == 1$, Pa_i' is empty
6:             $Pa_j' \leftarrow Pa_j[1...k-1]$
                    ▷ When $k == 1$, Pa_j' is empty
7:             **if** $Pa_i' == Pa_j'$ **then**
8:                 $Pa_{new} \leftarrow append(Pa_i, Pa_j[k])$
9:                 $Pa_{new.L} \leftarrow L(Pa_{new})$
10. $Pa_{new.IL} \leftarrow MaxIL(Pa_{new})$
11:                 **if** $Pa_{new.IL} < Pa_{new.L}$ **then**
12:                     $Pa_{new.IL} \leftarrow Pa_{new.L}$
13:                     $Pa_{new.adv} \leftarrow true$
14:                 **end if**
15:                 **if** $S(Pa_{new}) > S_T$ **then**
16:                     $\mathbf{PA}_{k+1} \leftarrow \mathbf{PA}_{k+1} \cup Pa_{new}$
17:                 **end if**
18:             **end if**
19:         **end for**
20:     **end for**
21:     **return** $\mathbf{PA}_{k+1}$
22: **end procedure**

The proposed algorithm starts with the risk pattern set $\mathbf{R}$ obtained by Algorithm 1, which has no redundant pattern left. Then for each malware, we find all candidate risk patterns requested by it in $\mathbf{R}$ and keep the top k patterns in terms of likelihood. If there are candidate patterns with the same value of likelihood that make the size of pattern group larger than k, patterns are randomly chosen from them to make the group size to be k.

4   Performance Evaluation

4.1   Application Datasets

The datasets used for evaluating the performance of ARP-Miner are shown in the following.

Benign App Dataset. Benign app dataset is used for calculating the likelihood of a permission request pattern. The same dataset used in [10] are adopted in this study which includes 27,274 apps collected from the official Android market, Google Play. Google considers the security issue as a big concern. Thus, an security layer called Bouncer was developed and deployed to protect the market. All uploaded Android apps are automatically scanned for known malware, as

Algorithm 3. Top-k Covering Risk Pattern Groups Discovery

1: **procedure** TOP-K(k, **M**, **R**)
2:      **TOP$_k$** $\leftarrow \emptyset$
3:      **for all** $m \in$ **M do**
4:          **R**$_m \leftarrow \emptyset$
5:          **for all** $Pa \in$ **R do**
6:              **if** Pa *reqested by* m **then**
7:                  **R**$_m \leftarrow$ **R**$_m \cup Pa$
8:              **end if**
9:          **end for**
10:         $Sort($**R**$_m)$
                                    ▷ Sort the **R**$_m$ by the descent order of likelihood
11:         **TOP$_k$** $\leftarrow$ **TOP$_k$** $\cup FirstK($**R**$_m, k)$
                                    ▷ Add the first k patterns from **R**$_m$ to **TOP$_k$**
12:     **end for**
13:     **return TOP$_k$**
14: **end procedure**

well as for potential malicious behavior through dynamic analysis. On the other hand, user reported malware on the market will be removed immediately when confirmed.

Malware Dataset. Android Malware Genome Project [12] published an Android malware dataset including 1,260 samples from 49 malware families found between August 2010 and October 2011. The malicious behaviors performed by these malware include but not limited to remotely controlling, private information leaking, root-level exploits, unauthorized messaging and phone calling. Many of these malware are repackaged versions of benign apps with malicious payload.

4.2   Results

The number of risk patterns extracted from the malware dataset are shown in Table 2. Without top-k covering risk pattern groups discovery ($k = \infty$), a large number of patterns are extracted. The number of extracted patterns is significantly reduced when only the representative k risk patterns are kept for

Table 2. The number of extracted risk patterns

	$L_T = 0.6$	$L_T = 0.7$	$L_T = 0.8$
$k = 1$	87	79	55
$k = 3$	238	209	144
$k = 5$	377	325	213
$k = \infty$	9,493	8,645	6,327

Table 3. Top 10 most frequently requested risk patterns

	1	2	3	4	5	6	7	8	9	10
ACCESS_COARSE_LOCATION							★			
ACCESS_NETWORK_STATE			★	★				★		
ACCESS_WIFI_STATE			★							
CALL_PHONE				★	★					
CHANGE_WIFI_STATE						★				
DISABLE_KEYGUARD	★				★					
INSTALL_PACKAGES		★					★	★		
INTERNET			★						★	★
KILL_BACKGROUND_PROCESSES					★				★	★
READ_CONTACTS	★			★						★
READ_LOGS	★									
READ_PHONE_STATE			★							
READ_SMS		★		★	★			★		
RECEIVE_BOOT_COMPLETED	★									
RECEIVE_SMS				★				★	★	
SEND_SMS								★	★	
VIBRATE										★
WRITE_APN_SETTINGS	★			★	★			★		★
WRITE_CONTACTS	★									★
WRITE_SMS			★							
Likelihood	0.989	0.989	0.989	0.989	0.989	0.938	0.897	0.907	0.806	0.807

each malware family. It can be observed that the number of extracted patterns decreases with the decrement of the value of k and the increment of the threshold of likelihood, L_T. The threshold of local support, S_T, is set to 0.1 for all cases.

The most frequently requested 10 risk patterns are listed in Table 3, where k, S_T and L_T are set to 1, 0.8 and 0.1, respectively. Under the same parameter setting, the top 10 patterns in terms of likelihood are listed in Table 4. These patterns indicate the following common behaviors of Android malware: (1) collecting private information including contacts, short messages, logs, location information etc.; (2) checking the network setting or automatically changing the network setting without users awareness for further attacks; (3) trying to kill other applications especially antivirus applications; (4) accessing short messages which may result in privacy leaking and sending premium rate messages; and (5) installing packages for update attacks.

Table 4. Top 10 risk patterns in terms of likelihood

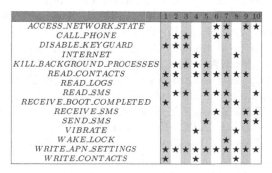

	1	2	3	4	5	6	7	8	9	10
ACCESS_NETWORK_STATE						★	★		★	★
CALL_PHONE		★	★			★	★			
DISABLE_KEYGUARD	★	★	★							
INTERNET				★			★			
KILL_BACKGROUND_PROCESSES		★	★	★	★					
READ_CONTACTS	★	★		★	★	★	★	★		
READ_LOGS	★									
READ_SMS		★	★		★	★	★			★
RECEIVE_BOOT_COMPLETED	★					★	★			
RECEIVE_SMS						★			★	★
SEND_SMS					★				★	★
VIBRATE				★			★			
WAKE_LOCK						★				
WRITE_APN_SETTINGS	★	★	★	★	★	★	★	★	★	★
WRITE_CONTACTS	★			★			★			

4.3   Typical Android Malware Families and Their Representative Risk Patterns

Risk pattern can indicate an app's malicious behavior. For example, if a calculator app requests permissions to read the contacts and send SMS, it can cause privacy leaks. In the following, we relate the malicious activities of typical Android malware families with their representative risk patterns. The patterns are extracted with $k = 1$, $L_T = 0.8$ and $S_T = 0.1$.

- **ADRD Family:** This threat may be injected into a repacked application and appears on some third party markets. It tries to steal some private information such as IMEI, IMSI and upload them to a remote server. If no network available, it will try to open network connection by itself. It may also send search requests to increase the site ranking. We found that 86.36 % of samples in ADRD family request the risk pattern {WRITE_APN_SETTINGS, READ_PHONE_STATE, ACCESS_NETWORK_STATE, INTERNET}. WRITE_APN_SETTINGS is used for changing the network connection; READ_PHONE_STATE can let the threat get the IMEI; ACCESS_NETWORK_STATE and INTERNET are used for opening network sockets. Obviously, this risk pattern clearly reveals the behavior of the threat.
- **YZHC family:** YZHC is a SMS trojan which may send SMS to premium rate number and incurs unexpected charges. This threat can retrieve the premium numbers from internet. We found that all samples in this family request the risk pattern {WRITE_SMS, READ_SMS, ACCESS_WIFI_STATE, ACCESS_NETWORK_STATE, INTERNET}. WRITE_SMS and READ_SMS are used to write and read SMS, respectively. ACCESS_WIFI_STATE and ACCESS_NETWORK_STATE are two permissions for checking the network status. INTERNET is used for internet access. When the network is available, this Trojan may send out the premium numbers.
- **DroidKungFu family:** DroidKungFu is a sophisticated Android malware found on many third party markets. It is a large malware family with many variants. Some variations try to elevate the privilege to get root access through certain exploits. Private information is also a target of this malware. They may send HTTP post to some hard coded servers to collect IMEI, phone model, Android operating system version and other information. DroidKungFu may try to install an app called **legacy** which pretends to be a benign Google Search app but is indeed a backdoor. {INSTALL_PACKAGES, CHANGE_WIFI_STATE} is the risk pattern requested by 76.5 % of the samples in the sub-family DroidKungFu1, while 79.2 % of the samples in the sub-family DroidKungFu4 request the risk pattern {WRITE_SMS, ACCESS_WIFI_STATE, READ_PHONE_STATE, ACCESS_NETWORK_STATE, INTERNET}. INSTALL_PACKAGES may be used for installing the backdoor, which may turn the host device into a bot. READ_PHONE_STATE is usually used for stealing private information. Other network related permissions are requested for uploading the sensitive data.

– **GoldDream family:** GoldDream was reported as a malware that spies SMS messages and phones calls. It may turn into a bot controlled by remote server. All the samples in this family are found to request the risk pattern {INSTALL_PACKAGES, READ_SMS}. The two permissions are used for installing backdoor and collect SMS messages.
– **Kmin family:** Kmin is a trojan that collects and uploads user data to a remote server. All the samples in this family request the pattern {WRITE_APN_SETTINGS, READ_SMS, RECEIVE_SMS, SEND_SMS, ACCESS_NETWORK_STATE}. Obviously, this malware has the capability of stealing and leaking sensitive information from SMS messages.

4.4 Malware Detection

Since the extracted risk patterns are related to malicious activities, they could be used to construct a classification system for malware detection. The malware detection performance of such classifier built using a similar approach as in [10] is shown in Table 5. The performance of the ARP-Miner classifier is compared with Droid Detective proposed in [7] which also uses permission request patterns for Android malware detection. A 5-fold cross validation was used in the experiment. The parameters of Droid Detective were set as $MIN_s = 10$, $MIN_{diff} = 90\%$ and $k = 6$, which were claimed to achieve the best performance. The metrics used for performance comparison are sensitivity, specificity, and F_1 score. Sensitivity is the true positive rate which measures the percentage of malware samples that are correctly detected. Specificity is the true negative rate which indicates the percentage of benign apps being correctly detected. F_1 score is the harmonic mean of precision and sensitivity as shown in Eq. 3, where precision is the number of correctly detected malware samples divided by the number of samples being detected as malware.

$$F_1 = 2 \times \frac{precision \times sensitivity}{precision + sensitivity} \tag{3}$$

It can be seen from Table 5 that ARP-Miner outperforms Droid Detective in terms of all performance metrics. The results demonstrate that the risk patterns extracted by ARP-Miner are highly correlated with the activities of Android malware. Thus, the classifier constructed with the extracted risk patterns achieves a good performance in malware detection.

Table 5. Performance of malware detection

	sensitivity	specificity	F_1 score
ARP-Miner	90.48 %	86.94 %	0.38
Droid Detective	85.08 %	84.99 %	0.33

5   Conclusion

In this paper, we proposed ARP-Miner, an algorithm for extracting risk patterns of Android malware, which is based on the popular association rule mining algorithm, Apriori, and the top-k covering rule groups extraction algorithm. Two interesting measures, local support and likelihood, are adopted in ARP-Miner to find risk patterns indicating malicious activities. It has been shown that the proposed algorithm can effectively discover risk patterns that can be related to the malicious activities of typical Android malware families. The classifier constructed with the extracted risk patterns can achieve a better performance in malware detection compared with existing method.

Acknowledgments. This work was supported in part by the U.S. Department of Homeland Security under Award Number: "2010-ST-062-000051" and the Institute of Complex Additive Systems Analysis (ICASA) of New Mexico Tech.

References

1. F-secure, threat report h2 (2013). http://www.f-secure.com/static/doc/labs_global/Research/Threat_Report_H2_2013.pdf
2. Agrawal, R., Srikant, R., Others: Fast algorithms for mining association rules. In: Proceeding 20th International Conference Very Large Data Bases, VLDB, vol. 1215, pp. 487–499 (1994)
3. Cong, G., Tan, K.L., Tung, A.K., Xu, X.: Mining top-k covering rule groups for gene expression data. In: Proceedings of the 2005 ACM SIGMOD international conference on Management of data, pp. 670–681. ACM (2005)
4. Enck, W., Ongtang, M., McDaniel, P.: On lightweight mobile phone application certification. In: Proceedings of the 16th ACM conference on Computer and Communications Security, CCS 2009 pp. 235–245 (2009)
5. Felt, A.P., Ha, E., Egelman, S., Haney, A., Chin, E., Wagner, D.: Android permissions: user attention, comprehension, and behavior. In: Proceedings of the Eighth Symposium on Usable Privacy and Security, SOUPS 2012, pp. 3:1–3:14 (2012)
6. Frank, M., Dong, B., Porter Felt, A., Song, D.: Mining permission request patterns from android and facebook applications. In: Proceedings of the 2012 IEEE 12th International Conference on Data Mining, ICDM 2012, pp. 870–875. IEEE Computer Society, Washington, DC (2012)
7. Liang, S., Du, X.: Permission-combination-based scheme for android mobile malware detection. In: IEEE International Conference on Communications (ICC), pp. 2301–2306. IEEE (2014)
8. Moonsamy, V., Rong, J., Liu, S.: Mining permission patterns for contrasting clean and malicious android applications. Future Gener. Comput. Syst. **36**, 122–132 (2014)
9. Sarma, B.P., Li, N., Gates, C., Potharaju, R., Nita-Rotaru, C., Molloy, I.: Android permissions: a perspective combining risks and benefits. In: Proceedings of the 17th ACM symposium on Access Control Models and Technologies, SACMAT 2012 pp. 13–22 (2012)

10. Wang, Y., Zheng, J., Sun, C., Mukkamala, S.: Quantitative security risk assessment of android permissions and applications. In: Wang, L., Shafiq, B. (eds.) DBSec 2013. LNCS, vol. 7964, pp. 226–241. Springer, Heidelberg (2013)
11. Xu, W., Zhang, F., Zhu, S.: Permlyzer: analyzing permission usage in android applications. In: ISSRE, pp. 400–410. IEEE (2013)
12. Zhou, Y., Jiang, X.: Dissecting android malware: characterization and evolution. In: Proceedings of the 33rd IEEE Symposium on Security and Privacy, Oakland 2012, pp. 95–109 (2012)
13. Zhou, Y., Wang, Z., Zhou, W., Jiang, X.: Hey, you, get off my market: detecting malicious apps in official and alternative android markets. In: Proceedings of the 19th Network and Distributed System Security Symposium, NDSS (2012)

Markov Based Social User Interest Prediction

Dongyun An[1,2(✉)] and Xianghan Zheng[1,2]

[1] College of Mathematics and Computer Science,
Fuzhou University, Fuzhou, China
dongyun_an@163.com, xianghan.zheng@fzu.edu.cn
[2] Fujian Key Laboratory of Network Computing and Intelligent Information
Processing, Fuzhou, China

Abstract. In this paper, we propose a new approach to predict users' interest eigenvalues based on multi-Markov chain model, which provides a better personalized service for the users timely. We first collect a dataset from Sina Weibo that includes 4613 users and more than 16 million messages; Then, preprocess data set to obtain users' interest eigenvalues. After that, divide users into several categories and establish multi-Markov chain to predict users' interest eigenvalues. Our experiments show that using multi-Markov model to predict users' interest eigenvalues is feasible and efficient, and could predicting both long-term and short-term user interests based on a suitable selection of the initial state distribution, λ.

Keywords: Social network · Sole–Markov chain · Enhanced–Markov chain · Interest eigen values

1 Introduction

With the development of Internet technology and the emerging forms of media, the Internet entered into a mass of information age [1]. At the same time a new type of information sharing and publishing platform (such as Weibo emergences, so the degree of Internet users' participation and active in China is showing explosive growth. Through this platform, users can express their views by posting some original essays or sharing information [2]. Therefore, effective feature learning and interest prediction [3] is significant not only for users (e.g., looking for users with similar interests [4]), etc.), but also for service providers in a set of application scenarios (e.g., user behavior analysis, personalized recommendation).

Most existing research considers user interest prediction from mainly three aspects: user registration information [5], browsing and posting history [6], and social interaction and relationship [7]. But prediction performance based on these aspects is unsatisfactory. In this paper, we investigate the social network environment and propose a Markov chain model that is feasible and efficient in predicting both long-term and short-term user interests. The contribution of this paper can be concluded that: first, develop specific data crawler to collect dataset from Sina Weibo. After features extraction through a set of operations, each user could be converted and represented as a feature vector; second, obtain user interest eigenvalue sequence by establishing a

© Springer International Publishing Switzerland 2015
A. Bikakis and X. Zheng (Eds.): MIWAI 2015, LNAI 9426, pp. 376–384, 2015.
DOI: 10.1007/978-3-319-26181-2_35

sole–Markov chain model; implement the SOM algorithm to find similarities among users and construct enhanced–Markov chain model that merges users into specific predefined interest categories; finally, conduct experiments to validate the feasibility and efficiency of the proposed solution; validate that the proposed solution can be implemented for both long-term and short-term user interest predictions.

The rest of the paper is organized as follows. Section 2 presents the background of social network and Markov Chain and reviews existing research on user interest prediction. Section 3 introduces dataset preprocess and feature vector extraction. Section 4 describes the construction of sole–Markov chain and enhanced–Markov chain models for user clustering and interest prediction. Experiments and evaluations are conducted in Sect. 5. Finally, conclusions are drawn in Sect. 6.

2 Related Work

In recent years, a lot of research on how to predict user interest has been undertaken. In the view of industry, Twitter and Facebook mention in their reports that they are using AI (deep learning) to understand the significance behind users' posting messages. However, the detailed techniques are proprietary and refuse to public. In academia, related works are: Attenberg et al. [8] propose a user interest prediction mechanism by analyzing the content of messages posted by users or by analyzing their interest eigenvalues. Xu et al. [9] propose a modified author-topic model to discover topics of interest on Twitter by filtering interest-unrelated tweets (noisy posts) from the aggregated user profiles. Yan et al. [10] establish a human dynamic model codriven by interest and social identity and show that user interest in sending posts is positively correlated with the number of comments on their previous posts. In the field of future interest prediction. Nori et al. [11] propose a new graphic representation (Action Graph) for modeling user multinomial with time-evolving actions and predict user interest by computing the similarity between each user and a set of resources. However, this ignores the influence of the user's friends on his or her interests.

In summary, existing research explores user interest from mainly three aspects: user registration information, browsing and posting history, social interaction and relationships. Compared to existing research, our work contains a few distinguished points: (a) investigate and consider the factor of time and influence among friends or similar users, and propose a possible solution that combines Markov model with clustering technology; (b) through the construction of the sole–Markov chain and enhanced–Markov chain models, the proposed solution is capable of providing excellent performance with the true positive rate of clustering reaching 88.5 %.

3 Dataset Collection and Analysis

We develop a specific data crawlers and feature collection mechanisms for dataset collection: Firstly, Hundred normal users (celebrity, company, and government that post, repost, and comment frequently) with 20 interest categories of Weibo messages are manually selected as the data source. Secondly, specific data crawlers are developed for the ordinary user and for the hot Weibo category. And finally, 4,612 Weibo users, and 20 categories of hot Weibo messages are extracted. After that, for each user, the basic user information (e.g., username) is retrieved by the Weibo API. Through the username, it is possible to retrieve a set of message IDs through which the text messages can be obtained. Finally, more than 16 million messages are crawled. Finally, for each user, a feature vector is constructed according to the crawled user and the message information with the operations in the following section.

As soon as the dataset is crawled, it is preprocessed [12] and each user's messages are converted into a vector which will be used in the establishment of the model.

1. Word Segment. This paper uses the Chinese Institute of Computing segmentation system (ICTCLAS) [13] divided user message content into separated words.
2. Frequency Statistics. The TF-IDF (term frequency–inverse document frequency) [14] algorithm to obtain the keywords appearing in 20 predefined interest categories is used. Finally, the top *50* keywords in each category are extracted. After de-duplication, *579* keywords as user interest eigenvalues are obtained.
3. Feature Vectors Generation. Based on these *579* keywords, it is possible to convert each user's messages into a feature vector.

4 Markov Based User Interest Modeling

Figure 1 illustrates the system framework of proposed interest prediction solution. The concept is: After generation of a series of feature vectors (described in previous section), Markov chain model is implemented to construct the prediction model, and generates a series of user interest categories.

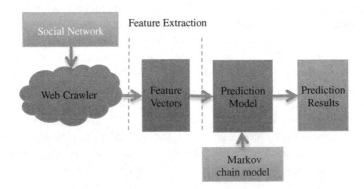

Fig. 1. Overview of interest prediction model

4.1 Sole-Markov Chain Based Interest Prediction (SMC)

According to user interest eigen values, a sole–Markov chain model [15] can be constructed.

A sole–Markov chain model can be represented as a triplet, $MC = <X, A, \lambda>$, where X is a discrete random variable in the range of $\{x_1, x_2, \ldots, x_n\}$, in which each x_i represents user interest eigenvalue. A is the transition rate matrix and λ is the initial state distribution represented as followed:

$$A = (p_{ij}) = \begin{bmatrix} P_{11} & P_{12} & \cdots & P_{1j} & & P_{1n} \\ P_{21} & P_{22} & \cdots & P_{2j} & & P_{2n} \\ \cdots & \cdots & \cdots & \cdots & \cdots & \cdots \\ P_{i1} & P_{i2} & \cdots & P_{ij} & \cdots & P_{in} \\ \cdots & \cdots & \cdots & \cdots & \cdots & \cdots \\ P_{n1} & P_{n2} & \cdots & P_{nj} & \cdots & P_{nn} \end{bmatrix} \tag{1}$$

$$\lambda = (p_i) = (p_1, p_2, \ldots, p_n) \tag{2}$$

Where $p_{ij} = P(X_t = x_j \mid X_{t-1} = x_i)$ refers to the transition probability from state x_i to state x_j; $p_i = P(X_{t=0} = x_i)$.

Through the collection of user messages in a certain time period, it is possible to extract a sequence of user interest eigenvalues (variables x). After that, with the maximum likelihood estimation function, it is possible to estimate the value of the parameters in a SMC model, referred to in the following formulas:

$$p_{ij} = \frac{S_{ij}}{\sum\limits_{j=1}^{n} S_{ij}} \qquad p_i = \frac{\sum\limits_{j=1}^{n} S_{ij}}{\sum\limits_{i=1}^{n} \sum\limits_{j=1}^{n} S_{ij}} \tag{3}$$

Where S_{ij} refers to the number of state pairs (x_i, x_j) appearing in users' messages posted.

Let vector $H(t) = [0, 0, \ldots, 1]$ refer to the user interest eigenvalue sequence in time point t and $V(t) = [P(X_t = x_1), P(X_t = x_2), \ldots, P(X_t = x_n)]$ refer to the probability of each eigenvalue. Therefore, it is possible to predict user interest eigenvalue with the formula 4. And the most related user interest eigenvalue is the highest probability value in vector $V(t)$. With the multistage weighted combination model (for considering historical user interest eigenvalues), prediction accuracy can be improved as represented in formulas 5 and 6:

$$V(t) = H(t-1) \times A \tag{4}$$

$$V(t) = w_1 H(t-1) \times A^1 + w_2 H(t-2) \times A^2 \\ + \cdots + w_h H(t-h) \times A^h \tag{5}$$

$$w_1 + w_2 + \cdots + w_h = 1 \tag{6}$$

Experimental results (Sect. 5.2) show that prediction accuracy increases with the higher value of h, until stabilized eventually.

4.2 Enhanced-Markov Chain Based Interest Prediction (EMC)

Based on the SMC, we further propose an Enhanced-Markov chain based interest prediction. First, two assumptions about user interest eigenvalue sequences are described:

Assume that there are K categories of interests, represented as $C = \{c_1, c_2, \ldots, c_k\}$, $P(C = c_k)$ refers to the probability of the *i-th* category the user belongs to, then, for each user:

$$\sum_{k=1}^{K} P(C = c_k) = 1 \tag{7}$$

Assume that users in the same interest category have similar behavior features and that the corresponding interest eigenvalues sequences are random process that follow the discrete homogeneous Markov chain.

With above two assumptions, it is possible to construct a user interest prediction classification model containing multiple Markov chains, known as the EMC model.

The EMC interest model is defined as a quaternion: $<X, K, P(C), MC>$, in which X is a discrete random variable in range $\{x_1, x_2, \ldots, x_n\}$, where each x_i represents an interest eigenvalue, $C = \{c_1, c_2, \ldots, c_k\}$ represents a group of user interest categories with the number k, $P(C = c_k)$ refers to the probability of the i-th category the user belongs to, $MC = \{mc_1, mc_1, \ldots, mc_k\}$ expresses a set of Markov chains and each element mc_k is the Markov eigenvalue chain that belongs to a specific category c_k. The transition rate matrix A_k of mc_k and the initial state distribution λ_k could be expressed as:

$$A = (p_{kij}) = \begin{bmatrix} P_{k11} & P_{k12} & \cdots & P_{k1j} & & P_{k1n} \\ P_{k21} & P_{k22} & \cdots & P_{k2j} & & P_{k2n} \\ \cdots & \cdots & \cdots & \cdots & \cdots & \cdots \\ P_{ki1} & P_{ki2} & \cdots & P_{kij} & \cdots & P_{kin} \\ \cdots & \cdots & \cdots & \cdots & \cdots & \cdots \\ P_{kn1} & P_{kn2} & \cdots & P_{knj} & \cdots & P_{knn} \end{bmatrix} \tag{8}$$

$$\lambda_k = (p_{ki}) = (p_{k1}, p_{k2}, \ldots, p_{kn}) \tag{9}$$

According to Definition 3, based on user interest eigenvalue sequences, it is possible to construct a set of Markov chains. Formula 10 are expressed for calculating p_{kij}, with p_{ki} belonging to A_k. Where k represents the number of interest categories; S_{kij} represents the number of status pairs (x_i, x_j) appearing in user content; a_{kij} is a super parameter as formula 11 shows:

$$p_{kij} = \frac{S_{kij} + \alpha_{kij}}{\sum_{j=1}^{n} (S_{kij} + \alpha_{kij})} \qquad p_{ki} = \frac{\sum_{j=1}^{n} S_{kij} + \alpha_{kij}}{\sum_{i=1}^{n} \sum_{j=1}^{n} (S_{kij} + \alpha_{kij})} \tag{10}$$

$$\alpha_{kij} = \frac{\beta}{n \times n} \tag{11}$$

Where β is the constant value of the problem domain size n.

In cases where the transfer matrixes of two users have a high degree of similarity, δ_{kl}, let the two matrixes merge together, with the calculation formulas:

$$CE(p_{ki}, p_{li}) = \sum_{j=1}^{n} p_{kij} \log \frac{p_{kij}}{p_{lij}} \tag{12}$$

$$Similarity\ (A_k, A_l) = \sum_{i=1}^{n} CE(p_{ki}, p_{li})/n \tag{13}$$

$$\delta_{kl} = Similarity\,(mc_k, mc_l)$$
$$= \frac{2}{Similarity\,(mc_k, mc_l) + Similarity\,(mc_l, mc_k)} \tag{14}$$

Where $CE(p_{ki}, p_{li})$ is the cross entropy of p_{ki} and p_{li}. In case the δ_{kl} value falls between mc_k and mc_l, the Markov chain is large enough or infinite and the corresponding two users are regarded to be in the same interest category, with merging formulas that follow:

$$P_{(k+l)ij} = \frac{S_{kij} + S_{lij} + \alpha_{(k+l)ij}}{\sum_{j=1}^{n} \left(S_{kij} + S_{lij} + \alpha_{(k+l)ij}\right)} \tag{15}$$

$$P_{(k+l)i} = \frac{\sum_{j=1}^{n} \left(S_{kij} + S_{lij} + \alpha_{(k+l)ij}\right)}{\sum_{i=1}^{n}\sum_{j=1}^{n} \left(S_{kij} + S_{lij} + \alpha_{(k+l)ij}\right)} \tag{16}$$

Step by step, user interest prediction with the EMC model can be generated.

5   Experiment Analysis

In this section, an experimental analysis of our proposed solution from four aspects is introduced: a user clustering experiment, prediction comparisons between the SMC and EMC models. Based on the collected dataset containing 4600 users, 3700 users are randomly selected with the Pareto principle as training data and the messages of the remaining 900 users are used as testing data. For simplification, the impact of the events that cause interruption is neglected. The experiment is carried out in MATLAB environment running in two Core i5-3470, 2 * 3.20 GHZ CPU.

5.1   User Clustering

The SOM neural network algorithm [16] is used to cluster users. The SOM algorithm reduces user n-dimensional original transfer matrixes into two-dimensional matrixes and keeps the original topology of the user transfer matrix. From the clustering result shown in Fig. 2, it can be seen that a set of clusters are formed, but these clusters have a lot of noise data. Further investigation reveals that this is caused by a group of spammers who might distribute spam messages in a lot of interest fields. These spam messages greatly reduce prediction accuracy. For noise filtering, the independent component analysis method (provided by Matlab) to remove spammers is imported and denoised user interest clusters are obtained, as shown in Fig. 3.

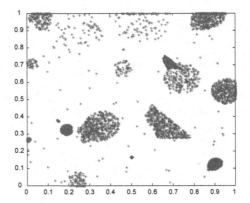

Fig. 2. User classification based on SOM neural network algorithm

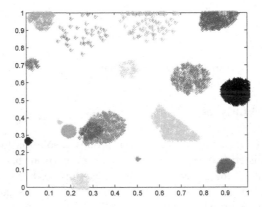

Fig. 3. User classification after denoised

5.2 Prediction Comparisons Between the SMC and EMC Models

In this experiment, the selected user in Sect. 5.1 was reused and the prediction of accuracy between the SMC and EMC models was compared. The results in Fig. 4 show that (1) the prediction is independent of the number of order h (similar to the result of the EMC model described in Sect. 4.2) and (2) the EMC model is capable of separating interest categories with bigger intervals (from 0.03 to 0.35) than the SMC model and, therefore, capable of obtaining the most suitable interest category classification result.

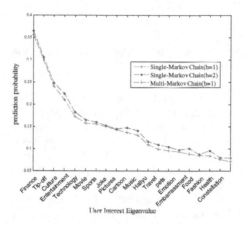

Fig. 4. Prediction accuracy of sole and enhanced-Markov chain model

From the training dataset, 20 users from each category, with 400 users in total, are randomly selected. After that, both SMC and EMC based approaches are implemented, with the average prediction of accuracy results listed in Table 1. The results show that the average value of the SMC model is only 0.5249, with a variance value of 0.0460, while the corresponding values of the EMC model are 0.8699 and 0.0007. This further proves that the EMC based interest prediction model is capable of achieving better accuracy of prediction.

Table 1. Predictive accuracy of sole and enhanced-Markov chain model

Category	SMC	EMC
Average	0.5249	0.8699
Variance	0.0460	0.0007

6 Conclusions

Social user interest prediction has become an important topic in the social network research field. In this paper, interest prediction based on the Markov chain modeling on clustered users is introduced. The solution considers user content feature and obtains user interest eigenvalue sequences to a establish SMC model; implement user clustering algorithms to construct a EMC model that classifies different users into specific predefined interest categories. Through a multitude of analyses, experiments and evaluations, it can be concluded that the proposed solution is feasible, efficient, and capable of achieving a much higher accuracy of prediction than any of the other existing approaches.

Acknowledgements. The authors would like to thank the support of the Technology Innovation Platform Project of Fujian Province under Grant No. 2009J1007, the Program of Fujian Key Project under Grant No. 2013H6011, the Natural Science Foundation of Fujian Province under Grant No. 2013J01228.

References

1. Wasserman, S.: Social network analysis: Methods and applications. Cambridge University Press, Cambridge (1994)
2. Statista, in: http://www.statista.com/
3. Bao, H., Li, Q., Liao, S.S., et al.: A new temporal and social PMF-based method to predict users' interests in micro-blogging. Decis. Support Syst. **55**(3), 698–709 (2013)
4. Chen, K.H., Han, P.P., Wu, J.: User clustering based social network recommendation. Jisuanji Xuebao(Chinese Journal of Computers) **36**(2), 349–359 (2013)
5. Yang, S.H., Long, B., Smola, A., et al.: Like like alike: joint friendship and interest propagation in social networks. In: Proceedings of the 20th International Conference on World Wide Web, pp. 537–546. ACM (2011)
6. Van Iddekinge, C.H., Putka, D.J., Campbell, J.P.: Reconsidering vocational interests for personnel selection: The validity of an interest-based selection test in relation to job knowledge, job performance, and continuance intentions. J. Appl. Psychol. **96**(1), 13 (2011)
7. La Greca, A.M., Harrison, H.M.: Adolescent peer relations, friendships, and romantic relationships: Do they predict social anxiety and depression? J. Clin. Child Adolesc. Psychol. **34**(1), 49–61 (2005)
8. Attenberg, J., Pandey, S., Suel, T.: Modeling and predicting user behavior in sponsored search. In: Proceedings of the 15th ACM SIGKDD International Conference on Knowledge Discovery and Data Mining, pp. 1067–1076. ACM (2009)
9. Xu, Z., Lu, R., Xiang, L., et al.: Discovering user interest on twitter with a modified author-topic model In: 2011 IEEE/WIC/ACM International Conference on IEEE Web Intelligence and Intelligent Agent Technology (WI-IAT), vol. 1, pp. 422–429 (2011)
10. Yan, Q., Yi, L., Wu, L.: Human dynamic model co-driven by interest and social identity in the Microblog community. Physica A **391**(4), 1540–1545 (2012)
11. Nori, N., Bollegala, D., Ishizuka, M.: Interest prediction on multinomial, time-evolving social graph. In: IJCAI, vol. 11, pp. 2507–2512 (2011)
12. Phan, X.H., Nguyen, C.T., Le, D.T., et al.: A hidden topic-based framework toward building applications with short Web documents. IEEE Trans. Knowl. Data Eng. **23**(7), 961–976 (2011)
13. Wang, C., Jin, C.: Based on the established vocabulary of yi automatic segmentation system design and implementation. Sci. Technol. Eng. **10**, 020 (2012)
14. Teevan, J., Ramage, D., Morris, M.R.: # TwitterSearch: a comparison of microblog search and web search. In: Proceedings of the Fourth ACM International Conference on Web Search and Data Mining, pp. 35–44. ACM (2011)
15. Liu, C.: Stochastic Process (fourth edition). Huazhong University of Science and Technology Press, Wuchang Yu Jiashan, vol. 8, pp. 1–113 (2008)
16. Ghaseminezhad, M.H., Karami, A.: A novel self-organizing map (SOM) neural network for discrete groups of data clustering. Appl. Soft Comput. **11**(4), 3771–3778 (2011)

Feature-Driven Formal Concept Analysis
for Malware Hierarchy Construction

Nguyen Thien Binh[1(✉)], Tran Cong Doi[2(✉)], Quan Thanh Tho[1(✉)],
and Nguyen Minh Hai[1]

[1] Ho Chi Minh City University of Technology, Ho Chi Minh City, Vietnam
551105019@stu.hcmut.edu.vn, qttho@cse.hcmut.edu.vn,
551307910@hcmut.edu.vn
[2] Dong Nai University, Dong Nai, Vietnam
tcdoi@dnpu.edu.vn

Abstract. As the number of computer viruses have rapidly been increasing
nowadays, automatic classification of viruses into a concept hierarchy is one of
the emerging issues of malware research community. Among various approaches, Formal Concept Analysis (FCA) is a well-known technique which is
capable of producing a concept lattice/hierarchy from a formal concept. However, the traditional approach of concept representation offered by FCA is not
enough to capture the semantics of virus behaviors.

In recent literature, the operational mechanism of virus has often been represented by temporal logic for formal analysis. This motivates us to extend FCA
into F-FCA (Feature-driven FCA) to overcome the discussed problem. In
F-FCA, each formal object and concept is associated with a temporal logic
formula. We also introduce an on-the-fly algorithm, known as FOCA, to generate a concept hierarchy on F-FCA by means of an object-joining operator.
Experiments on a real dataset of 3000 virus samples demonstrate the efficiency
of our approach, as compared to the traditional approach.

Keywords: Computer virus · Malicious software · Malware detection · Formal
concept analysis · Feature-driven FCA · FOCA

1 Introduction

Computer virus (from now we call *virus*), or *malware*, is a segment of computer
programs which executes actions to harm to a computer system potentially. When
infecting a file, virus also copies a unique syntactic pattern, known as *signature*, to the
file. When a virus discovers this pattern from a file, it recognizes that this file is infected
and does not replicate itself. Based on this characteristic, most of industry anti-virus
programs detect virus by scanning whether signature appears or not. However, this
method has difficulty in dealing with advanced viruses such as polymorphic and
metamorphic virus [1, 2] because these viruses virtually create different signature after
each infection.

To solve this problem, recent studies have suggested a method of virus detection
based on determining hazardous behavior instead of matching pattern [3–5]. We
consider an example in Fig. 1a, which illustrates harmful behavior of the well-known

© Springer International Publishing Switzerland 2015
A. Bikakis and X. Zheng (Eds.): MIWAI 2015, LNAI 9426, pp. 385–396, 2015.
DOI: 10.1007/978-3-319-26181-2_36

Avron virus. This behavior includes pushing *zero* to the top of stack (by assigning zero for register *ABX* and pushing it to the top of stack), then the virus body will be executed to invoke *GetModuleFileNameA* in order to get the name and path of the victim file. Then the virus will proceed to replace the original code in the victim by the malicious code of the virus itself. Thus, each time the victim file is executed on a computer, it will infect virus on the whole system. This process will be ongoing and the number of infected computers will increase rapidly. Hence, even though Avron has several variants which have different signatures, the *behavior* of finding the name and path of the victim by means of *GetModuleFileNameA* is still always remained. However, this behavior can also be executed by different code execution as subsequently discussed.

ID	Sample Pattern	Logic Formulas	Meaning
A	mov ebx,0 push ebx call GetModuleFileNameA	**F**(mov(ebx,0)∧**X**push(ebx)∧ **X**call(GetModuleFileNameA)	*Avron virus*
B	mov ecx,0 push ecx call GetModuleFileNameA	**F**(mov(ecx,0∧**X**push(ecx)∧ **X**call(GetModuleFileNameA)	*Avron variant*
C	xor ebx,ebx push ebx inc a call GetModuleFileNameA	**F**(xor(ebx,ebx)∧**X**push(ebx)∧**X**inc(a)∧ **X**call(GetModuleFileNameA)	*Avron variant with junk code*
D	mov ebx,0 push ebx push 1 call GetModuleFileNameA	**F**(mov(ebx,0)∧**X**push(ebx)∧ **X**push(1)∧ **X**call(GetModuleFileNameA))	*Not a virus*
E	call GetModuleFileNameA push ebx mov ebx,0	**F**(mov(ebx,0)∧ **X**call(GetModuleFileNameA) ∧**X**push(ebx))	*Not a virus*

F	sub esp, 4
	mov [esp], 0
	dec b
	jmp GetModuleFileNameA
		F(sub(esp,4)∧**X**mov([esp],a)∧**X**dec(b)∧
		Xjmp(GetModuleFileNameA))
			Complex Avron variant with junk code

Fig. 1. Some code segments illustrating the viral behaviors

One of the interesting approaches of describing the virus' behavior is using logical formula. *Temporal Logic* (TL) [6] is used commonly due to its capability of describing correctly the execution sequence of virus behaviors. For example, the viral code in Fig. 1a will be represented by a TL formula of *F(mov(ebx,0)∧Xpush(ebx)∧Xcall (GetModuleFileNameA)*, in which the operator *F* is understood as *Eventually*, and the operator *X Next*. This logic formula can be interpreted as: "*in a binary code, if eventually there is an instruction assigning 0 for register ebx, and then value of ebx is pushed to stack and subsequently GetModuleFileNameA function is called, this binary code is infected by Avron virus*".

However, using logic formula to represent virus' behaviors leads to difficulty in classifying virus, because viruses are no longer syntactically distinguished as in signature-based approach. Thus, we need to distinguish virus semantically based on the semantic meaning of the logical formulas used to represent them.

To overcome this problem, we propose an approach based on Formal Concept Analysis (FCA) [7], because this method supports to construct a concept lattice representing the hierarchical relationships among the sets of objects in data set. Since then, FCA has strong support for conceptual clustering, in which objects are grouped to form real-life concepts. To support representing the logic formula for these concepts, we extend the traditional FCA to F-FCA (*Feature-driven Formal Concept Analysis*), in which each concept has a *featured property* represented as a logic formula. The contributions of our research include:

- Proposing the F-FCA model in which a formal concept created by FCA is represent as a logic-based featured property, so that the concept can capture more semantic information.
- Proposing *Feature-driven On-the-fly Conceptual Clustering* (FOCA) algorithm to create concept hierarchy from F-FCA efficiently for large data sets.
- Using FOCA to create Malware Hierarchy from the data set consisting of 3000 actual viruses.

The rest of this paper is organized as follows. Section 2 introduced a motivating example. Section 3 reviews some related works. In Sects. 4 and 5, we present F-FCA and FOCA. Section 6 discusses our experiments. Finally, Sect. 7 concludes the paper.

2 Motivating Example

To illustrate our contribution, let us consider again some virus behavior given in Fig. 1. Figure 1a shows a sample of the Avron virus as described. Figure 1b shows a variant of this virus, which uses *ebx* instead of *eax*. Figure 1c shows another variant of this virus, in which XOR is executed to assign zero for *eax*. Figure 1f shows a sophisticated variant, in which the assignment instruction and call stack are replaced with instruction access via pointer and instruction jump to entry address of function. Particularly, in Fig. 1c and f, the virus use *obfuscation techniques* of *junk code*, which insert some instructions that do not make sense, such as *inc a* or *dec b*, where *a* and *b* are two dumping variables[1].

Meanwhile, there are no virus found in Fig. 1d and e, but the instructions involved on those pieces of code are also quite similar. The code in Fig. 1d contains assigning zero for stack and then calling *GetModuleFileNameA*. However, after executing instructions between two calling instructions, the value of stack has been changed, so this is a harmless program. In Fig. 1e, instructions of assigning 0 for call stack and calling function are also executed but in inversed order, so this also makes no harm. If

---

[1] Detailed discussion on obfuscation technique is beyond the scope of this paper.

Table 1. Formal context created from the segment of programs in Fig. 1.

	mov	ebx	push	call	moduleA	ecx	xor	inc	a	sub	esp	[]	dec	b	jmp	1	0	4
A	x	x	x	x	x												x	
B	x		x	x	x	x											x	
C		x	x	x	x		x	x	x									
D	x	x	x	x	x											x	x	
E	x	x	x	x	x												x	
F	x				x					x	x	x	x	x	x			x

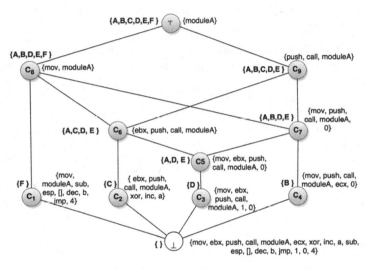

Fig. 2. The concept lattice of virus is generated by FCA method

naively applying the traditional FCA approach[2] on six routines illustrated in Fig. 1, we would obtain formal context and concept lattice as shown in Table 1 and Fig. 2.

As shown in Fig. 2, three sample programs A, D and E will be grouped into a concept because they have the same attributes. It is not reasonable because only A is a virus in three programs. Moreover, the intent representation of each concept is not precise enough to determine a virus set. For example, in the *C8* concept, its intent of {*mov, call ModuleA*} is too general since a normal program can also invoke those instructions as illustrated in the segment programs of D and E.

With our new proposed F-FCA method, we represent each concept by a *featured property*, denoted as a *temporal logic* (TL) formula, as described in Fig. 3. In the meantime, each concept set of virus is represented by a logical formula TL capturing correctly the corresponding virus behaviors. Thus, our concept lattice can be divided into two sets virus and non-virus respectably. To keep the figure better readable, we

---

[2] In this paper, we do not discuss on details the traditional FCA technique, of which interested readers can refer to [7].

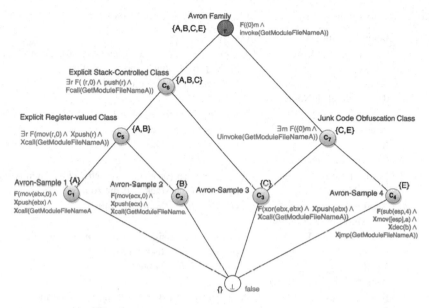

Fig. 3. The concept lattice generated by F-FCA

only display concepts corresponding to virus samples. In Table 2, we describe logical formulas representing for the concepts appearing in Fig. 3 and their meaning.

3 Related Works

3.1 Formal Concept Analysis

Formal Concept Analysis [7] is a data analysis technique aiming at recognizing formal concepts of a formal context and constructing a concept lattice accordingly. In data mining, this technique has many diverse applications, such as bioinformatics [8], social network analysis [9], semantic web search [10], technology trend monitoring [11], discovery association rules of Web pages [12], terrorist threat assessment [13], text adaptation [14] and software engineering [15]. More specifically, [16] used the FCA technique to analyze the state of publishing paper in this field.

In order to construct concept lattice more efficient and meaningful, many enhancements have been proposed. Zhang (2013) [17] combined FCA, Chu space and Domain Theory to analyze the dependency among the attributes. [18] proposed using closure operator to analyze the dependency among the attributes. Concept location method was proposed by [19]. In work of [20], an algebraic structure was proposed to build the concept hierarchy and ontology merging from concept lattice.

When using the FCA technique for conceptual clustering, a challenge is the contextual implication of the attributes to represent the concept [21]. [10] proposed analyzing the independent sub-contexts. [22] used the set of attributes to make preference models, hence introducing the *ceteris paribus preferences* [9]. Using linguistic hedges have been proposed by [23].

Table 2. Virus concepts in Fig. 3 are described by logical formulas and interpretations

Concept	Logical formulas	Description
{A}	**F**(mov(ebx,0)∧**X**push (ebx)∧ **X**call (GetModuleFileNameA))	Assigning zero for *ebx* by *mov(ebx, 0)*. Then *ebx* is pushed to the stack and the routine *GetModuleFileNameA* is called
{B}	**F**(mov(ecx,0)∧**X**push (ecx)∧ **X**call (GetModuleFileNameA))	Assigning zero for *ecx* by *mov(ecx, 0)*. Then *ecx* is pushed to the stack and the routine *GetModuleFileNameA* is called
{C}	**F**(xor(ebx,ebx)∧**X**push (ebx)∧ **X**call (GetModuleFileNameA))	Assigning zero for *ebx* by *xor(ebx, ebx)*. Then *ebx* is · pushed to the stack and the routine *GetModuleFileNameA* is called
{A,B}	∃r **F**(mov(r,0)∧**X**push (r)∧ **X**call (GetModuleFileNameA))	A register *r* is assigned zero by *mov(r,0)*. Then *r* is pushed to the stack and the routine *GetModuleFileNameA* is called
{A,B, C}	∃r **F**((r,0)∧push(r)∧ **F**call (GetModuleFileNameA))	A register *r* is pushed to stack while its value is zero, then the routine *GetModuleFileNameA* is called
{C,E}	∃m **F**({0}m∧ **U**ivoke (GetModuleFileNameA))	Exists a movement that the value of stack is zero, and unchanged until the routine GetModuleFileNameA is called
{A,B,C, E}	**F**({0}m∧ invoke (GetModuleFileNameA))	The routine *GetModuleFileNameA* is called when value of top of stack is zero

However, while the domain concepts are represented by the complex logic behaviors such as the malicious actions of virus, the concept representation requires techniques to analyze and handle the relationship between the complex attributes, which remained unresolved by existing works.

3.2 Malware Classification and Specification

In industry, the current common technique handling malware is based on virus signature, i.e. using syntactic pattern matching. Due to the rapidly increased amount of virus, malware signatures are classify into classes to speed up processing. Table 3 illustrated some common viruses. However, for some advance virus, the signature-based analysis is sometime easily cheated by obfuscation techniques. To tackle this, the formal methods propose using logic formulas, especially temporal logic to describe the behavior of the virus [3–5]. Since then, virus classification becomes more difficult, because (i) the classification must investigate the logic relationship of formulas describing the behavior virus and (ii) a class of virus must be represented by an appropriate formula.

Table 3. Some common virus groups [2]

No	Description	Characteristic harmful behavior	Representative
1	Change API addresses	Read Hast Table to find out API addresses	Aztec, Cabanas, Highway.
2	Use interrupts (SHE)	Actively create interrupt then use exception handler mechanism to execute the desired code.	Anon, Assault, Alcaul, Agent, Adson, Plexus, Vapsup, SDBot, Sasser.
3	Access to PEB structure to detect debugging mode.	Check the 2nd byte of PEB to detect whether system is running in debugging mode if this value is non-zero	Abul, Agent, Cycler, Migotrup, Gaobot, Zlob.
4	Hide entry point	Access to PE of infected files to change a harmless calling function by a function call the virus code.	Amon, Cjdisease, Elfinfect, Dissolution, Egypt32.
5	Self-Encryption	Hide harmful code by encoding method.	Adson, Aldebaran, Belial, Cefet, Sasser, Yoddos.

4 Feature-Driven Formal Concept Analysis for Concept Hierarchy Construction

Definition 1 (Featured Formal Context). A featured formal context is a triple $K = (G, M, I = \varphi(G \times f(M))$ where G is a set of objects, M is a set of attributes, $f(M)$ is a logical formula over subset M' of M, denoted as $M' = f\_domain(f)$ and I is a relation on domain $G \times f(M)$.

Thus, in comparing to the formal context on FCA, each object $o \in G$ in featured formal context is represented by a logical formula f_o. We say the relation $i \in I$ is relevant to f_o, denoted as φ_rel (i, f_o) if $i \in f\_domain$ (f_o).

For example, the formal context in Fig. 1 can be expressed in a featured formal context as follows. We only present the information of $f(Object)$ and $f\_domain(f)$ since the set G and M still remain the same.

Definition 2 (Featured Formal Concept). Given a featured formal context $K = (G, M, I)$ we define $A' = \{m \in M | \forall g \in A: m \in f\_domain(f(g))\}$ for $A \subseteq G$ and $B' = \{g \in G | \forall m \in B: m \in f\_domain(f(g))\}$ for $B \subseteq M$. A *featured formal concept* (or *featured concept*) of a featured formal context (G, M, I) is a triple (A, B, α) where $A \subseteq G$, $B \subseteq M$, $A' = B$ and $B' = A$; α is a logical formula that $\forall g \in A$, $\alpha \vDash \varphi(g)$.

For example, in Fig. 3, concept C5 has two objects A and B. We have $\varphi(A) = F(mov(ebx,0) \wedge Xpush(ebx) \wedge Xcall(GetModuleFileNameA)$ and φ $(B) = F(mov(ecx,0 \wedge Xpush(ecx) \wedge Xcall(GetModuleFileNameA)$; meanwhile C5 is represented as $\alpha(C5) = \exists r\ F(mov(r,0) \wedge Xpush(r) \wedge Xcall(GetModuleFileNameA))$. Then, the following propositions hold: $\alpha(C5) \vDash \varphi(A)$ and $\alpha(C5) \vDash \varphi(B)$.

Definition 3 (Featured Subconcept and Superconcept). Let (A_1, B_1, α_1) and (A_2, B_2, α_2) be two featured concepts of a featured formal context (G, M, I). (A_1, B_1, α_1) is the *subconcept* of (A_2, B_2, α_2), denoted as $(A_1, B_1, \alpha_1) \leq (A_2, B_2, \alpha_2)$, if and only if

$(A_1 \subseteq A_2)$ $(\Leftrightarrow B_2 \subseteq B_1)$ and $\alpha_1 \models \alpha_2$. Equivalently, (A_2, B_2, α_2) is the superconcept of (A_1, B_1, α_1).

For example, in Fig. 3, concept C6 is superconcept of concept C5, we have $\alpha(C6) = \exists r \, F((r,0) \wedge push(r) \wedge Fcall(GetModuleFileNameA))$. Thus, one can observe that $\alpha(C5) \models \alpha(C6)$.

Definition 4 (Featured Concept Lattice). A featured concept lattice of a featured formal context K is a set F(K) of all featured formal concepts of K with the partial order $\leq$.

Figure 3 is a simple featured concept lattice of virus generated by F-FCA applied on the featured formal context given in Fig. 4. On this lattice, all of viruses are group correctly as descendants of the superconcepts of *Avron family*.

Object	f(Object)	f_domain(f)
A	**F**(mov(ebx,0)∧**X**push(ebx)∧ **X**call(GetModuleFileNameA)	{mov, ebx, push, call, GetModuleFileNameA}
B	**F**(mov(ecx,0∧**X**push(ecx)∧ **X**call(GetModuleFileNameA)	{mov, ecx, push, call, GetModuleFileNameA}
C	**F**(xor(ebx,ebx) ∧**X**push(ebx)∧ **X**inc(a) ∧ **X**call(GetModuleFileNameA)	{xor, ebx, push, inc, call, GetModuleFileNameA}
D	**F**(mov(ebx,0)∧**X**push(ebx)∧ **X**push(1)∧**X**call(GetModuleFileNameA))	{mov, ebx, push, call, GetModuleFileNameA}
E	**F**(mov(ebx,0)∧ **X**call(GetModuleFileNameA) ∧**X**push(ebx))	{mov, ebx, push, call, GetModuleFileNameA}
F	**F**(sub(esp,4)∧**X**mov([esp],a) ∧**X**dec(b)∧**X**jmp(GetModuleFileNameA))	{sub, esp, mov, ⊔, dec,jmp, GetModuleFileNameA}

Fig. 4. A featured formal context

5   FOCA: Feature-Driven on-the-Fly Conceptual Clustering Algorithm

In order to construct the final featured concept lattice from a given featured formal context, obviously one needs to consider "grouping" objects into concepts. To do so, we introduce the object-joining operator as follows.

Definition 1 (Object-joining Operator). Object-joining operator ⋈ on two featured concepts $C_1 = (A_1, B_1, \alpha_1)$ and $C_2 = (A_2, B_2, \alpha_2)$ is defined as follows:
$C_1 \bowtie C_2 = \{C^*, B^*, \alpha^* = \alpha_1 \uplus \alpha_2\}$
where:

- $\uplus$ is *widening operator* two formulas α_1 and α_2. The simplest widening operator is the operator $\vee$.
- $C^* = \{x| \, x \in (C_1 \cup C_2) \text{ and } \alpha * \models \varphi(x)\}$
- $B^* = \{i| \, \exists \, g \in C * : g \circ i\}$

For example, when one performs C5 = C1 ⋈ C2 as illustrated in Fig. 3, we have φ(C5) = f(C1) ⊎ f(C2) = **F**(mov(ebx,0)∧**X**push(ebx)∧**X**call(GetModuleFileNameA)) ⊎ **F**(mov(ecx,0)∧**X**push(ecx)∧ **X**call(GetModuleFileNameA)) = ∃r (mov(r,0)∧**X**push (r)∧**X**call(GetModuleFileNameA)).

In order to perform the widening as described in above example, one needs a mathematical prover. To develop a general prover is an NP-hard problem (equivalent to solving SAT). However, in this study, we have developed a simple prover to work with simple logical expressions corresponding to popular instructions appear in the virus samples.

Based on object-joining operator, we developed FOCA algorithm to generate a concept hierarchy from a set of objects of a formal context (i.e. generating the final featured concept lattice). This algorithm is equivalent to the conceptual clustering algorithm described in [24]. However, FOCA comply on-the-fly approach, so it results in better efficiency.

Detail of FOCA can be described as follows.

```
For each object o in the formal context
    For each concept c on current concept lattice
        Add o as a concept of the current lattice
        If o ⋈ c = o' ≠ ε (empty) then
            If o' and o has the same featured formula then
    add o to c
        Else introducing new concept c'
```

Thus, the complexity of FOCA is N × M, where N is the number of objects and M is the number of concepts. If M ≪ N, the complexity is linear to N. It means that this algorithm can not perform in real-time, but can be done with a rather large number N.

6   Experiments

We performed experiments by generating malware concept hierarchy from a dataset of 3000 virus samples downloaded from VXHeaven[3]. From this dataset, we constructed the concept hierarchy using baseline method of traditional FCA-based conceptual clustering and our proposed F-FCA-based conceptual clustering.

To measure the efficiency of two methods, we evaluated the quality of the conceptual clusters created, using the following metrics.

Conceptual Cluster Goodness

Relaxation error [25] implies dissimilarities of items in a cluster based on attributes' values. Since conceptual clustering techniques typically use a set of attributes for concept generation, relaxation error is quite a commonly used measure for evaluating the goodness of conceptual clusters. The relaxation error RE of a cluster C is defined as

---

[3] http://vxheaven.org/.

$$RE(C) = \sum_{a \in A} \sum_{i=1}^{n} \sum_{j=1}^{n} P(x_i)P(x_j)d^a(x_i, x_j)$$

where A is the set of the attributes of items in C, $P(x_i)$ and $P(x_j)$ are the probabilities of items x_i and x_j occurring in C respectively, and $d_a(x_i; x_j)$ is the distance of x_i and x_j on attribute a. The cluster goodness G of cluster C is defined as

$$G(C) = 1 - RE(C)$$

Obviously, smaller cluster relaxation error implies better cluster goodness.

Hierarchical Relation Goodness

Average Uninterpolated Precision (AUP) [26] is defined as the sum of the *precision* value at each point (or node) in a hierarchical structure where a relevant item appears, divided by the total number of relevant items. Typically, AUP implies the goodness of a concept hierarchical structure. For evaluating AUP, we have manually classified the virus samples into classes. For each class, we extract 5 most frequent patterns. Then, we use these patterns as inputs to form retrieval queries and evaluate the retrieval performance using AUP.

Tables 4 and 5 give the performance results for G(C) and AUP measures. From Table 4, we have found that F-FCA has achieved better cluster goodness than FCA. This has shown the advantage of using logic-based features for representing object attributes. We have also found that when the number of samples gets larger, the performance on AUP gets better. In addition, the performance on AUP of F-FCA is generally better than that of FCA. It means that the logic formulas generated for conceptual clusters are appropriate formulas for representing the concept hierarchical structure.

Table 4. Performance results based on cluster goodness

Size of sample	150	200	250	500	1000	1500	2000	2500	3000
FCA	0.760	0.720	0.655	0.665	0.670	0.690	0.680	0.671	0.650
F-FCA	0.891	0.870	0.819	0.710	0.785	0.790	0.813	0.807	0.803

Table 5. Performance results based on AUP

Size of sample	150	200	250	500	1000	1500	2000	2500	3000
FCA	0.036	0.0509	0.095	0.139	0.219	0.265	0.265	0.281	0.285
F-FCA	0.049	0.099	0.148	0.182	0.273	0.314	0.379	0.388	0.395

7   Conclusion

In this paper, we extend the traditional FCA (Formal Concept Analysis) technique into F-FCA (Feature-driven FCA) technique, in which each object and formal concept is featured by a logic formula. This formalism allows us to capture and present precisely behaviors of virus when constructing a concept hierarchy of malware. As results, we successfully generated a Malware Hierarchy from a dataset of 3000 real virus samples. Experimental results show that the concept hierarchy developed by our proposed FFCA gained better quality than that from the traditional FCA.

References

1. Muttik, I.: Silicon implants. Virus Bulletin, pp. 8–10 (1997)
2. Szor, P.: Advanced code evolution techniques and computer virus generator kits. The Art of Computer Virus Research and Defense (2005)
3. Kinder, J., Katzenbeisser, S., Schallhart, C., Veith, H.: Detecting Malicious Code by Model Checking In: Julisch, K., Kruegel, C. (eds.) DIMVA 2005. LNCS, vol. 3548, pp. 174–187. Springer, Heidelberg (2005)
4. Song, F., Touili, T.: Efficient malware detection using model-checking. In: Giannakopoulou, D., Méry, D. (eds.) FM 2012. LNCS, vol. 7436, pp. 418–433. Springer, Heidelberg (2012)
5. Song, F., Touili, T.: Pushdown model checking for malware detection. Int. J. Softw. Tools Technol. Transfer 16(2), 147–173 (2014)
6. Huth, M., Ryan, M.: Logic in Computer Science: Modelling and reasoning about systems. Cambridge University Press (2004)
7. Ganter, B., Wille, R., Wille, R.: Formal concept analysis, vol. 284. Springer, Berlin (1999)
8. Coste, F., Garet, G., Groisillier, A., Nicolas, J., Tonon, T.: Automated enzyme classification by formal concept analysis. In: Glodeanu, C.V., Kaytoue, M., Sacarca, C. (eds.) ICFCA 2014. LNCS, vol. 8478, pp. 235–250. Springer, Heidelberg (2014)
9. Obiedkov, S.: Modeling Ceteris Paribus preferences in formal concept analysis. In: Cellier, P., Distel, F., Ganter, B. (eds.) ICFCA 2013. LNCS, vol. 7880, pp. 188–202. Springer, Heidelberg (2013)
10. Dubois, D., Prade, H.: Possibility theory and formal concept analysis: characterizing independent sub-contexts. Fuzzy Sets Syst. 196, 4–16 (2012)
11. Lee, C., Jeon, J., Park, Y.: Monitoring trends of technological changes based on the dynamic patent lattice: a modified formal concept analysis approach. Technol. Forecast. Soc. Chang. 78(4), 690–702 (2011)
12. Du, Y., Li, H.: Strategy for mining association rules for web pages based on formal concept analysis. Appl. Soft Comput. 10(3), 772–783 (2010)
13. Elzinga, P., Poelmans, J., Viaene, S., Dedene, G., Morsing, S.: Terrorist threat assessment with formal concept analysis. In: IEEE International Conference on Intelligence and Security Informatics (ISI), pp. 77–82. IEEE (2010)
14. Dufour-Lussier, V., Lieber, J., Nauer, E., Toussaint, Y.: Text adaptation using formal concept analysis. In: Bichindaritz, I., Montani, S. (eds.) ICCBR 2010. LNCS, vol. 6176, pp. 96–110. Springer, Heidelberg (2010)
15. He, N., Rümmer, P., Kroening, D.: Test-case generation for embedded simulink via formal concept analysis. In: Proceedings of the 48th Design Automation Conference, pp. 224–229. ACM (2011)
16. Doerfel, S., Jäschke, R., Stumme, G.: Publication analysis of the formal concept analysis community. In: Domenach, F., Ignatov, D.I., Poelmans, J. (eds.) ICFCA 2012. LNCS, vol. 7278, pp. 77–95. Springer, Heidelberg (2012)
17. Zhang, G.-Q.: Chu spaces, concept lattices, and domains. Electron. Notes Theor. Comput. Sci. 83, 287–302 (2013)

18. Ganter, B.: Two basic algorithms in concept analysis. In: Kwuida, L., Sertkaya, B. (eds.) ICFCA 2010. LNCS, vol. 5986, pp. 312–340. Springer, Heidelberg (2010)

19. Poshyvanyk, D., Gethers, M., Marcus, A.: Concept location using formal concept analysis and information retrieval. ACM Trans. Software Eng. Methodol. (TOSEM) **21**(4), 23 (2012)

20. Wang, L., Liu, X., Cao, J.: A new algebraic structure for formal concept analysis. Inf. Sci. **180**(24), 4865–4876 (2010)

21. Duquenne, V.: Contextual implications between attributes and some representation properties for finite lattices. In: Cellier, P., Distel, F., Ganter, B. (eds.) ICFCA 2013. LNCS, vol. 7880, pp. 1–27. Springer, Heidelberg (2013)

22. Obiedkov, S.: Modeling preferences over attribute sets in formal concept analysis. In: Domenach, F., Ignatov, D.I., Poelmans, J. (eds.) ICFCA 2012. LNCS, vol. 7278, pp. 227–243. Springer, Heidelberg (2012)

23. Belohlavek, R., Vychodil, V.: Formal concept analysis and linguistic hedges. Int. J. Gen Syst. **41**(5), 503–532 (2012)

24. Quan, T.T., Hui, S.C., Cao, T.H.: A Fuzzy FCA-based Approach to Conceptual Clustering for Automatic Generation of Concept Hierarchy on Uncertainty Data. In: CLA, pp. 1–12 (2004)

25. Chu, W.W., Chiang, K.: Abstraction of High Level Concepts from Numerical Values in Databases. In: KDD Workshop, pp. 133–144. Citeseer (1994)

26. Nanas, N., Uren, V., De Roeck, A.: Building and applying a concept hierarchy representation of a user profile. In: Proceedings of the 26th annual international ACM SIGIR conference on Research and development in informaion retrieval, pp. 198–204. ACM (2003)

Machine Learning Based Scalable and Adaptive Network Function Virtualization

Kun Li[1,2(✉)], Xianghan Zheng[1,2], and Chunming Rong[1,2]

[1] College of Mathematics and Computer Science, Fuzhou University, Fuzhou, China
kunlio@163.com, xianghan.zheng@fzu.edu.cn, chunming.rong@uis.no
[2] Fujian Key Laboratory of Network Computing and Intelligent Information Processing,
Fuzhou, China

Abstract. Due to the continuous development of SDN and NFV technology in recent years, it is important to improve the network performance to the users. But the traditional technology is not mature and has many shortcomings. In order to apply these two technologies (Adaptive and Autoscaling) in computer networks, we use SDN not only to separate the forwarding plane and control plane, but has the nature of the programmability also. Based on the actual business requirements for automatic deployment, NFV technology has the resources of virtualization and the characteristics of flexibility and fault isolation. Two kinds of technology are different, but they can work cooperatively very well. In this paper, we apply the algorithms of machine learning, combine the SDN and NFV technology, and build NFV dynamic control system architecture on the CloudStack cloud platform to provide users with customized service. Furthermore, in the fourth part, we added the feasibility of architecture to the home network and mobile core network.

Keywords: SDN · NFV · Machine learning · Network architecture

1 Introduction

Nowadays, network virtualization has already involved in many aspects in our life. In the background of rapid generation of virtualization technologies, NFV and SDN [1], these two high-profile virtualization technologies have received more and more attention all over the world. Both academia and industry have seen NFV and SDN as the development direction of computer network in the future.

In the era of big data, the demands of users are increasing rapidly on the performance of the network, instead of merely basic network performance requirements. Such as network speed, stability, delay, flexibility, programmability and customized and so on [2]. While on the one hand we enjoy the benefits of NFV and SDN, on the other hand, researchers are also facing many technical challenges. The first problem is how to implement the network in NFV ways, achieving the function of in hardware implementation of network. Second, we pay attention on how to manage NFV to meet dynamic control function of the network. It makes network operators to consider at the lowest cost to meet the functional requirements of growing network users. Though the study of NFV and SDN are gradually deepening, the two technologies still have defects in

© Springer International Publishing Switzerland 2015
A. Bikakis and X. Zheng (Eds.): MIWAI 2015, LNAI 9426, pp. 397–404, 2015.
DOI: 10.1007/978-3-319-26181-2_37

many of aspects [3]. As a result, the demand for user's network function is not a efficient solution in order to meet user's demand on network function of dynamic, automated deployment, configuration, and manage network services [4].

In this paper, the main work is to set up the framework of NFV dynamic control system which based on machine learning, and to focus on its applicability and feasibility and to introduce the relevant theory of SDN [5] and NFV. Then, the Sect. 2 introduces the concepts of SDN and NFV. The Sect. 3 proposes the structure of NFV dynamic control system. The Sect. 4 introduces the architecture used in the case of home network and mobile core network, as the examples to explain the feasibility and applicability. The paper conclusions are given in Sect. 5.

2   Related Work

2.1   NFV

NFV is a dynamic creation and management function of network. On the one hand, we can bind different network functions together to reduce the complexity of the network management, on the other hand, we can also use the function of the network function decomposition to smaller modules, which is a helpful way to improve the reusability and speed up the response time. NFV architecture has four main function modules: the coordinator, VNF manager, virtualization layer and virtualization infrastructure manager [6] respectively.

NFV's benefits can be summarized as the following: first, it can be formed through virtualization network to reduce the network infrastructure investment and energy consumption, such as virtual switches, routers, etc. Second, it improves the function of network deployment flexibility, because without the limitation of hardware facilities, deploy and configure of the network become easily. Moreover, it separates the network function and hardware to strengthen the programmability of network. In the end, it's easier to meet the user's custom service, because operators can dynamically provide service for the users, and do not need to add additional hardware devices.

2.2   SDN

SDN originated from the OpenFlow project at Stanford University. After many years of hard work, SDN is usually defined as the control plane separation from the underlying data plane, and control functions are integrated into the controller of the network architecture [7].

SDN's core concept is to decouple the data plane and control plane of traditional network equipment, through a centralized controller in a standardized interface to manage and configure the various network devices. SDN abstract the network intelligent from the hardware device, it not only simplifies the operation of the users to update the network function, greatly reducing the number of hardware facilities bought by operators, but also reduces the complexity of network management and configuration. Due to the characteristics of SDN, it will have a significant impact on the development of future network.

Based on OpenFlow, SDN basic system structure is mainly divided into three layers. Infrastructure layer is composed of all kinds of support OpenFlow protocol network equipment, mainly responsible for data forwarding operations. Control layer mainly includes OpenFlow controller and network operating system. The application layer is composed of many application softwares, the software provides the specific algorithm for controller, and through the controller into traffic control command, issued to the infrastructure layer in practical devices.

3   Design and Architecture Framework

The aim of system is to realize the dynamic control of the VNF, including VNF automatic deployment, delete, configuration, and other functions. As Fig. 1 shown, we put forward the VNF dynamic control system framework. The whole system framework is implemented based on CloudStack cloud platform, combining the SDN thoughts and NFV characteristics of centralized control, introducing theory of machine learning algorithms [8] processing network performance data.

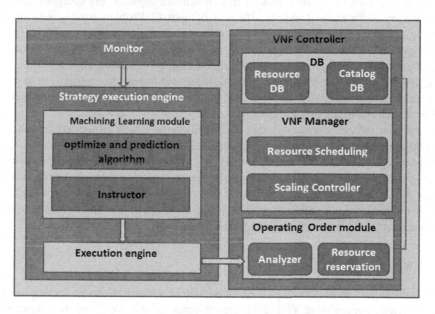

Fig. 1. VNF dynamic control system framework

The three main parts of the system is the strategy execution engine, VNF controller and monitor. Strategy execution engine, which is the core of the system, is mainly responsible for predict the state of the network of the next moment and generate operation instruction. VNF controller system is an important part of the system. It receives instructions from the strategy execution engine, and then perform corresponding operations according to the resources. Listener is mainly used to monitor network, collect useful data to form the parameters. And execution engine use these parameters to analysis strategy and process.

3.1 Strategy Execution Engine

Strategy execution engine is the core of the system architecture. The network performance parameters are mainly concentrated in the processing of this part. Its function is to use the machine learning algorithm to deal with network parameters, generate network forecast of the next moment and form the order code to realize the dynamic control of the network and the automatic configuration. Strategy execution engine is divided into machine learning module and execution engine module two parts.

In the machine learning module, we introduce the theory of machine learning algorithm. Probability theory, statistics, machine learning theory fusion algorithms such as multidisciplinary knowledge, according to the existing data and past experience, combined with a variety of learning methods, to carry out the forecast and problem solving. From the point of task type, application of machine learning research mainly concentrated on the following three categories [8–10]: (1) Unsupervised clustering, where similar object by static classification method for different groups or subset, makes the same subset of members of the object with similar properties. (2) The supervised classification, namely, on the basis of the known sample data, the category attribute unknown model for analysis and classification of input. (3) Problem solving, i.e., through learning to obtain knowledge can improve the efficiency of problem solving. In this architecture it will be an important tool to resolve the network performance prediction in application. Optimal prediction algorithm of strategy execution engine will be used to deal with parameters obtained from the listener, and formed by instructions to adjust the operating instructions of the network. Instruction implement relevant instructions will be forwarded to the execution engine.

Execution engine works to judge the instructions which generated by machine learning module and decide whether to send the instruction to VNF controller. These instructions include the VNF delete, modify, add, and so on. Considering the robustness of the system, we formulate that execution engine must receive the same instruction two consecutive times from instruction implement until operating instructions will change to VNF controller. At the same time, it gives the feedback of machine learning module, and confirms the instruction has been forwarded.

3.2 VNF Controller

VNF controller as an important part of the system is responsible for executing instructions. It is used for realizing VNF deployment, deleting and modifying the action. VNF controller contains instructions module, VNF manager as well as Database (DB).

Operation instruction module will receive instructions from execution engine, and through the analyzer to analyze reading instruction. Then, it forms the corresponding operation code. Finally, it send to the resource reservation module. The resource reservation module calculate the required resources for operation, according to the instructions in the query DB catalog DB.

The resource Scheduling of VNF manager, access to the resource DB in the DB request to obtain the required physical and logical resources. Scaling controller module is an important part which according to the result of the request to decide whether perform VNF deployment, modify or delete operation. If resources permit, it will perform the corresponding operation, if insufficient resources, it will refuse to perform this operation. In this architecture, we temporarily do not consider the limited resources, and assume that operations are in the range of allowable resources to ensure that the Scaling controller to make the corresponding operation.

DB as VNF controller resource manager, stores the information resources of the entire network. It consists of two parts, one is the catalog DB and the other is resource DB. The Catalog is used to store online and offline DB directory. Entities directory respectively is virtual router, virtual switches and other network equipment. Deposit Resource DB is the actual system with the physical resources (such as the number of CPU cores, storage capacity, etc.) and logic resources (such as IP address, etc.).

3.3 Monitor

The status of the network system change is very complex, and due to such many parameters, it is difficult to capture all of them. So we need to set a Monitor to monitor the status of the network, which can regularly feedback network status. The role of the Monitor is to monitor the user's network, collect data network parameter, and carry on the preliminary processing, get the useful parameters. The Monitor only monitors the network latency, the network traffic and network speed in the system. Monitor will feedback state parameters of the transmission network to the policy execution engine to process.

4 Case Analysis

4.1 The Application of the Home Network

As shown in Fig. 2(A), the average family network provided by the service provider, through a family gateway access. Home network must support a multitude of increasingly sophisticated network operator and user services, from remote lighting activation for user to communicate with smart appliances, while remaining affordable. However, they are struggling to keep pace with traffic-flow variety and the demand for always-on connectivity. Another complication is the home user's desire to change services but not the gateway hardware—in light of continual service innovation, the gateway hardware will quickly become obsolete [11].

We put forward the solution to set up NFV dynamic control framework which is applied to the home network and use the virtualization technology to provide family access network, as shown in Fig. 2(B). The strategy combines advantage of SDN and NFV, and fusion of machine learning algorithm. It can not only meet user demands for dynamic network function, but also provides the convenience for the management of the network function. The framework is based on the user state automatically to adjust to the network. Network service providers do not have to upgrade hardware facilities, but only need to provide virtualization capabilities needed resources, and ensure that sufficient resources. Therefore, the framework of NFV dynamic control system still apply in home network.

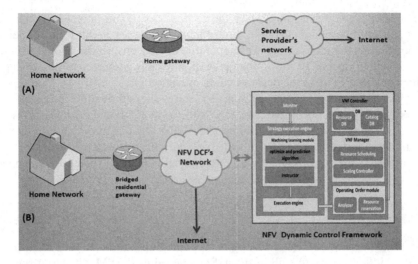

Fig. 2. Home network application

4.2   The Application of the Mobile Core Network

The last few years have witnessed a tremendous growth of mobile data traffic, in large part due to increasing popularity of smart phones, tablet computers and Machine-to-Machine (M2M) devices. This trend is expected to continue in the future [12]. Also, protocols in today's wireless networks are tightly coupled with the hardware in the form of ASICs (Application-Specific Integrated Circuits) designed for each of them, so any important protocol update requires upgrade of base stations or even their replacement [13].

In this article, we introduced in mobile core network NFV dynamic control system framework to replace the traditional distributed mobile core network. As shown in Fig. 3, the mobile core network composed of NFV dynamic control system framework. The framework of the listener is responsible for monitoring all the feedback from the base station network parameters, and it conveys the parameters to machine learning module processing to form the forecast of the wireless network status and adjust the function of function of the network to meet the different needs of different users. At the same time, the service provider can be applied to the management of mobile core network system framework. As Fig. 3 shown a good NFV example in mobile core network dynamic control system framework, the further application illustrates the framework that can be applicable to the different network architecture, which has the very high practical value.

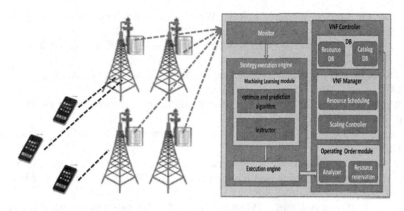

Fig. 3. The application of the mobile core network

5   Conclusion

In this paper, based on principle of SDN, we put forward an application on the Cloud-Stack cloud platforms to realize VNF dynamic control system which can be achieved under the network status to monitor and predict the network in the next moment. The system provides the corresponding network resources, services or configuration. The function of dynamic is to meet user network service requirements, and guarantee the SLA requirements. In this paper, I support two cases, family core network and mobile Internet application, to illustrate the NFV dynamic control framework. It can be good for different structure of the network system based on machine learning system.

Not specified in this article, however, the application of machine learning algorithm only describes the design idea and structure of the system framework. In the follow-up study, I will study further to conclude the network parameters and predict the state of the network of machine learning algorithm.

References

1. Jain, R., Paul, S.: Network virtualization and software defined networking for cloud computing: a survey. IEEE Commun. Mag. **51**, 24–31 (2013)
2. Han, B., Gopalakrishnan, V., Ji, L., Lee, S.: Network function virtualization: challenges and opportunities for innovations. IEEE Commun. Mag. **53**, 90–97 (2015)
3. Evangelos, H., Jamal, H.S., Spyros, D., Odysseas, K.: Towards a network abstraction model for SDN. J. Netw. Syst. Manage. **23**, 309–327 (2015)
4. Von Lehmen, A., Doverspike, R., Clapp, G., Freimuth, D.M., Gannett, J., Kolarov, A., Kobrinski, H., Makaya, C., Mavrogiorgis, E., Pastor, J.: Coronet: testbeds, demonstration, and lessons learned [invited]. IEEE/OSA J. Opt. Commun. Netw. **7**, A447–A458 (2015)
5. Raza, M.H., Sivakumar, S.C., Nafarieh, A., Robertson, B.: A comparison of software defined network (SDN) implementation strategies. Procedia Comput. Sci. **32**, 1050–1055 (2014)
6. Veitch, P., Mcgrath, M.J., Bayon, V.: An instrumentation and analytics framework for optimal and robust NFV Deployment. IEEE Commun. Mag. **53**, 126–133 (2015)

7. Akyildiz, I.F., Lee, A., Wang, P., Luo, M., Wu, C.: A roadmap for traffic engineering in SDN-openflow networks. Comput. Netw. **71**, 1–30 (2014)

8. 何清, 李宁, **罗**文娟, and 史忠植, 大数据下的机器学习算法**综述**, 模式**识别**与人工智能 **4**, 9 (2014)

9. Sun, S.: A survey of multi-view machine learning. Neural Comput. Appl. **23**, 2031–2038 (2013)

10. Vink, J.P., de Haan, G.: Comparison of machine learning techniques for target detection. Artif. Intell. Rev. **43**, 125–139 (2015)

11. Dillon, M., Winters, T.: Virtualization of home network gateways. Computer **47**, 62–65 (2014)

12. Sama, M.R., Contreras, L.M., Kaippallimalil, J., Akiyoshi, I., Qian, H., Ni, H.: Software-defined control of the virtualized mobile packet core. IEEE Commun. Mag. **53**, 107–115 (2015)

13. Tomovic, S., Pejanovic-Djurisic, M., Radusinovic, I.: SDN based mobile networks: concepts and benefits. Wirel. Pers. Commun. **78**, 1629–1644 (2014)

Fault Localization by Imperialist Competitive Algorithm

Afshin Shahriari[1], Farhad Rad[2], and Hamid Parvin[1,3(✉)]

[1] Nourabad Mamasani Branch, Islamic Azad University, Nourabad, Mamasani, Iran
a.shahriari@mamasaniiau.ac.ir
[2] Department of Computer Science, Yasuj Branch, Islamic Azad University, Yasuj, Iran
[3] Young Researchers and Elite Club, Nourabad Mamasani Branch, Islamic Azad University,
Nourabad, Mamasani, Iran

Abstract. Faults in computer networks may result in millions of dollars in cost. Faults in a network need to be localized and repaired to keep the health of the network. Fault management systems are used to keep today's complex networks running without significant cost, either by using active techniques or passive techniques. In this paper, we propose a novel approach based on imperialist competitive algorithm using passive techniques to localize faults in computer networks. The presented approach using end-to-end data detect that there are faults on the network, and then we use imperialist competitive algorithm (ICA) to localize faults on the network. The aim of proposed approach is to minimize the cost of localization of faults in the network. According to simulation results, our algorithm is better than other state-of-the-art approaches that localize and repair all faults in a network.

Keywords: Fault management system · Imperialist competitive algorithm · Normalized testing cost · End-to-end data

1 Introduction

Fault localization techniques must be designed and implemented to localize faults in the network with minimum cost test and have not a negative impact on network traffic. So we proposed a passive-base algorithm using imperialist competitive algorithm for fault localization in computer networks. The presented approach using end-to-end data detect that there are faults on the network, and then we use imperialist competitive algorithm to localize faults on the network. The contents of this article are organized as follows. Network modeling is in Sect. 2. Section 3 shows proposed approach in detail. The results are analyzed in Sects. 4 and 5 presents the concluding results.

2 Problem Setting and Assumptions

Generally, computer networks are composed of nodes and links [1, 2]. Therefore, the network components can be represented as a tree or a graph. Imagine a network that is composed of several components (links, routers and etc.). The failure of any of these components can disrupt the communication between the client and one (or both) of the

© Springer International Publishing Switzerland 2015
A. Bikakis and X. Zheng (Eds.): MIWAI 2015, LNAI 9426, pp. 405–412, 2015.
DOI: 10.1007/978-3-319-26181-2_38

servers [1]. For fault localization in a computer network, we need to make a physical network into logical form. We first map each potentially faulty physical component (topology) to a *logical topology*.

We assume can recognize the existence faults according to end-to-end data on the network. The amount of end-to-end data can be used to detect faults in the network: insufficient amount of data indicates faults, while sufficient amount of data indicates that the network is operating normally. The status of a component (i.e., whether faulty or not) can be tested to determine status it. So, test of each component in network have costs [3]. Therefore choice candid component for test is very important to reduce test costs [4].

When the known network has abnormal behavior, therefore are faulty components in the network and the network has been disrupted. But the main problem is that we do not know exactly which component (or components) is faulty [4]. So network components should be tested to find the exact location of faulty components of the network. But all the network components cannot be tested, because testing of components has costs. So it should be a minimum number of tests to identify the exact location of the faults to reduce costs testing. The main problem in this paper is to minimize the cost of testing. Therefore, we have chosen to test the node that has the highest probability of being faulty. This problem is NP-Complete [3, 4].

3   Proposed Method

Fault detection is the first step of network fault management (not fault localization). In other words, first step of network fault management is fault detection, and then second step is fault localization. In this paper, we used end-to-end data for fault detection in first step. In fact, if the data is sent from the source to server do not fit, then we can recognize that the network is functioning abnormal. In second step the exact location of faults in the network must diagnosed and repaired them. Therefore, we introduced an algorithm based on the imperialist competitive algorithm for finding the exact location of faulty components.

3.1   Fault Localization by Imperialist Competitive Algorithm

We use binary coding for represent countries. In fact, each country represents a path, and each element of the country refers to a node. Each element can have two values, 0 or 1. If the element value of country is 1, the node corresponding to the element of the path do not uses for data transfer or routing. If the element value of country is 0, the node corresponding to the element uses in routing. Length of each country is equal to the total number of nodes in the network. Table 1 shows an example of the paths. According to Table 1 the total number of nodes are three, so clients these paths for transmit data to the server. According to Table 1 can be produced three countries. Figure 1 shows binary representation of Table 1.

Table 1. Paths from client to server.

Path 1	Node 1	Node 2	Node 3
Path 1	Node 2	–	–
Path 1	Node 1	Node 2	–

	Node 1	Node 2	Node 3
Country 1	0	0	0
Country 2	1	0	1
Country 3	0	0	1

Fig. 1. Binary representation of Table 1.

In the example of Fig. 1, the path 1 uses of the three nodes for data transmission. The path 2 uses of one node and path 3 uses of two nodes for data transfer to server (or servers). We consider bad or faulty paths as an initial set of countries.

In this paper, we use two parameters to evaluate fitness of countries:

- The number of nodes sharing the different paths (degree of node)
- The number of nodes used in a path.

First, we calculate the weight of each node using Eq. (1). N is number of total countries. W_i is weight of $node_i$. n_{ik} shows amount of each $element_i$ in $country_i$. We use Eq. (2) for calculate the number of nodes used in a path. n_i is the $element_i$ in the $country_i$. N is number of total countries.

$$w_i = N - \sum_{k=1}^{N} n_{ik} \tag{1}$$

$$A = \sum_{i=1}^{N} n_i \tag{2}$$

According to the above description, fitness function to evaluate the fitness of each country is based on Eq. (3). α is a value between 0 and 1 ($\alpha = [0,1]$). If $\alpha = 1$, then the evaluation is based on the number of nodes used in a path. If $\alpha = 0$, then the evaluation is based on degree of node. α can have a value between 1 and 0, this value determines importance of W_i, and importance of A. More value of the f_i corresponds to a better fitness value for the country.

$$f_i = (1 - \alpha)A + \alpha w_i \tag{3}$$

There is an exception in evaluating fitness of countries that if the all elements if country have value of 1, therefore fitness country will be equal to zero ($f_i = 0$).

To start the optimisation algorithm, initial countries of size $N_{Country}$ is produced ($N_{Country}$ = number of bad paths). We select N_{imp} of the most powerful countries (countries with high fitness) to form the empires. The remaining N_{col} of the initial countries will be the colonies each of which belongs to an empire. To form the initial empires, the colonies are divided among imperialists based on their fitness. The colonies are randomly chosen and given to the nth imperialist. For calculate power of empires uses Eq. (4). Where P_n is the total cost of empire and α is a positive number that is between 0 and 1, and f shows fitness.

$$P_n = f(imperialist_n) + \alpha \left(f \left(colonies\ of\ empire_n \right) \right)$$ (4)

To define a policy of assimilation, we use the same operator with crossover operator in genetic algorithm. Each country is subjected to crossover with probability P_c. One country is selected from the population, and a random number ($RN = [0,1]$) is generated for it. If $RN < P_c$, these country is subjected to the crossover operation with empire using single point crossover. Otherwise, these countries are not changed. The pseudo code of the crossover function is as follows.

1. *Select one country*
2. *Let RN a random real number between 0 and 1*
3. *If RN < 0:5/* operators probability*
4. *Crossover (country, empire).*

Revolution is a fundamental change in power that takes place in a relatively short period of time. In the terminology of ICA, revolution causes a country to suddenly change its socio-political characteristics. That is, instead of being assimilated by an imperialist, the colony randomly changes its position in the socio-political axis [5].

While moving toward the imperialist, a colony might reach to a position with lower fitness than the imperialist. In this case, the imperialist and the colony change their positions. Then the algorithm will continue by the imperialist in the new position and the colonies will be assimilated by the imperialist in its new position.

All empires try to take the possession of colonies of other empires and control them. The imperialistic competition gradually brings about a decrease in the power of weaker empires and an increase in the power of more powerful ones. The imperialistic competition is modelled by just picking some (usually one) of the weakest colonies of the weakest empire and making a competition among all empires to possess these (this) colonies [5].

To start the competition, first a colony of the weakest empire is chosen and then the possession probability of each empire is found. The possession probability PP is proportionate to the total power of the empire. The process of selecting an empire is similar to the roulette wheel process which is used in selecting parents in GA.

If only one element value of each country is 0, and the rest of the elements values are 1, therefore the algorithm should be terminate. Element whose value is 0, node corresponding to the element tested. After the test node, if the node is faulty, the node is repaired else the algorithm is repeated. The Proposed algorithm will be repeated until all faulty nodes are localized.

4 Experimental Study

We evaluate the performance of our algorithm through extensive simulation (using a MATLAB) in a general network. The proposed method compared with the methods in [2, 6, 7]. Several scenarios have been considered for simulations. In general, three different scenarios have been considered for simulation that is shown in Table 2. The performance metrics we use are the normalized testing cost, false positive and total test cost.

Table 2. Scenarios for simulation.

Scenarios	Number of Nodes	Number of clients
Scenario 1	60	40
Scenario 2	60	80
Scenario 3	150	80

Figure 2 shows the results of total test cost for algorithms in three scenarios. The results of proposed algorithm, Greedy, Ordering and NFDM are plotted in the Fig. 2. We observe that proposed algorithm has better total test cost compared to other algorithms.

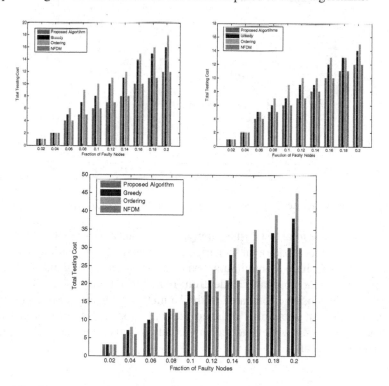

Fig. 2. Total test cost in scenario 1 (left top), 2 (right top) and 3 (down).

Figure 3 shows the results of normalized testing cost for algorithms in three scenarios. According to the results shown in Fig. 3, we observe that proposed algorithm has better normalized testing cost compared to other algorithms, and it has minimum normalized testing cost compared to other three algorithms.

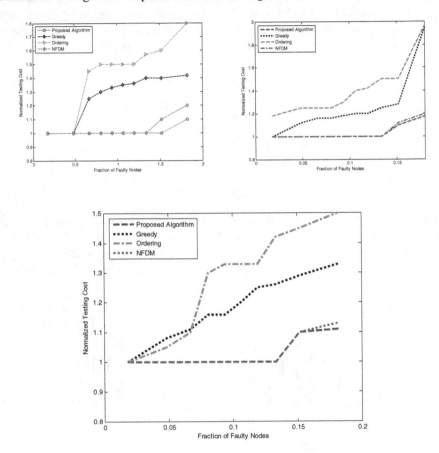

Fig. 3. Normalized testing cost in scenario 1 (left top), 2 (right top) and 3 (down).

Figure 4 shows the total test cost by applying proposed algorithm, Greedy, Ordering and NFDM for fault localization in scenarios 1–3 respectively. According to Fig. 4 total test cost of our algorithm is minimum compared to other algorithms. According to simulation results, our approach aside from reducing the cost of fault localization provides good false positive compared to similar methods. Obtained results in large and small scales (different scenarios) indicate that our proposed algorithm can provide similar results in different scales and scenarios, and prove the robustness of the proposed method in different scales.

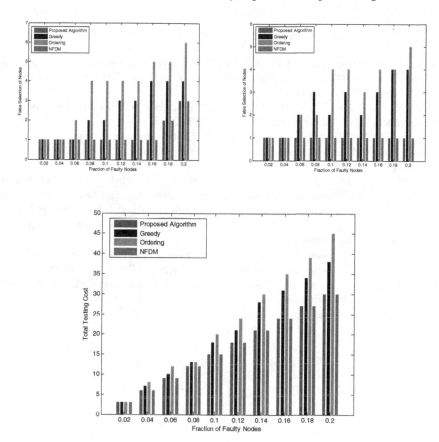

Fig. 4. Total test cost in scenario 1 (left top), 2 (right top) and 3 (down).

5   Conclusion

In this paper a new algorithm based on imperialist competitive algorithm was implemented for fault localization in computer networks. The presented approach used end-to-end data detected that there are faults on the network, and then used imperialist competitive algorithm (ICA) to localized faults on the network. Extensive simulation demonstrated that our algorithm is outperforms the other algorithm for fault localization in computer networks. As future work, we are pursuing in the following two directions: (1) evaluating the performance of our approach under other scenarios, for instance, when the location of faulty components follows a more clustered distribution instead of uniform random distribution and (2) developing proposed algorithm for temporary fault localization.

References

1. Patrick, L., Vishal, M., Rubenstein, D.: Toward optimal network fault correction via end-to-end inference. In: 26th IEEE International Conference on Computer Communications, New York, October 25–26, pp. 165–172 (2007)
2. Yijiao, Y., Qin, L., Liansheng, T.: A graph-based proactive fault identification approach in computer networks. Comput. Commun. **28**, 366–378 (2005)
3. Bing, W., Wei, W., Hieu, D., Wei, Z., Krishna, R.: Fault localization using passive end-to-end measurements and sequential testing for wireless sensor networks. IEEE Trans. Mob. Comput. **11**, 439–452 (2012)
4. Garshasbi, M.S., Jamali, Sh.: A new fault detection method using end-to-end data and sequential testing for computer networks. Int. J. Inf. Technol. Comput. Sci. **1**, 93–100 (2014)
5. Atashpaz-Gargari, E.: Imperialist competitive algorithm: an algorithm for optimization inspired by imperialistic competition. In: 2007 IEEE Congress on Evolutionary Computation, CEC 2007, 25–28 September 2007
6. Hung, N., Patrick, T.: Using end-to-end data to infer lossy links in sensor networks. In: 25th IEEE International Conference on Computer Communications Proceedings (2006)
7. Balaji, P., Shakti, K., Vinay, K.P.: Fault management of computer networks based on probe station and probe set selection algorithms. In: International Conference on Advances in Electronics, Qinhuangdao, China, June 25–27, pp. 504–508 (2012)

Other AI Applications

A Recommender System for Mobile Commerce Based on Relational Learning

Shengnan Chen[1(✉)], Hongyan Qian[1,2], and Jiayi Gu[1]

[1] College of Computer Science and Technology,
Nanjing University of Aeronautics and Astronautics,
Nanjing 211106, China
{csn0423,qihua1990}@aliyun.com, qhy98@nuaa.edu.cn
[2] Information Technology Research Base of Civil Aviation Administration
of China, Civil Aviation University of China, Tianjin 300300, China

Abstract. Recommender systems are intelligent tools to extract useful information from a large collection of online data. They have been widely used in various fields, including the recommendation of music, movies, documents, tourism attraction, e-learning and e-commerce. Many approaches, such as content-based filtering and collaborative filtering, have been proposed to run the recommender system, but they are not completely compatible with the m-commerce context. Therefore, this paper focuses on how to develop a recommender model that can be applied to the mobile environment. In addition, this paper also presents the methods to preprocess the data. Through applying the model to a real-world data supported by *Alibaba* Group, it is shown that our model works effectively in m-commerce.

Keywords: m-commerce · Recommender system · Relational learning

1 Introduction

With the convenience of mobile phones nowadays, online shopping has become increasingly popular among Internet users. It is worthy to build an intelligent recommender system on mobile phones because this system can identify users' needs in various circumstances. As one of the most celebrated corporations in China, Alibaba group has many business groups. The task of Alibaba Competition 2015 Track1 aims to develop such a system in which behavior records of mobile clients are analyzed to capture their interests and predict their consumption preference in the near future. This prediction task is not only a challenge to researchers, but also is of more practical value for the accurate prediction can increase the sales of vendors and improve users' experience on the trade platform.

We choose to propose our own model instead of utilizing the existing ones, because most conventional models are based on the similarity of users or items [1–4]. However, in m-commerce (or e-commerce), similar users may hold different attitudes toward the same attributes of an item. Besides, users and items alter continuously in this context, which makes it difficult to apply traditional approaches.

A. Bikakis and X. Zheng (Eds.): MIWAI 2015, LNAI 9426, pp. 415–428, 2015.
DOI: 10.1007/978-3-319-26181-2_39

The contributions of our paper are as follows:

(1) A novel strategy for missing data is presented to help extract the positioning information of consumers to identify their geo-features. Besides, to solve the imbalance of training set, an effective method is performed to select negative samples.

(2) The prediction task is regarded as a binary classification, which simplifies our problem into a relational learning problem, namely the prediction of relationship between users and items (buy or not buy). Various features including geo-features are extracted from users' previous behaviors and then integrated into the proposed model. In addition, a two-stage training process is used so that the resulting model can gain a high accuracy with good performance in the mobile shopping environment.

The rest of the paper is organized as follows. In Sect. 2, the framework of our system and the preparation of the data set are proposed. In Sect. 3, two individual models for classification are introduced and we show how to combine them effectively for a better performance. The evaluation metric and analysis of experimental results are discussed in Sect. 4 and prior works on personal systems are presented in Sect. 5. Finally, conclusions are made about our work in Sect. 6.

2   Framework

In this section, a concise overview of our system is first provided. Then three key parts of the preparation are discussed: how to fill the missing data, how to solve the imbalance of the training set and how to construct our feature engineering.

2.1   The Overview of the System

The recommender system includes three stages: Generating individual model, aggregating models with validation set and Ensemble learning with the test set.

Firstly, different approaches are adopted to capture different concepts and then train a diverse set of individual models. Diversity reflects different aspects of our modeling. In the second stage, the results from stage 1 are combined through non-linear blending methods, and the validation set is used to tune the weight and decide the contribution of each model to our final result. The blending model can greatly improve the prediction performance and the generalization ability. In the final stage, the blending model is applied to the test set and the final prediction result is gained (Fig. 1).

2.2   Data and Task Description

The data set is offered by *Alibaba* Group, consisting of complete 31 days' 5,000,000 users' records, ranging from November 18, 2014 to December 18, 2014. The data set is much larger than other released data sets. In sake of users' privacy, the data set is masked. The task attempts to predict which people would purchase items in December 19, 2014 and recommend items to them according to these records.

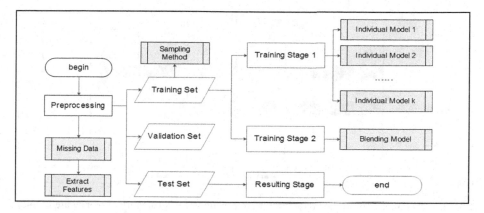

Fig. 1. The illustration of our recommender system

The data set has two parts, user table and item table. The user table has 6 attributes, including *user_id, item_id, behavior_type, item_category, user_geohash, time*, which constructs relational information between users and items.

Four kinds of behaviors that people can conduct when using this platform on mobile clients. *'click'* means people click an item. *'store'* means people add the item to their wish list. *'cart'* means people add the item to the cart. *'buy'* means people buy it.

The item table has 3 attributes, including item_id, item_category, item_geohash, which constructs supplemental information.

Researchers can use the data set on the web site www.yushanfang.com.

2.3   Filling Missing Location Data

In the data set, there are two kinds of missing data for location: users and items. We need to incorporate the given information to infer other unknown information. Since the items that we predicted are primarily catering service in our life, we think that the characteristics of the dataset like time and location may be more obvious. We do some survey based on the dataset to verify our assumption. For simplicity, we divide a day (24 h) into 3 parts. 0 a.m. ∼ 8 a.m. stands for period 1, when most people sleep or on their way to work. 9 a.m. ∼ 16 p.m. stands for period 2, which can be working hours in most circumstances. 17 p.m. ∼ 23 p.m. stands for period 3, when people come back home and enjoy family time. Figure 2 depicts how many people visits a category of items in different periods. The color stands for different operations.

We randomly select behavior records from 500 people with perfect location data and verify our assumption below.

(1) The preference of positioning. People frequently visit fixed places so some places are more frequently visited than other places. This happens because people prefer familiar places and some places may provide more entertainment venues than others.

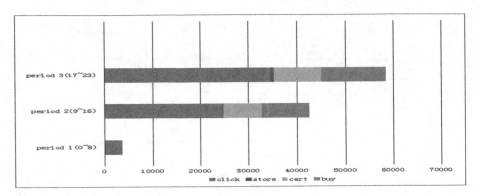

Fig. 2. The counts of four behaviors in different periods for a category (Color figure online)

(2) The preference of traveling distance. People are reluctant to travel a long distance to entertain them or enjoy the service. Thus, it makes no sense to recommend them items that are far beyond them.

We find that about 400 people have a fixed active range, which accounts 80 % of the total. Table 1 lists part of our result.

Table 1. Some people's active range

User id	Location	The counts of their behaviors in 3 period			Total
		0–8	9–16	17–23	
64310783	955m***	4	17	21	42
58812216	94nv***	5	8	10	23
	94nu***	5	28	2	35
	9r7d***	0	1	5	6

We find that about 80 % of people tend to have a fixed active range. Table 1 lists part of our result: the counts that people visit the place in 3 periods. People may sometimes leave away for a while, but we do not take in to consideration.

Besides, the other assumption we extract from life experience. When we add it to our feature engineering, we glad to find it work effectively to improve our result.

There is a mutual effect between users and items, which means users' location can partly effect items' location, vice versa. The algorithm for filling missing location data can be divided into two steps. For those users' with imperfect location information, we utilize their personal location information to fill it and use the location information to infer items' location. For those user with no location data, we infer the users' possible location by using items' location according to *u-i* relationship.

Step1

For user in user_list:

 If user's records has location information

  1) Count users' behavior in three periods

  2) Find the maximum in three period, which can be the most possible place people visit.

  3) Use it to fill the missing data for user

  4) According to u-i relationship, fill the missing item's location data

 Else Add the users' record to remain_list

Step2

For item in item_list:

 If item's records has location information related to users

  1) Count the behaviors from different location

  2) Find the maximum as the items' location

For user in remain_list;

 Use the u-i relationship to fill their location data

Algorithm 1. Fill the missing location data

Table 2. The example for Step 1

user_id	Detailed information				
	User's location	(0–8)	(9–16)	(17–23)	(count, time_period)
14397660	99ck***	1	1	**15**	(15, 3)
	99cl***	0	1	3	

Table 3. The example for Step 2

item_id	Detailed information		
	User's location	Count	(count, time_period)
14397660	99ck***	20	(97mk, 20)
	99cl***	1	

A typical example that we give to illustrate this process.

From Table 2, we know that user *14397660* is more likely to be in *99ck* during 17–23. We use *99ck* to fill the missing location of user *14397660* when records' time is in range 17 to 23.

From Table 3, we choose the location *97mk**** as the items' location because it contains more records than others. And we use the items' location to supplement other uses' records based on *u-i* relationship (Table 3).

2.4   Imbalance of Training Set and a Novel Sampling Method

The number of positive and negative samples can have an impact on our training. Positive samples refers to purchasing records and negative samples refers to others. In the training set, it is observed that only 39,279 records are positive and 21,600,829 are negative. We keep all positive samples. However, it remains a key task in the pre-processing stage to tackle such a large quantity of negative samples.

A common method, the random sub-sampling, is used to tackle samples, but the random sampling has limitations: the loss of important information. Thus, we combine K-means clustering with the weighted sampling method to conduct sub-sampling.

(1) We first use K-means clustering to classify data negative samples into different groups.
(2) Then groups are weighted by different value ω according to the number of samples they contain.

$$\omega = log_2 \left(\frac{n_{all\_negative\_samples}}{n_{group\_negative\_samples}} \right) \tag{1}$$

(3) We adopt weighted random sampling based on previous results and the small groups have more opportunities to be chosen.

In Fig. 3, we choose 10 initial centers for clustering process, because the large value of initial centers can result in a high cost of computation and time comparing to its common effect. Besides, we find that 10 centers have already satisfied our need in this dataset after we tested other values. We further discuss how it influences our result in Sect. 5.

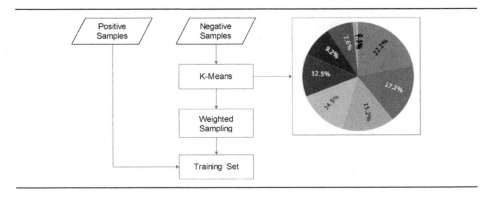

Fig. 3. Result of clustering and process for creating training set

2.5   Effective Features

Feature engineering is an informal term but it plays a vital role in applied machine learning. The features can directly influence the predictive models. Here we show how we conduct feature engineering.

(1) Basic Features

Features which can be directly extracted from the data set are regarded as Basic Features. These features are constructed from the following six aspects: user, item, category, the relationship of user-item, user-category and item-category. For instance, how many times does the user click on a certain item? Through calculating the data or sorting out the data, some features are obtained and other extended features are combined with previous features. In general, basic features mainly reflect users' preference, purchasing power, items' popularity and so on.

(2) Location Features and Time Feature

We put emphasis on how to create location features and time features, which can mainly decide whether a user will buy the item. Users have preference on specific places and categories (items) may be sold in specific time. And recent records of users have more effects on our modeling. For these reasons, we create location and time features as following.

In terms of time (*yyyy-MM-dd hh*), We have divided a day (24 h) into 3 parts and count the number of users' operations in each period for the whole month. We also consider that recent dates have more effect on current behaviors, so we count the number of operations during the last 1 day, the last 2 days, the last 3 days and the whole month separately. For locations, we categorize the places into 10 classes according to the number of visiting users in different periods by using K-means. In addition, the cosine method is adopted to calculate the similarity between each location. The distance between the user and the item is also taken into consideration because the shorter distance between the item and the user makes it more likely for the user to buy it.

3 Our Model

For individual model, we exploit Random Forest and Gradient Boosting Decision Tree to train the classifier. We also tried SVM and Naïve Bayes, however they do not perform well on the data set. The main reason is that we can hardly guarantee the independence between every two features. Then parameters are tuned based on validation set. Also, we use validation to ensemble individual classifier in blending stage for better performance and generalization ability.

3.1 Random Forest

Random Forest is a popular algorithm and widely used to solve classification problems since it combines two useful techniques: bagging ideas proposed by Breiman [5, 6] and random features selection introduced by Ho et al. [7].

Bootstrap or bagging method adopt sampling with replacement from original data to estimate the distribution of the estimator. Recent researches [20, 21] have witnessed that bootstrap is applied to generate subsets of data when training base models. Given a training data set with N samples, m new training sets are generated through bagging methods. It is possible to use sampling with replacement to keep the size of training set unchanged, but contain different observations. Then newly generated m training sets are used to train k base models. The final result is combined by voting.

Random feature selection is also a technique based on bootstrapped sampling. At every splitting node of tree, it chooses to randomly use subset of features instead of all features, which results diversity between each base model.

Random forest improves prediction accuracy and reduces variance by averaging noisy but unbiased tree. The variance of random forest with K-number trees is:

$$\text{var} = \rho\sigma^2 + \frac{1-\rho}{K}\sigma^2 \tag{2}$$

where σ^2 is the variance of the individual tree, ρ is correlation between trees and K is the number of trees. Hence, the variance of random forest is primarily determined by three factors.

3.2   Gradient Boosting Decision Tree

Unlike bagging methods, boosting methods sequentially generate base models. Boosting methods strategically re-sample the training data and provide consecutive models with the most efficient information. For each step, the distribution is adjusted according to the error produced by previous steps [8, 9]. Samples which are not classified have more chance to be chosen. The gradient boosting decision builds models in a stage-wise fashion and updates it by reducing the expected value of predefined loss function. The updating process can be described through:

$$f_m(x) = f_{m-1}(x) + \rho_m g_m(x) \tag{3}$$

where m is the number of trees, J is learning rate and C is variable interaction. The more trees exploited in the model, the smaller training error fitted models can achieve.

3.3   Blending Models

The performance of blending model is decided by three factors, the diversity of training set, features and individual models.

The basic models that we exploit are Random Forest (RF) and Gradient Boosting Decision Tree (GBDT). We further construct individual models based on these models and aggregate the results from individual models.

(1)   Random GBDT Forest

We select different negative samples and make up different training set. Then randomly select features from training set and applied them to GBDT models. The classified results are average in order to gain the final result (Fig. 4).

(2)   Random Forest with GBDT

We first train a classifier by using Random Forest and the initial target value is defined as y. Then we gain the prediction result y_1 and $\Delta y = y - y_1$. We regard Δy as

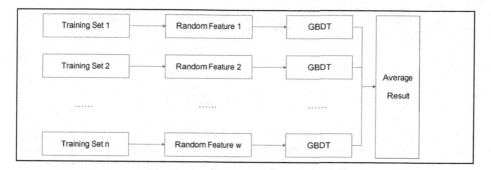

Fig. 4. Illustration of random GBDT Forest

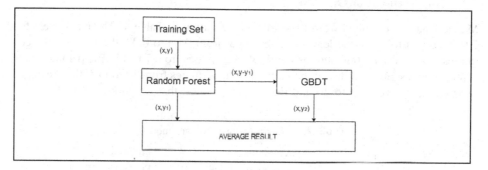

Fig. 5. Illustration of Random Forest with GBDT

our new target and use GBDT to train the model. The prediction result is defined as y_2. The final result is the average of y_1 and y_2 (Fig. 5).

(3) Blending models

There are mainly two methods used to blend model. One is to vote results from individual model and the highest number of votes is the final result. Another is to select top k results from individual model and combine them. However, they are both time-consuming. In blending stage, we use validation set and logistic regression to select the individual model and decide the contribution of each model to final result. Finally, we apply the model to the test set and gain the result.

4 Evaluation

4.1 Metrics

We expect that not only the precision rate of the predicted items will be as high as possible but also it will cover more users and items. Therefore, we take the precision rate and the recall as the indicators given in (3) and (4).

$$precision = \frac{|\cap(predictionset, referenceset)|}{|predictionset|}, \quad recall = \frac{|\cap(predictionset, referenceset)|}{|referenceset|}$$

$$\tag{4}$$

$$F1 = \frac{2 \times precision \times recall}{precision + recall} \tag{5}$$

predictionset is predicted and provided by us. *recall* is provided by *Alibaba* Group to evaluate the performance of our model, which contains the real relationship between users and their buying items. The higher of F1 score the better of our model.

4.2   Improvement of Our Model

Metrics that are obtained from experiments are shown in Table 4. The parameters for GBDT are as follows: the learning rate ranges from 0.05 to 0.08, the number of trees ranges from 1500 to 3000 and the depth of trees ranges from 7 to 11. The parameter for Random Forest are as follows: the number of trees ranges from 1000 to 3000, the depth of trees ranges from 13 to 20 and the random method that we choose is $N/3$.

Table 4. The improvement of our model

	Model	Validation			Public test		
		Precision %	Recall %	F1 score (%)	Precision %	Recall %	F1 score (%)
1	Base Model (random sampling + Basic features + GBDT Model)	7.6001	8.6702	7.6350	7.6306	7.6612	7.6459
2	1 + novel weighted sampling	7.6824	7.7108	7.7108	7.7714	7.6819	7.7264
3	2 + information of time and location	8.1205	8.1536	8.1536	8.1470	8.1825	8.1647
4	3 + Random GBDT Forest	8.0520	8.1807	8.1807	8.0640	8.3345	8.1970
5	3 + RF combined with GBDT	8.0677	8.1744	8.1744	8.0750	8.2992	8.1856

We can conclude from the table that information of location and time, the most important information in this context give a great improvement to individual model. The novel sampling method mentioned in Sect. 2 is also helpful.

In addition to this, the two individual models that we design significantly enhance the performance of prediction, though the random GBDT performs a bit better.

Here, we do not apply our system to other standard data sets, e.g. MovieLens, because they lack information in terms of users' preference towards items' attributes and decision factors. Besides, it is not necessary to compare our system with other existing recommender systems, since we do not construct uses' model and thus can not integrate users' preference into our model.

4.3   Blending Models

We use 5 individual models, including 3 Random GBDT Forest models and 2 Forest combined with GBDT models, for blending resulting model. Table 5 shows the performance of different ensemble methods, where No. 3 model can perform the best. This is mainly because the utilization of logistic regression can help us remove the less useful model and tune the parameters for blending.

Table 5. The blending result of our model

	Model	Validation			Public test		
		Precision %	Recall %	F1 score %	Precision %	Recall %	F1 score %
1	Select top k (k = 85000)	8.1057	8.2967	8.2001	8.1174	8.3058	8.2105
2	Voting technique	8.2144	8.2667	8.2405	8.2217	8.3501	8.2687
3	Logistic Regression	8.2446	8.4462	8.3442	8.2504	8.4522	8.3501

5   Related Work

Most existing recommender models have been successfully applied in different domains [1–4]. Widely used recommender techniques can be generally categorized into collaborative filtering (CF) [10], content based filtering (CB), and hybrid techniques. Each technique has its advantages and limitations; for example, collaborative filtering relies on users' previous behavior to make recommendations. However, it also can hardly be applied to new systems because the information of users' interaction is not enough to capture their interest. Content-based filtering approaches make effort to solve this problem, *cold start problem*. It focuses on using information about items and users to recommend items to users [10]. However, it brings about a problem, *over-specialization*. Currently, many advanced recommendation approaches have been proposed to avoid these limitations, such as knowledge-based recommender systems, context-aware based recommender systems and social network-based recommender systems [11].

Knowledge-based (KB) recommender systems provide users with items based on knowledge related to users, items or the interaction between them [12]. They keep a detailed knowledge base which can help them know whether the item meets the user's need. Martínez Luis et al. [13] used linguistic information to manage and model the uncertainty of users' preference. This system was able to complete imperfect users' linguistic preference relations and make recommendation.

Context aware-based recommender systems incorporate contextual information to create relatively more accurate and relevant recommendations for users [18, 19]. Duenkai Chen [14] adapted recommender systems to web service. The paper presented how to use contextual information to compose necessary web services and meet users' request. Mehdi et al. [15] mined the data from users' search records and other sources, which helped to enhance the presentation of content on visited pages.

Social network-based recommender systems combine social network analysis (SNA) with recommender systems [17–19]. Zhoubao et al. [16] proposed a social regularization method that combined social relationship to benefit recommender systems. The proposed system utilized friendships to calculate the similarity between users and it also concern about the correlation between users and items.

Recent year, some researchers have concerned about how to apply the recommender techniques to mobile environment [22, 23]. Li-Hua Li et al. [22] proposed their approach based on Collaborative Filter. They first discover mobile users' moving patterns by clustering and provide recommendations based on their preferences in mobile context. However, they do not focus time and location at the same time and capture the pattern inside their records. Manuel J. Baranco et al. [23] incorporated location based service (LBS) into mobile tourism guides. They utilized speed and trajectory to recommend interesting points to them. But in our cases, the data set do not contain that much information and their purpose are far from us.

The recommendation approaches mentioned above mainly concern about the similarity between users or items. However, in real world, especially for e-commerce, the similar users may show different attitudes towards the attributes of items. This problem has not been addressed in these approaches and mobility of users also a challenge to recommender systems, both of which are the main focus of the current paper.

6   Conclusion and Future Work

Recent years has witnessed a tremendous progress in recommender systems. Many approaches have been designed to address recommendation but few have considered about relational learning. In this paper, we simplify the recommender problem as the prediction of the relationship between users and items. We further utilize the information of location and time to construct our recommender model for mobile commerce. The real world data set has demonstrated that we are able to get an effective performance through our model.

Acknowledgments. This work is partly supported by Open Project Foundation of Information Technology Research Base of Civil Aviation Administration of China (NO. CAAC-ITRB-201301).

References

1. Lu, J., Wu, D., Mao, M., Wang, W., Zhang, G.: Recommender system application developments: a survey. Decis. Support Syst. **74**, 22–32 (2015)
2. Park, D.H., Kim, H.K., Choi, I.Y.: A literature review and classification of recommender system. Expert Syst. Appl. **39**(11), 10059–10072 (2012)

3. Choi, K., Suh, Y.: A new similarity function for selecting neighbors for each target item in collaborative filtering. J. Knowl-Based Syst. **37**, 146–153 (2013)
4. Ortega, F., SáNchez, J.L., Bobadilla, J., GutiéRrez, A.: Improving collaborative filtering-based recommender systems results using Pareto dominance. J. Inf. Sci. **239**, 50–61 (2013)
5. Zhang, Y., Haghaini, A.: A gradient boosting method to improve travel time prediction. Transp. Res. Part C: Emerg. Technol. **58**(B), 308–324 (2015)
6. Li, C., Hua, X.-L.: Towards positive unlabeled learning for parallel data mining: a random forest framework. In: Luo, X., Yu, J.X., Li, Z. (eds.) ADMA 2014. LNCS, vol. 8933, pp. 573–587. Springer, Heidelberg (2014)
7. Wu, Q., Ye, Y., Zhang, H., Ng, M.K., Ho, S.S.: ForesTexter: an efficient random forest algorithm for imbalanced text categorization. J. Knowl.-Based Syst. **67**, 105–116 (2014)
8. Khot, T., Natarajan, S., Kersting, K., Shavlik, J.: Gradient-based boosting for statistical relational learning: the Markov logic network and missing data cases. Mach. Learn. **100**, 75–100 (2015)
9. Kocsis, L., Gyorgy, A., Ban, A.N.: Boosting tree: parallel selection of weak learners in boosting, with application to ranking. Mach. Learn. **93**(2), 293–320 (2013)
10. Kaklauskas, A., Zavadskas, E., Seniut, M., et al.: Recommender system to analyze students' academic performance. Expert Syst. Appl. **40**(15), 6150–6165 (2015)
11. Diaby, M., Viennet, E., Launay, T.: Toward the next generation of recruitment tools: an online social network-based job recommender system. In: Proceedings of the 2013 IEEE/ACM International Conference on Advances in Social Networks Analysis and Mining, pp. 821–828. ACM (2013)
12. Fernandez, Y.B., Pazos, J.J., Cabrer, M.R., Nores, M.L., Duque, J.G., Vilas, A.F., et al.: A flexible semantic inference methodology to reason about user preferences in knowledge-based recommender system. Knowl.-Based Syst. **21**(4), 305–320 (2008)
13. Luis, M., Pérez, L.G., Manuel, B., Macarena, E.: Improving the effectiveness of knowledge based recommender systems using incomplete linguistic preference relations. Int. J. Uncertainty Fuzziness Knowl.-Based Syst. **16**, 33–56 (2008)
14. Chen, D.-K.: A context-aware recommender system for web service composition. In: Proceeding of the Eighth International Conference on Intelligent Information Hiding and Multimedia Signal Processing, pp. 227–229. IEEE Press (2012)
15. Elahi, M.: Context-aware intelligent recommender system. In: Proceedings of the 15th International Conference on Intelligent User Interfaces, pp. 407–408. ACM (2010)
16. Sun, Z., Han, L., Huang, W., Wang, X., Wang, M., et al.: Recommender systems based on social network. J. Syst. Softw. **99**, 109–119 (2014). Springer
17. Knijnenburg, B.P., Kobsa, A.: Making decision about privacy: information disclosure in context-aware recommender system. ACM Trans. Interact. Intell. Syst. **2**(3), (2013). Article No. 20. ACM. 50(01),730-731
18. Zhao, L., Huang, J., Zhong, N.: A context-aware recommender system with a cognition inspired model. In: Miao, D., Pedrycz, W., Ślęzak, D., Peters, G., Hu, Q., Wang, R. (eds.) RSKT 2014. LNCS, vol. 8818, pp. 613–622. Springer, Heidelberg (2014)
19. Liu, X., Aberer, K.: A social network aided context-aware recommender system. In: Proceedings of the 22nd International Conference on World Wide Web, pp. 781–802. ACM (2013)
20. Liu, F., Zhang, X., Ye, Y., Zhao, Y., Li, Y.: MLRF: multi-label classification through random forest with label-set partition. In: Huang, D.-S., Han, K. (eds.) ICIC 2015. LNCS, vol. 9227, pp. 407–418. Springer, Heidelberg (2015)
21. del Río, S., López, V., Benítez, J., Herrera, F.: On the use of MapReduce for imbalanced big data using random forest. Inf. Sci. **285**, 112–137 (2014)

22. Li, L.H., Lee, F.M., Chen, Y.C., Cheng, C.Y.: A multi-stage collaborative filtering approach for mobile recommendation. In: Proceedings of the 3rd International Conference on Ubiquitous Information Management and Communication, pp. 88–97. ACM (2009)
23. Barranco, M.J., Noguera, J.M., Castro, J., Martínez, L.: A context-aware mobile recommender system based on location and trajectory. In: Casillas, J., Martínez-López, F. J., Rodríguez, J.M.C. (eds.) Management Intelligent Systems. AISC, vol. 171, pp. 153–162. Springer, Heidelberg (2012)

An Agent-Oriented Data Sharing and Decision Support Service for Hubei Provincial Care Platform

Liang Xiao[✉]

Hubei University of Technology, Wuhan, Hubei, China
lx@mail.hbut.edu.cn

Abstract. Research today is often dedicated in isolation to the fields of regional clinical data sharing and clinical decision support with closed boundary. A framework has been proposed in this paper for integrating agent-oriented data sharing and agent-oriented argumentation upon shared data, for the Hubei Provincial Care Platform. This is built upon the LCC technology and CDA standard, demonstrated with a hypertension management example, and in compliant with IHE XDS standard. The agent-oriented platform services will support, in the entire province, a regional collaborative health service paradigm where the right clinical data will be available at the right place at the right time, for making the right decision.

Keywords: Agent · Clinical data sharing · Clinical decision support · CDA · LCC

1 Introduction and Motivation

Healthcare service is becoming increasingly complex and collaborative and this drives a continuous need of clinical data sharing and decision-making in a distributed environment. National or regional level data sharing paradigms are emerging in many countries, with mixed fortunes [1], e.g. England's Summary Care Record [2], Scotland's Emergency Care Summary [1], Denmark's National Patient Index [3], which have specific implementation mechanisms and contrasting socio-technical or socio-political characteristics on shared content, driving force and aim, population coverage, patient and stakeholder consensus, uptake and effective hit rate, and so on. Others propose the use of multi-agent systems in deploying multilateral exchange agreements among clinical sites for data sharing [4].

Clinical decision support will improve clinical practice when recommendations are given at the time and location of decision-making [5, 6]. This relies heavily upon two factors: (1) the availability of specific patient data against which the decision support system or platform can apply and, (2) the structuring of generic clinical guideline knowledge or evidence from which the applicable pieces can be retrieved and then customised advices generated. Traditional clinical decision support systems are developed in pre-determined hospital settings and integrated with local hospital information systems with an aim of solving particular decision problems. Although such integration makes local patient data easily available, systems as such have closed boundary and are inappropriate anymore when applied in the open data-sharing scope.

© Springer International Publishing Switzerland 2015
A. Bikakis and X. Zheng (Eds.): MIWAI 2015, LNAI 9426, pp. 429–440, 2015.
DOI: 10.1007/978-3-319-26181-2_40

An opportunity will be missed if a great effort has already been put upon the development of national or regional data sharing paradigms involving great resource commitment, but with no integration of a systematic and generic decision support service. It is often this service that will magnify the value of clinical data shared among healthcare organisations in the region, when dynamically incorporated.

This paper reports the latest progress in the programme of Hubei Provincial Care Platform in China by us as participants in the specification and design of its clinical data-sharing infrastructure. A framework has been proposed for integrating the agent-oriented data sharing and agent-oriented argumentation upon shared data. The integrated service will support, in the entire province, a regional collaborative health service paradigm where relevant data will be available at anytime for anyone towards improved care. Previous experience will be drawn upon: LCC-based data and knowledge sharing specification in healthcare [7] through the EU Framework 6 projects of HealthAgents and OpenKnowledge, and PROforma-based argumentation scheme and its extension [8] through a collaborative research programme with Oxford University.

2   The Requirements of an Integrated Service in Hubei

Healthcare services are often delivered to the same patient at different places along lifetime, and data generated from these sources are usually associated and when made available, they can support clinicians to make improved decisions on complicated conditions. In the view of Hubei province in Central China, around 90 % of population are reported to have most of their clinical services delivered within their municipal regions and less than 5 % outside the province. However, the proportion of cross-boundary service may be well magnified in foreseeable future, as 36 among 72 top-ranked 3-A level hospitals are located in the capital city of Wuhan, which has a population of around 8 million out of 60 million in the entire province. Many are seeking more accurate diagnosis and better treatment in Wuhan from countryside everyday due to the unbalanced clinical service distribution.

A provincial platform is therefore required to aggregate data scattered among clinical organisations for the whole population and along everyone's lifetime. An added value would be to take advantages of the shared datasets and continuously inform the concerned parties a patient's allergies, lab results, current medication and alike, and even better, automatically produced recommendations under the current circumstance.

3   The Design of a Data Sharing Service Across the Province

The Design Principle and Overall Structure of the Data Sharing Service. As the platform shall systematically support the entire province of any general purpose of data exchange, it is crucial to specify where a patient's shared electronic health record should be available and when the scattered data pieces should be aggregated into it, what it may be used for under what circumstances, and how clinical organisations should liaise across regional boundary through platforms for both the contribution and retrieval of records. Centered on these is a key design principle of localised healthcare record management. Everyone will be registered, at birth time, a birth certificate that is the start of healthcare data recording. From then on, all their health-related records will be continuously accumulated, all located in their initially registered administrative county/ district-level data centre. While a service is delivered at a remote clinical site, a copy of the record will be aggregated to the same data centre to become part of the complete lifelong dataset. Likewise, historical records can be accessed, with approval, by local or remote clinicians via the same data centre.

A hierarchical platform structure is proposed as a solution of scalability, including a provincial-level platform, a dozen or so municipal-level platforms, and over a hundred county-level data centres. (1) The *provincial-level platform* has connection with it, directly, all municipal-level platforms in this province as well as provincial-level clinical organisations. (2) A *municipal-level platform* has connection with it, directly, all county-level data centres in this city as well as municipal-level clinical organisations. (3) A *county-level data centre* has connection with it, directly, all clinical organisations in this county. Platforms at each level shall have the complete indexes to all data resources under their domains as well as the summary records (full records in case of data centres) for clinical organisations in direct connection. They will also be responsible, through collaboration, for aggregating datasets into complete records for all patients who fall within their jurisdiction, and indexing new data resources when they become available.

The Analysis and Design of Actors/Roles, Role Interactions, and CDA Messages.
The aforementioned platforms and organisations can be associated with roles largely in accordance with IHE Cross-Enterprise Document Sharing (XDS) standard [9] in its Integration Profile of XDS.b, where a provincial/municipal platform acts as a *Registry* role, a county data centre as a *Repository*, and a clinical organisation as either a *Data Consumer*, a *Data Source*, or a *Repository*. Although role interactions can be flexible in serving clinical data sharing for all kinds of purposes in the platform, we will describe from a perspective of efficient healthcare resource deployment in society. This is a very crucial goal to be achieved through role interactions and demonstrated in the rest of the paper, in line with XDS service recommendations.

It is recently reported that, some of our major hospitals have up to 50 % of their outpatient service provided to patients with chronic diseases, of which 50 % are on hypertension or diabetes. The inefficient medical resource use, in relation with the deficiency of a Primary Care infrastructure in China, may be partly alleviated by a more rigid reimbursement policy along with the data sharing function of the platform. It is envisioned in near future patients will have initial diagnosis of chronic diseases and medical intervention in specialised hospitals. On stabilisation and discharge, they will be managed and followed-up at local community care in a more routine basis. Diagnosis at hospitals will be informed, via the platform, and prescribing adjusted as necessary until deterioration, where patients are referred again to specialised services. It is hoped that patient reimbursement will strictly adhere to this service pattern, and the platform will support data sharing on referrals and discharges towards an integrated healthcare service in Hubei.

Assuming this paradigm with a cross-boundary (municipal) service, two typical use scenarios are put forward whereas the platform's technical metrics and social implication can be sensed in a concrete manner. One is termed as *Remote Service Delivery followed by Localised Data Management*, with two components:

(1) A specialised clinical service at a municipal level organisation (*Document Source*) is followed by a XDS service of *RegisterDocumentSet (ITI-42)*: uploading to the municipal platform (*Registry*) the generated clinical data's summary, and to the provincial platform (*Registry*) the data index, along with the full clinical data whereas service is delivered to patients across municipal boundary.

(2) Once the provincial platform receives the full clinical data passed by a municipal platform other than the concerned patient's administrative domain, it distributes the data to the appropriate municipal platform which follows a XDS service of *ProvideAndRegisterDocumentSet (ITI-41)*: passing the detailed data to the local data centre for storage (*Repository*) and keeping the data summary for itself (*Registry*).

The other use scenario is termed as *Local Community Service Delivery on the basis of Latest Record followed by Index Update on Platforms*, with two components:

(1) A community service (*Document Consumer*) starts with a XDS service of *RegistryStoredQuery* (ITI-18): enquiring the platforms (*Registry*) for appropriate dataset and *RetrieveDocumentSet* (ITI-43): retrieval of the latest dataset from the suggested data centre (*Repository*). Current diagnosis or treatment in hospital will be returned, the patient conditions followed-up, and medication or any other intervention adjusted whatsoever necessary, accompanied with data update.

(2) When new data is generated by the community service (*Document Source*) and becomes available at patient's local site (*Repository*), it is followed by a XDS service of *RegisterDocumentSet (ITI-42)*: data summary registered on municipal platform (*Registry*) and index on the provincial platform (*Registry*), for later retrieval.

In the interest of conciseness, it is illustrated in Fig. 1 an interaction protocol of just the former use scenario, with 7 participants acting as platforms (4), clinical service deliverers (2), and patient (1), data sharing realised via message passing among them and a formal interaction specification towards service delivery will be described next. Equally important, a generic and standard message format will accompany each specific interaction. The HL7 Clinical Document Architecture (CDA) [10] is a document exchange standard that enables interoperability among disparate systems. Together with terminologies such as LOINC, ICD, or SNOMED, they can enable a syntactic and semantic agreement between communicating parties, where both the clinical data structures can be understood and concepts of interest can be unambiguously identified and their instance values retrieved.

Role Interaction Specification in LCC – A Multi-agent Solution. Our role interaction specification follows the Lightweight Coordination Calculus (LCC) language developed in the OpenKnowledge project [11–13], which emphases the explicit knowledge modelling of role interactions in a heterogeneous environment towards agent goals. A major advantage of this framework is that the declarative specification can be transmitted and interpreted by agents dynamically when joining and involving in interactions.

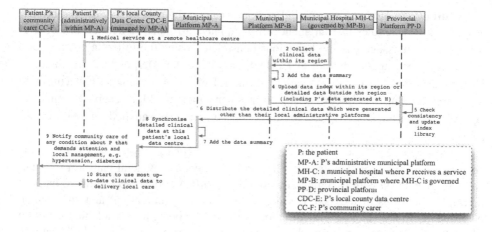

Fig. 1. A role interaction protocol for cross-boundary data sharing in the platform

The language is designed explicitly to separate the higher-level knowledge sharing among an agent group (specified by knowledge engineers) from the lower-level operation of individual agents (specified by software engineers). Sitting in a distributed environment, agents can be dynamically bound with components or services under local arrangement for computational need, while interoperability achieved at the agent interaction level. The LCC specification for our scenario is shown in Fig. 2, referencing the actors and messages (in particular message 2, 4, 6, and 8 which will be abbreviated accordingly) identified in Fig. 1.

Agent MP-B: receives a message 2 (abbreviated as M2) and sends M4, to keep a summary and in different conditions, upload to provincial platform an index of the record or its detail.

$a(R_{municipal-platform}, MP-B)::$

    $uploadRecord(Record-p, RecordSummary-p, RecordIndex-p, ID-p) \Leftarrow a(R_H, MH-C)$ then

    $null \leftarrow addRecordSummary(RecordSummary-p, ID-p)$ then

        $uploadRecordIndex(RecordIndex-p, ID-p) \Rightarrow a(R_{provincial-platform}, PP-D)\leftarrow getMunicipalDomain(ID-p) \in MP-B$

        or $uploadRecord(Record-p, RecordSummary-p, RecordIndex-p, ID-p) \Rightarrow a(R_{provincial-platform}, PP-D)\leftarrow$
$getMunicipalDomain(ID-p) \notin MP-B$

Agent PP-D: receives M4 and sends M6, to add an index passed up by municipal platform, and pass down the full record to where it should be under localised management.

$a(R_{provincial-platform}, PP-D)::$

    $uploadRecord(Record-p, RecordSummary-p, RecordIndex-p, ID-p) \Leftarrow a(R_{municipal-platform}, MP-B)$ then

    $null \leftarrow addRecordIndex(RecordIndex-p, ID-p)$ then

        $localMgnt(Record-p, RecordSummary-p, ID-p) \Rightarrow a(R_{municipal-platform}, MP-A)\leftarrow findMunicipalDomain(ID-p, MP-A)$

Agent MP-A: receives M6 and sends M8, to keep a summary record of the data passed down by provincial platform and stores at the local county data centre the detailed record.

$a(R_{municipal-platform}, MP-A)::$

    $localMgnt(Record-p, RecordSummary-p, ID-p) \Leftarrow a(R_{provincial-platform}, PP-D)$ then

    $null \leftarrow addRecordSummary(RecordSummary-p, ID-p)$ then

        $storeRecord(Record-p, ID-p) \Rightarrow a(R_{county-datacentre}, CDC-E)\leftarrow findDataCentre(ID-p, CDC-E)$

Fig. 2. A service specification for municipal platform A, B and provincial platform D

A full LCC dialogue framework description can be referred in [11, 12] and briefly, $a(R_i, A_i)$ :: Def denotes that an agent (type) A_i plays a role R_i as defined in Def. Def describes the message passing behaviour constructed using the following forms: Def_j then Def_k (Def_j satisfied before Def_k), Def_j or Def_k (either Def_j or Def_k satisfied), or Def_j par Def_k (both Def_j and Def_k satisfied). In the Def, $M_l \Rightarrow A_m$ denotes that a message M_l is sent to agent A_m while $M_l \Leftarrow A_m$ denotes that a message M_l is received from agent A_m. Also in the Def, $\leftarrow Cons_n$ denotes that a constraint must be satisfied before the clause prior to it.

Following this precise syntax of LCC, the construct $a(R_{provincial-platform}, PP\text{-}D)$ and its behaviour specification denotes that PP-D which plays a provincial platform role will have an interaction with two municipal platforms MP-A and MP-B (with two pairs of highlighted clauses of message passing in the overall specification) and an eventual outcome of clinical data generated remotely (at MP-B) being also available locally (at MP-A). This declares a pattern where three (or more) agents' shared goal of localised data management (our design principle) may be achieved, and a similar pattern can be defined on data access across municipal boundary for clinical investigation or other purposes, together making up a complete data-sharing infrastructure.

The execution of LCC specification is fully supported by the OpenKnowledge kernel, hence the interactions among platform and organisation agents can fulfill the services recommended by the IHE XDS standard. The only additional requirement left is the customised implementation of agent constraint solving facilities, including *getMunicipal-Domain()*, *findMunicipalDomain()*, and *findDataCentre()*, where the identity of a patient's administrative data centre or municipal domain should be cross-referenced from a patient identity (refer to the IHE Patient Identifier Cross-referencing (PIX) standard). Other key functions include *addRecordIndex()*, *addRecordSummary()*, and *storeRecord()*, where platforms at different levels should manage data at different levels of granularity. It is worth noting that message passing among platforms (and organisations) are abstracted in the above specification but may be instantiated for all kinds of actual clinical need (as coded in CDA), e.g. recent diagnosis or prescribing at a remote site shared for current decision-making.

4   The Design of a Rule-Driven Argumentation Service

A rule-based argumentation scheme has been designed on top of PROforma language in describing clinical guidelines and extended towards dynamic agent interpretation and multi-agent choreography [8], shown in Fig. 3. Briefly, each argumentation rule specifies multiple decision options (branches) and each option is in turn associated with multiple arguments either in support or against these options. Being a key element in the scheme, an argument has criteria for judging if this argument holds or not. The criteria may be logically linked statements for predicating the satisfaction of patient symptoms, signs or lab rest results, returning a Boolean value of true or false. A ranked preference of decision options will be offered on the basis of the overall weight of successfully justified arguments, via a separate recommendation rule.

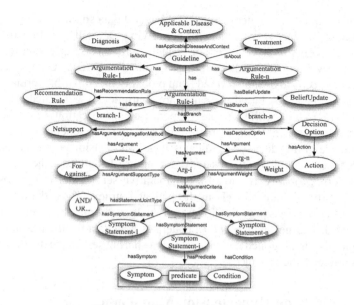

Fig. 3. An ontological representation of the argumentation scheme

A rule schema is specified further towards agent-interpretation, with a simplified example shown in Fig. 4. The rule is defined for judging whether a patient has flu or cold, with an (strong) argument in confirming flu (with a weight of 100) on patients with "fever" and arguments in support of flu (with a weight of 10) on patients with "severe cough", "exhaustion", among others. While an argument against cold is specified on patients with "headache", it can be specified equivalently as an argument in support of fever. A recommendation rule can be further defined as an algorithm of aggregating arguments and provision of suggested diagnosis. Argumentation rules as such are generic enough for clinical decision knowledge representation and specific enough for agent interpretation and execution.

```
<xs:element name="argumentationrule">          <argumentationrule>
  <xs:complexType>                               <candidate name = "Flu">
    <xs:sequence>                                  <arguments support = "confirming">
      <xs:element name="candidate" maxOccurs="unbounded">   <argument weight = "100">
        <xs:complexType>                               <obj att = "fever" predicate = "present" value = "true"/>
          <xs:sequence>                              </argument>
            <xs:element name="name" type="xs:string"/>    </arguments>
            <xs:element name="arguments" maxOccurs="unbounded">   <arguments support = "for">
              <xs:complexType>                             <argument weight = "10">
                <xs:sequence>                                <obj att = "cough" predicate = "present" value = "severe"/>
                  <xs:element name="support" type="support-type"/>   </argument>
                  <xs:simpleType name="support-type">        <argument weight = "10">
                    <xs:restriction base="xs:string">          <obj att = "exhaustion" predicate = "present" value = "true"/>
                      <xs:enumeration value="for"/>            </argument>
                      <xs:enumeration value="against"/>      </arguments>
                    </xs:restriction>                      </candidate >
                  </xs:simpleType>                       <candidate name = "Cold">
                  <xs:element name="argument" maxOccurs="unbounded">   <arguments support = "against">
                    <xs:complexType>                           <argument weight = "10">
                      <xs:sequence>                              <obj att = "headache" predicate = "present" value = "true"/>
                        <xs:element name="weight" type="xs:decimal"/>   </argument>
                        <xs:element name="symptom" type="xs:string"/>   </arguments>
                        <xs:element name="predicate" type="xs:string"/>   </candidate>
                        <xs:element name="value" type="xs:string"/>   </argumentationrule>
                      </xs:sequence>
```

Fig. 4. The rule schema and a simplified example (fragmented)

Intending a semantically interoperable and integral data sharing and argumentation service, the same standard set of data items defined in CDA will constitute rule criteria, e.g. symptom concepts such as "fever" or "cough" will consistently refer to the corresponding ones (with an identical coding mechanism) in CDA. Otherwise, a rule's prerequisite data needs to be transformed from probably incompatible formats provided by local vendors. As CDA-based messages are continuously generated and passed in the platform in compliant with LCC-based interaction specification, up-to-date clinical data can be extracted straightaway from the universal message scheme whilst evaluating arguments and weighting decision candidates at various decision points. In addition, the receipt of such messages also indicates that decisions are demanded at the current clinical service providers, probably as part of a collaborative service delivery across several regions and/or administrative levels in the patient journey. Thus, the availability of shared data from a previous provider drives current decision-making (accompanied with patient referral/discharge), which in turn generates new data for the next provider and this keeps going – an integrated data and decision sharing service enables an integrated healthcare service in the province.

5   A Case Study on Hypertension Management

A case study on hypertension management is used for demonstration. Regarding the data sharing service, a total of 31 classified datasets have been published under the title of population and public health along with their compliant CDA documents or message structures for data sharing in Hubei's platform. Our example is among these and corresponds to a paper-based form currently used in data recording. It includes a total of 116 data items, many from a Chinese national recommendation and compliant with an accompanied coding system. Regarding the argumentation service, the National Institute for Health and Clinical Excellence (NICE) provides national clinical guidelines in England, a sound and reliable source of clinical knowledge as we reflect upon our previous studies of referral and triple assessment for breast cancer [8]. The NICE guideline of "antihypertensive drug treatment" [14] is borrowed here for demonstration, with argumentation rules developed from it, since no equivalent national guideline is currently available in China.

Case description.
   A patient aged 40 from a remote countryside was diagnosed of hypertension at a large hospital in the city of Wuhan. She was first prescribed by the doctor with a medication of an ACE inhibitor, which was later found not tolerated because of cough. Then a low-cost ARB was offered instead (Step 1). After a period of time, the patient is stabilised and later discharged for community care. The medication works fine in the first months but deteriorated after some time. The community caregiver is now confronted with a decision to make (Step 2).

The Data-Sharing Service with CDA-Based Message Passing for Hypertension.
A fragment of the role interaction protocol in Fig. 1 and its LCC specification in Fig. 2 has been extracted in Fig. 5, with a focus on the provincial platform. The abstract messages (M4/M6) passed between the provincial platform and two municipal platforms, as in

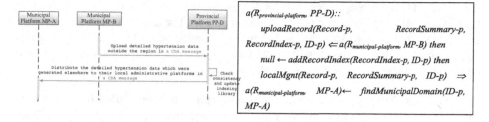

$a(R_{provincial\text{-}platform}, PP\text{-}D)::$

$uploadRecord(Record\text{-}p, \qquad RecordSummary\text{-}p,$

$RecordIndex\text{-}p, ID\text{-}p) \Leftarrow a(R_{municipal\text{-}platform}, MP\text{-}B)$ then

$null \leftarrow addRecordIndex(RecordIndex\text{-}p, ID\text{-}p)$ then

$localMgnt(Record\text{-}p, RecordSummary\text{-}p, ID\text{-}p) \Rightarrow$

$a(R_{municipal\text{-}platform}, MP\text{-}A) \leftarrow findMunicipalDomain(ID\text{-}p,$

$MP\text{-}A)$

Fig. 5. A concentrated view on the provincial platform's service specification

```
<ClinicalDocument xmlns="urn:hl7-org:v3" xmlns:mif="urn:hl7-org:v3/mif" xmlns:xsi="http://www.w3.org/2001/XMLSchema-instance">
    <id root="2.16.156.10011.1.1.1.4" extension="D2011000001" code="DE01.00.008.00"/>
    <code code="HOSB04.02" codeSystem="2.16.156.10011.2.4" codeSystemName="卫生信息共享文档规范编码体系"/>
    <effectiveTime value="20150624154823"/>
    <recordTarget typeCode="RCT" contextControlCode="OP">
        <patientRole classCode="PAT">
            <id root="2.16.156.10011.1.2" extension="HR201405273366666" code="DC01.00.009.00"/>
            <patient classCode="PSN" determinerCode="INSTANCE">
                <id root="2.16.156.10011.1.3" extension="ID420106195001011919"/>
                <name>翼小明</name>
                <age value="40"/>
                <ethnicGroupCode code="15" displayName="土家族" codeSystem="2.16.156.10011.2.3.3.3" codeSystemName="民族类别代码表 (GB 3304)"/>
            <author typeCode="AUT" contextControlCode="OP">
                <assignedAuthor classCode="ASSIGNED">
                    <id root="2.16.156.10011.1.7" extension="430006119471201"/>
                    <name>楼医生</name>
                <representedOrganization>
                    <id root="2.16.156.10011.1.5" extension="0187565656"/>
                    <name>湖北省中医院</name>
                    <addr>武汉市武昌区昙福路粮道街花园山4号</addr>
<structuredBody>
    <component>
        <section>
            <code code="11450-4" displayName="症状/Symptoms" codeSystem="2.16.840.1.113883.6.1" codeSystemName="LOINC"/>
            <entry>
                <observation classCode="OBS" moodCode="EVN">
                    <code code="DE04.01.118.00" codeSystem="2.16.156.10011.2.2.1" codeSystemName="卫生信息数据元目录" displayName="症状名称"/>
                    <value xsi:type="ST">眩花耳鸣/vertigo and tinnitus</value>
    <component>
        <section>
            <code code="8716-3" displayName="生命体征/Signs" codeSystem="2.16.840.1.113883.6.1" codeSystemName="LOINC"/>
            <entry>
                <observation classCode="OBS" moodCode="EVN">
                    <code code="DE04.10.174.00" codeSystem="2.16.156.10011.2.2.1" displayName="收缩压/systolic"/>
                    <value xsi:type="PQ" value="140" unit="mmHg"/>
                <observation classCode="OBS" moodCode="EVN">
                    <code code="DE04.10.176.00" codeSystem="2.16.156.10011.2.2.1" displayName="舒张压/diastolic"/>
                    <value xsi:type="PQ" value="90" unit="mmHg"/>
                <observation classCode="OBS" moodCode="EVN">
                    <code code="DE04.10.188.00" codeSystem="2.16.156.10011.2.2.1" displayName="体重/weight"/>
                    <value xsi:type="PQ" value="85" unit="kg"/>
    <component>
        <section>
            <code code="10160-0" codeSystem="2.16.840.1.113883.6.1" displayName="用药/Medication" codeSystemName="LOINC"/>
            <entry>
                <substanceAdministration classCode="SBADM" moodCode="EVN">
                    <routeCode code="1" codeSystem="2.16.156.10011.2.3.1.158" codeSystemName="用药途径代码表"/>
                    <doseQuantity unit="mg" unit="mg"/>
                    <rateQuantity value="1-2" unit="per day"/>
                    <consumable name="ARB"/>
                    <entryRelationship typeCode="COMP">
                        <observation classCode="OBS" moodCode="EVN">
                            <code code="DE06.00.130.00" codeSystem="2.16.156.10011.2.2.1" codeSystemName="卫生信息数据元目录" displayName="药物不良反应标志"/>
                            <value xsi:type="ST">ACE inhibitor Not Tolerated/ACE抑制剂伴随有咳嗽等不良反应</value>
```

Fig. 6. A CDA message being passed on hypertension discharge (referral)

previous specification, are now in a concrete format of CDA, shown in Fig. 6. Its structure can be split into two parts serving very distinct goals. First, the key administrative data (the header part) indicates that about which patient the data is concerned and the patient's demographic information (<recordTarget> and its embedded <patient> sections), when and where the data is generated and by whom (<effectiveTime>, <author>, and <representedOrganization>), and for what purpose (document <code>) it is produced. These will be used to construct a data index for platforms to keep, for tracking the data source, and forwarding the data if necessary, towards localised data management.

Second, key clinical data (the body part) generated at a remote hospital on investigations, diagnosis, interventions and so on fit into hierarchically annotated <component>-<section>-<entry>-<observation>/<substanceAdministration> parts. Each entry is a clinical statement typically with a uniquely and semantically identified <code> and corresponding <value>, describing precisely about this patient's symptoms (vertigo and tinnitus) and signs (systolic 140, diastolic 90 and weighted 85 kg, etc.), the doctor's diagnosis and prescribing (ARB with its dosage and frequency), and adverse drug reactions if any (ACE inhibitor not tolerated) (as in Step 1). The same type of CDA message being passed twice (*uploadRecord* and *localMgnt*) as specified in LCC, already observed patient conditions or administrated substances will be available to local data centre and community care as soon as these clinical statements are extracted.

Hypertension Management Guideline and Argumentation Rule. It is shown in Fig. 7 the clinical guideline on hypertension management recommended by the Nice Pathway, from which a rule is deduced in compliant with our argumentation scheme, shown in Fig. 8. The clinical data shared through the previous service indicates a Step 1 prescribing on ARB instead of ACE inhibitor due to accompanied cough, compliant with the guideline and a pre-condition of Step 2.

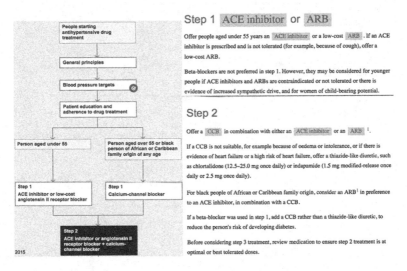

Fig. 7. Nice Pathway for hypertension prescribing (a partial careflow with narrative guidelines)

```
<ArgumentationRule-Ht2ndStep>
  <Ht1stStep><decision>ARB</decision></Ht1stStep>
  <candidate name = "Offer a CCB in addition to ARB">
    <arguments support = "against">
      <argument weight = "10"><obj att = "oedema or intolerance" predicate = "present " value = "true"/></argument>
      <argument weight = "10"><obj att = "heart failure" predicate = "present" value = "high risk level"/></argument>
    </arguments>
  </candidate >
  <candidate name = "Offer a thiazide-like diuretic"></candidate>
    <recommendation-rule>net-support('Ht2ndStep', 'Offer a CCB in addition to ARB') > -10</recommendation-rule>
</ArgumentationRule-Ht2ndStep>
```

Fig. 8. The argumentation rule deduced from the guideline

Informed observation on symptoms, adverse drug reactions and so on may be valuable when taken into account in rule execution but most likely, new relevant findings need to be established and data collected for weighting the prescribing options. Assuming the patient has deteriorated conditions, the argumentation rule will fire with an initial prescribing of an additional CCB. Evidence needs to be established on whether there is any argument against this – if patient has "CCB intolerance/oedema" or "high risk of heart failure" criteria satisfied then an aggregated weight of −10 or −20 would be assigned to this option. The recommendation rule will then indicate a shift from the default decision option of "CCB in addition to ARB" to "a thiazide-like diuretic" as the current most appropriate medication.

6 Discussions and Conclusions

An integrated agent-oriented data sharing and decision-support framework has been proposed in this paper, built upon the LCC technology and CDA standard, demonstrated with a hypertension management example, and in compliant with IHE XDS standard. An agent engine is being implemented for LCC constraint solving functions, CDA message processing, and argumentation rule execution. In future work, we will look deeper at the generality of rule representation and consistency of rule execution, the completeness and effectiveness of CDA-based message passing for all scenarios of data sharing, and the accuracy of argumentation in conventional sense.

In our previous work, clinical decision support systems have been developed in various multi-national or national contexts [7, 8, 15] and recently, a standard protocol for provincial level clinical data exchange [16]. Like us, most current research is dedicated to the fields of regional clinical data sharing or clinical decision support (with closed boundary), in isolation. Nevertheless, a CDA wrapper with dual direction of mappings has been proposed for standardising input (patient data) and output (decision result) of decision support [17]. It's claimed that decision support with this wrapper may not be constrained by specific information systems any more. Work as such is still limited and a tangible gap exists between two communities aiming at two most critical IT services for healthcare, and unless an integrated service infrastructure is systematically designed and developed, their fruits cannot be fully enjoyed and devoted efforts compromised. The agent-oriented platform services as proposed in this paper will enable the right clinical data to be available at the right place at the right time, for making the right decision. Work so far indicates it is promising and will contribute in a novel and substantial way to the Hubei Provincial Care Platform.

Acknowledgment. This work is supported by National Natural Science Foundation of China (61202101) & Dept. of Health on Data Exchange Standard for Hubei Provincial Care Platform.

References

1. Greenhalgha, T., et al.: Introducing a nationally shared electronic patient record: case study comparison of Scotland, England, Wales and Northern Ireland. Int. J. Med. Inf. **82**(5), e125–e138 (2013)
2. Greenhalgh, T.: Adoption and non-adoption of a shared electronic summary record in England. BMJ **340**, c3111 (2010)

3. Bernstein, K., Andersen, U.: Managing care pathways combining SNOMED CT, archetypes and an electronic guideline system. Stud. Health Technol. Inf. **136**, 353–358 (2008)
4. Lluch-Ariet, M., et al.: Knowledge sharing in the health scenario. J. Transl. Med. **12**(Suppl. 2), S8 (2014)
5. Kawamoto, K., Houlihan, C.A., Balas, E.A., Lobach, D.F.: Improving clinical practice using clinical decision support systems: a systematic review of trials to identify features critical to success. BMJ **330**, 765 (2005)
6. Garg, A.X., et al.: Effects of computerized clinical decision support systems on practitioner performance and patient outcomes: a systematic review. JAMA **293**(10), 1223–1238 (2005)
7. Xiao, L., Lewis, P., Gibb, A.: Developing a security protocol for a distributed decision support system in a healthcare environment. In: Proceedings of the 30th International Conference on Software Engineering (ICSE 2008), pp. 673–682. ACM (2008)
8. Xiao, L., Fox, J., Zhu, H.: An agent-oriented approach to support multidisciplinary care decisions. In: Proceedings of the 3rd Eastern European Regional Conference on the Engineering of Computer Based Systems (ECBS 2013), pp. 8–17. IEEE (2013)
9. Integrating the Healthcare Enterprise (IHE): IT Infrastructure Technical Framework, vol. 1–3 (and its Supplement on XDS.b) (2007–2013)
10. Dolin, R.H., et al.: The HL7 clinical document architecture. J. Am. Med. Inf. Assoc. **8**(6), 552–569 (2001)
11. Robertson, D.: A lightweight method for coordination of agent oriented web services. In: Proceedings of AAAI Spring Symposium on Semantic Web Services, Stanford (2004)
12. Robertson, D.: A lightweight coordination calculus for agent systems. In: Leite, J., Omicini, A., Torroni, P., Yolum, P (eds.) DALT 2004. LNCS (LNAI), vol. 3476, pp. 183–197. Springer, Heidelberg (2005)
13. Xiao, L., et al.: Adaptive agent model: an agent interaction and computation model. In: Proceedings of the 31st IEEE Annual International Computer Software and Applications Conference, pp. 153–158. IEEE Press (2007)
14. National Institute for Health and Clinical Excellence (NICE): Antihypertensive drug treatment, p. 2 (2015)
15. Xiao, L., et al.: Developing an electronic health record for methadone treatment recording and decision support. BMC Med. Inf. Decis. Making **11**, 5 (2011)
16. Xiao, L., Wei, Q.: Developing a standard protocol for clinical data exchange and analysis. In: Proceedings of the 6th IEEE International Conference on Software Engineering and Service Science (ICSESS 2015) (2015) (in press)
17. Sáez, C., Bresó, A., Vicente, J., Robles, M., García-Gómez, J.M.: An HL7-CDA wrapper for facilitating semantic interoperability to rule-based clinical decision support systems. Comput. Meth. Prog. Biomed. **109**(3), 239–249 (2013)

Data Gathering with Compressive Sensing for Urban Traffic Sensing in Vehicular Networks

Dan Wang, Haifeng Zheng$^{(\boxtimes)}$, Xin Chen, and Zhonghui Chen

College of Physics and Information Engineering,
Fuzhou University, Fuzhou, Fujian, China
{nl31127016,zhenghf,chen-xin,czh}@fzu.edu.cn

Abstract. Vehicular networks have become as an important platform to monitor metropolitan-scale traffic information. However, it is a challenge to deliver and process the huge amount of data from vehicular devices to a data center. By studying a large number of taxi data collected from around 3,000 taxis from Shenzhen city in China, we find that the data readings collected by vehicular devices have a strong spatial correlation. In this paper, we propose a novel scheme based on compressive sensing for traffic monitoring in vehicular networks. In this scheme, we construct a new type of random matrix with only one nonzero element of each row, which can significantly reduce the number of data needed to be transmitted while guaranteeing good reconstruction quality at the data center. Simulation results demonstrate that our scheme can achieve high reconstruction accuracy at a much lower sampling rate.

Keywords: Vehicular networks · Compressive sensing (CS) · Data gathering

1 Introduction

With the number of vehicles increasing in the world, more and more cities suffer from traffic congestion. Road traffic monitoring provides traffic conditions of different roads which make congestion control possible. A number of applications benefit from traffic estimation such as traffic management, road engineering and trip planning. To alleviate the burden of traffic, efficient traffic management and metropolitan-scale traffic estimation is necessary.

Traditional approaches to monitor traffic conditions are based on static traffic sensors. However, the high cost of deployment and maintenance limits the coverage of those traditional approaches. With more vehicles equipped with various sensors, especially Global Positioning System (GPS) receivers, vehicles are becoming powerful mobile sensors which can support mobile information for monitoring traffic. In this paper, we estimate metropolitan-scale traffic condition in vehicular networks by recovering the original GPS readings accurately with few samples.

The mission of monitoring applications in a vehicular network is to collect accurate data from all the vehicles. It is straightforward to make all the vehicles send their GPS data to monitoring center independently. Although this approach may provide a large

© Springer International Publishing Switzerland 2015
A. Bikakis and X. Zheng (Eds.): MIWAI 2015, LNAI 9426, pp. 441–448, 2015.
DOI: 10.1007/978-3-319-26181-2_41

coverage, it faces several challenges. Firstly, the amount of data readings which will be sent to a monitoring center is large, thus causing high communication and storage cost. Secondly, it takes too much time for all the vehicles to deliver their readings for one monitoring round. Besides, the coverage of vehicles is not completed which may lead to data missing. To solve the problems above, we analyze GPS readings collected from taxis in Shenzhen city and find that there exhibits hidden structures which demonstrate that there exists a strong spatial correlation among the data readings. Inspired by this important observation, we propose a compressive sensing based scheme, where we construct a new type of random measurement matrix with containing only one nonzero element in each row.

As an effective theory for recovery with sparse samples, compressive sensing has drawn more and more attention by researchers. In [1], an algorithm based on compressive sensing is proposed to get the best estimate traffic condition matrix for a metropolitan-scale traffic sensing system. In [2], Wang et al. calculate the minimum number of seeds and the transmission hop length for compressive measurements in the network and propose CSM for reducing communication cost by using compressive sensing. In [3], Zhu et al. find the hidden structures with real data readings and apply compressive sensing to solve the missing data problem. Differently from the existing works, we adopt a novel representation matrix based on Gaussian joint distribution model. For instance, in traditional compressive sensing schemes [4], when the network size is N, O(log N) nonzero elements are required in each row of the measurement matrix. It has been also proved that both the representation matrix and measurement matrix we adopt in this paper satisfy restricted isometric property (RIP) [5], which guarantees that the data collected in the monitoring center can be recovered with a high accuracy.

We have made the following technical contributions in this paper:

- By analyzing a large amount of GPS readings on vehicles, we discover that there exists a strong spatial correlation among data readings.
- We propose a novel compressive sensing based scheme to monitor with vehicles, which is able to reduce the communication cost while guarantees the recovery accuracy.
- Comprehensive simulations with real GPS data demonstrate that our scheme outperforms the other existing schemes.

The rest of this paper is organized as follows. Section 2 presents the preliminaries of compressive sensing and the network model. The detailed design is described in Sect. 3. Section 4 discusses our experimental results. Finally, we make a conclusion in Sect. 5.

2 Preliminary of Compressive Sensing

Compressive sensing (CS) is an effective technique for data compression and sampling. CS theory can recover an original n-dimensional data with only m measurements (m < n) since signals usually contain redundant information. Assuming an original signal x, it can be represented by a k-sparse data set d which contains only k nonzero components in some domain with the representation basis Ψ, i.e., $d = \psi X$. Then, the signal can be compressed by $y = \Phi d$, where Φ is an m × n measurement matrix and y is an m-dimensional measurement vector (m ≪ n). Thus, to recover the original signal x, three problems must be considered: (1) representation matrix Ψ which transforms a signal into a sparse one, (2) measurement matrix Φ which compresses the sparse signal, (3) recovery algorithm which reconstruct the original signal x from y. To design the two matrices, Φ and $\Phi\Psi$ should satisfy the following restricted isometric property (RIP) [6]:

Definition 1 ([6]). A matrix Φ satisfies the restricted isometric property of order k if there exists a $\delta_k \in (0, 1)$ such that

$$(1 - \delta_k)\|s\|_2^2 \le \|\Phi s\|_2^2 \le (1 + \delta_k)\|s\|_2^2 \tag{1}$$

for all k-sparse vectors $s \in R^N$.

It has been shown in [7, 8] that several random matrices satisfy the RIP. Among those, ±1 Bernoulli matrix and Gaussian distribution matrix are used frequently. Many representation matrices can make the original data set sparse such as discrete Fourier transformation (DFT), discrete cosine transformation (DCT). To recover the signals, $\ell 1$ optimization algorithm is applied as:

$$\bar{s} = \arg\min_s \|s\|_1 \text{ s.t. } y = \Phi s \tag{2}$$

Many efficient algorithms are proposed to solve the above problems such as basis pursuit [6], orthogonal matching pursuit (OMP) algorithm [9], CoSaMP [10].

3   The Proposed Scheme with Compressive Sensing

3.1   System Model

Consider a vehicle network with N probe vehicles. Each vehicle moves on its will along the roads. The GPS receivers embedded in vehicles continuously detect instant traffic information and send the data report to the monitoring center via the wireless network. A GPS reading including vehicle ID, time, longitude, latitude and speed is instantaneous, which is shown as follows.

Table 1. GPS readings

Vehicle ID	Time	Longitude	Latitude	Speed

The instant data set at a certain time is denoted as $C_t(ID) = \{Lon_t(ID), Lat_t(ID), V_t(ID)|t = T\}$, where ID is vehicle identification, Lon and Lat is defined as the vehicle longitude and latitude of the vehicle respectively and V denotes the speed, respectively. T is the time at which vehicles send their GPS readings. The set of $C_t(ID)$ is different for different vehicles.

We consider the speed of probe vehicles, which depicts the traffic condition of roads, denoted as $V_t(ID) = \{v_t(id_0), v_t(id_1), v_t(id_2), \ldots, v_t(id_{N-1}) | t = T\}$. It is costly to send and receive all the data directly. Many CS algorithms are proposed in previous studies to make transmission more effective. Since the measurement matrices used in those CS algorithms are Gaussian random matrix or Bernoulli random matrix, of which there are more than one nonzero element in each row, it cannot reduce the communication cost obviously. To use the least measurements for recovering the original data set, we design a new representation matrix and the sparsest measurement matrix with the finding that the data has a strong spatial correlation. In this paper, the recovery error can be represented by normalized mean absolute error.

$$\in = \sum |x' - x| \, / \, |x| \tag{3}$$

3.2   Design of Representation Matrix

A few representation bases such as DFT, DCT, DWT are often used in the existing studies. Since the GPS data exhibits strong spatial correlation, it is reasonable to adopt Gaussian joint distribution model and this model performs well in WSN [5].

Define $K(x_i, x_j)$ as a Gaussian kernel function, which denotes the correlation between x_i and x_j. Denote d_{ij} as the distance between each two vehicles, and coefficient σ determines the effect between x_i and x_j. In [11], it has been shown how to estimate the parameter σ via maximum likelihood or Bayesian framework from training data. Based on the above assumptions, we can obtain the correlation matrix G as

$$G = \begin{bmatrix} e^{\frac{-d_{11}^2}{2\sigma^2}} & e^{\frac{-d_{12}^2}{2\sigma^2}} & \cdots & e^{\frac{-d_{1N}^2}{2\sigma^2}} \\ e^{\frac{-d_{21}^2}{2\sigma^2}} & e^{\frac{-d_{22}^2}{2\sigma^2}} & \cdots & e^{\frac{-d_{2N}^2}{2\sigma^2}} \\ \vdots & \vdots & \vdots & \vdots \\ e^{\frac{-d_{N1}^2}{2\sigma^2}} & e^{\frac{-d_{N2}^2}{2\sigma^2}} & \cdots & e^{\frac{-d_{NN}^2}{2\sigma^2}} \end{bmatrix} \tag{4}$$

Then we diagonalize G as $G = \Psi \Lambda \Psi^{-1}$, where Ψ is an orthogonal eigenvector basis generated from the Gaussian kernel function (GKB), Λ is a diagonal matrix and its diagonal entries are the eigenvalues of G. Ψ is used as an orthogonal representation matrix. Thus we can transform x into a sparse signal by $s = \Psi^{-1}x$. If s is sparse and $\Phi\Psi G$ satisfy RIP, Ψ would be considered as a representation basis of sparsest random projections.

3.3   Design of Measurement Matrix

In our scheme, we use a sparsest random matrix (SERM), where each row contains only one CS measurement which represents one sampling value. The rationale of the design is provided by [5]. We define the sparsest measurement as

$$\Phi(i,j) = \begin{cases} 1 & j = r_i \\ 0 & \text{otherwise} \end{cases} \tag{5}$$

where the range of i is from 1 to M, the range of j is from 1 to N, r_i represent the independent and identically distributed random indices, and $r_i < r_{i+1}, r_i \in [1, N]$. Then, the data received can be expressed as

$$x_r = \begin{bmatrix} x_{r_1} \\ x_{r_2} \\ \vdots \\ x_{r_M} \end{bmatrix} = \begin{bmatrix} \Phi_1 \\ \Phi_2 \\ \vdots \\ \Phi_M \end{bmatrix} \begin{bmatrix} x_1 \\ x_2 \\ \vdots \\ x_N \end{bmatrix} \tag{6}$$

4 Simulations and Analysis

In this section we first show that the GPS readings are spatial correlated in a vehicular network. Then we perform simulations and present evaluation results to demonstrate the effectiveness of our scheme.

4.1 Revealing Spatial Correlation

To demonstrate spatial correlation with the vehicle sensing data, we use real vehicular data to perform simulations. All the data is collected from vehicles which move along the roads in Shenzhen city, China. The formatted GPS readings are summarized in Table 1 and the collecting time we choose is 8 o'clock on 18th, April, 2011.

We transform the original data into a sparse vector with the proposed representation basis. The coefficients vector we get is defined as $C = \{c_1, c_2, c_3, \ldots, c_n\}$. In fact, few real datasets are strictly K-sparse. So the following method is adopted commonly by previous works to obtain an approximate coefficients vector. The coefficients are sorted in the descending order and denoted as $|\theta_1| \geq |\theta_2| \geq \cdots \geq |\theta_n|$. Then, we keep the largest K coefficients and ignore the other small entities. We find that the value of the transform coefficients follows power law as

$$|\theta_i| \leq R_i^{-1/p} \tag{7}$$

where R_i is a constant and $0 < p < 2$.

We sort all the transform coefficients in a descending order and normalize them. Figure 1 depicts the fast decaying property of the values. It reveals that the vehicle speed approximately follows the power law.

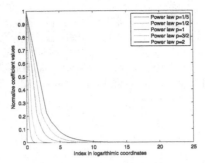

Fig. 1. Normalized coefficient values of real taxi speed data with auxiliary curves $x^{-1/p}$

4.2 Simulation Setup

We carry out simulations with a data set of GPS readings collected from 1,000 taxis at 8 o'clock. A comparative study is conducted to compare our scheme with the other competing schemes which are introduced in this section.

We divide the data sample task into two parts according to the theory of CS. In the first part, the readings are transformed into a sparse vector by the representation basis. In the other part, the n-dimension sparse vector is compressed into an m-dimension sample vector. Then, we recover the original data from the sample values and compare the estimation error of all the approaches. The representation bases we use include the Laplacian eigenvector basis [12], DCT, DFT and GKB proposed in our paper. We adopt Gaussian random matrix, Bernoulli random matrix, Sparse Random Projections (SRP) [13] with k = 80 non-zero elements in each row and our sparsest matrix as (5). For comparison, we combine different representation bases with the same measurement matrix. In addition, we also combine our GKB with different measurement matrices.

4.3 Impact of Representation Bases

We study the impact of representation bases on recovery accuracy. The number of measurements is varied from 200 to 500 at a step of 50. The comparison of the four representation bases for all the four measurement matrices is reported in Figs. 2, 3, 4 and 5, respectively.

We can find that the estimation error of DCT and DFT representation bases is much higher than the other two representation bases. Besides, we cannot recover the original data when DCT or DFT is combined with our SERM. It is because vehicular networks have an irregular topology due to random deployments while DCT and DFT are used for regular deployment networks. The recovery accuracy of the Laplacian eigenvector basis and our GKB is approximate. However, as shown in Fig. 2, the estimation error of our scheme using GKB is smaller compared to that of the Laplacian eigenvector basis. This is mainly because our GKB gets a better sparse degree.

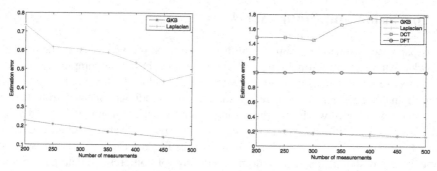

Fig. 2. Our sparsest random matrix (SERM)   **Fig. 3.** SRP with k = 80 nonzero measurements

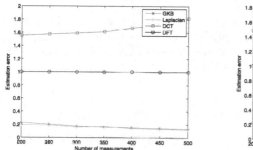

Fig. 4. Gaussian random matrix   **Fig. 5.** Bernoulli random matrix

4.4 Impact of Random Measurement Matrices

To study the impact of representation matrices on the recovery quality, we compare our scheme using SERM with those using Gaussian random matrix, Bernoulli random matrix and SRP, respectively. As shown in Fig. 6, it is shown that SERM works as well as the other measurement matrices, which demonstrates our scheme can reduce the communication cost while remaining comparative recovery accuracy.

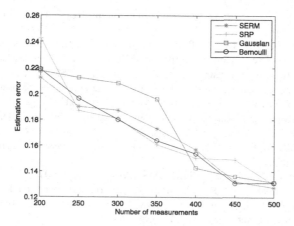

Fig. 6. Recovery quality for different four measurement matrices

5 Conclusion and Future Work

In this paper, we studied the traffic monitoring using compressive sensing for vehicular networks. Although traditional data gathering methods based on compressive sensing can recover original data accurately, these techniques induce high communication cost. Through analyzing a large set of real vehicular GPS readings collected from taxis in Shenzhen city, China, we found the strong spatial correlation among vehicular data readings. Based on this discovery, we proposed a novel CS scheme. Our scheme can reduce the communication cost with guaranteeing estimation accuracy. In this work, we only considered spatial sparsest random scheduling for compressive data gathering. In the future work, we will extend our work to the case where both spatial and temporal correlations for vehicular data are considered.

Acknowledgements. This work is supported by the Cross-strait joint fund of NSF China (No. U1405251); NSF China (No. 61571129); NSF of Fujian Province (No. 2013J01235, 2015J01250), Foundation of Fujian Educational Committee (No. JA12024), and Research Fund of Fuzhou University (No. 2013-XY-27, 2014-XQ-37, XRC-1460).

References

1. Li, Z., Zhu, Y., Zhu, H., Li, M.: Compressive sensing approach to urban traffic sensing. In: Proceedings of the IEEE ICDCS (2011)
2. Wang, H., Zhu, Y., Zhang, Q.: Compressive sensing based monitoring with vehicular networks. In: Proceedings of the IEEE INFOCOM, pp. 2923–2931 (2013)
3. Zhu, Y., Li, Z., Zhu, H., Li, M., Zhang, Q.: A compressive sensing approach to urban traffic estimation with probe vehicles. IEEE Trans. Mobile Comput. **12**(2), 2289–2302 (2013)
4. Wang, W., Garofalakis, M., Ramchandran, K.: Distributed sparse random projections for refined approximation. In: 6th International Symposium on Information Processing in Sensor Networks, IPSN 2007, pp. 331–339 (2007)
5. Wu, X., Yang, P., Jung, T., Xiong, Y., Zheng, X.: Compressive sensing meets unreliable link: sparsest random scheduling for compressive data gathering in lossy WSNs. In: MobiHoc, pp. 13–22 (2014)
6. Candes, E., Tao, T.: Decoding by linear programming. IEEE Trans. Inf. Theory **51**(12), 4203–4215 (2005)
7. Candes, E., Romberg, J., Tao, T.: Robust uncertainty principles: Exact signal reconstruction from highly incomplete frequency information. IEEE Trans. Inf. Theory **52**(2), 489–509 (2006)
8. Donoho, D.: Compressed sensing. IEEE Trans. Inf. Theory **52**(4), 1289–1306 (2006)
9. Tropp, J., Gilbert, A.: Signal recovery from random measurements via orthogonal matching pursuit. IEEE Trans. Form. Theory **53**(12), 4655–4666 (2007)
10. Needell, D., Tropp, J.: Cosamp: iterative signal recovery from incomplete and inaccurate samples. Appl. Comput. Harmon. Anal. **26**(3), 301–321 (2009)
11. Rasmussen, C.E., Williams, C.K.I.: Gaussian Processes for Machine Learning. MIT Press, Cambridge (2006)
12. Chung, F.: Spectral graph theory. In: CBMS-AMS, vol. 92 (1997)
13. Wang, W., Garofalakis, M., Ramchandran, K.: Distributed sparse random projections for refinable approximation. In: Proceedings of the 6th International Symposium on Information Processing in Sensor Networks, pp. 331–339 (2007)

A System Architecture for Smart Health Services and Applications

Jiangyong Chen[1,2] and Xianghan Zheng[1,2(✉)]

[1] College of Mathematics and Computer Science, Fuzhou University, Fuzhou 350108, China
xianghan.zheng@fzu.edu.cn
[2] Fujian Key Laboratory of Network Computing and Intelligent Information Processing,
Fuzhou 350108, China

Abstract. Given the increasing social needs for high-quality health and medical services at a low cost, smart health has gained significant attention as the leader in achieving national happiness and next-generation growth engine based on ICT convergence technology. To meet the needs of home and primary healthcare, this paper proposes an application scheme based on machine learning and similarity calculation algorithm for home and primary healthcare. Users can move freely at home at any time and obtain accurate human physiological parameters, good medical services, and personalized doctor recommendations. The scheme can be used for home and primary healthcare, and has good practical value.

Keywords: Home healthcare · Primary healthcare · Internet of things

1 Introduction

Information and communication technologies are transforming our social interactions, lifestyles, and work places. One of the most promising applications of information technology is healthcare and wellness management. Healthcare is moving from a reactive approach to a proactive approach, which is characterized by early detection, prevention, and long-term management of health conditions. The current trend contributes to health condition monitoring and wellness management for individual healthcare and well-being. This technology is particularly important in developed countries with a significant number of aging populations, where information technology can significantly improve the management of chronic conditions and improve quality of life.

Currently, family physical sign detection equipment (e.g., Sphygmomanometer, ECG detector, glucose meters, weight body fat analyzer, oxygen detector, pulse monitors) are being widely used. However, existing equipment and services are limited by the following problems:

1. Insufficient data collection, transmission sharing, and storage capacity. The single detection function of the family physical sign detection equipment only provides the function of "namely, the test is to see", which cannot perform continuous data acquisition. Thus, transmission test data cannot be achieved, as well as sharing and persistent storage.

A. Bikakis and X. Zheng (Eds.): MIWAI 2015, LNAI 9426, pp. 449–456, 2015.
DOI: 10.1007/978-3-319-26181-2_42

2. Missing data analysis mechanism. Existing sign detection services cannot be introduced into the mechanism of data analysis and conduct analysis of user-oriented health service such as disease forecasting.
3. Lack of value-added services. Traditional instruments only have a detection function, which cannot provide personalized, value-added services based on user demand scalability or physical signs data analysis results (e.g. doctor recommendations, registration services, and so on).

With the advancement in cloud computing in recent years, big data technology and hardware capabilities have continued to improve (for example, Bluetooth protocol). For home health testing fields, our system provides physical signs data collection, transmission, persistent storage, doctor recommendations, and other services. The key functions of the proposed system are summarized in three points:

1. The use of Bluetooth protocol to perform physical signs data collection and periodic transfer to the back-end cloud storage platform. Cloud platform with high scalability can provide huge amounts of data storage services.
2. The efficient and accurate method of the system design to predict disease by introducing the theory of machine learning, clustering, and causal relationship model based on customer symptom description and user characteristic data.
3. Collecting doctor's data of 12 big disease departments of authoritative hospitals (about 1500) to design personalized user-oriented physician referral service.

The rest of the paper is organized as follows: Sect. 2 introduces the background information related to smart health services and applications, extreme learning machine, and data set collection. Section 3 introduces the system architecture and cloud-based storage. Section 4 introduces the main system approaches. Section 5 describes the prototype implementation and corresponding performance evaluation. The conclusion and future works are given in Sect. 6.

2 Related Works

2.1 Smart Health Services and Applications

After the emergence of smart phone, various creative health applications have been released for patients, clinicians, and other health service consumers.

A design approach for smart health monitoring [2] has been proposed in this study. This project deals with a reliable health monitoring system designed for affordable wireless patient monitoring system. The monitoring signals can be obtained in PC and Android mobile devices. This project shows the patient's vital parameters, such as ECG, heart rate, Sp02, pulse rate, and temperature, which are measured using the patient's monitoring system. However, this system does not introduce the persistent stores, disease early warning and forecasting, and other applications.

Paper [3] proposed a smart health monitoring and evaluation system. This article describes the design and implementation of a health monitoring and evaluation system based on wireless technologies. The system architecture encompasses the abilities to collect multiple types of data from static and mobile sensors, transfer such data over a homogeneous or hybrid wireless network to a central server, and perform data fusion to extract vital information about the health status of the person being monitored.

The system architecture and corresponding approaches proposed in this paper is a combination of similarity-based prediction approach and ELM-based service recommendation approach.

2.2   Extreme Learning Machine

Extreme learning machine (ELM) [4] is based on the empirical risk minimization theory and makes use of a single layer feedforward network for the training of single hidden layer feedforward neural networks (SLFN). The learning process needs only one single iteration and avoids multiple iterations and local minimization. Compared with conventional neural network algorithms, ELMs are capable of achieving faster training speeds and can overcome the problem of over-fitting.

2.3   Data Set Collection

The doctors' information is crawled from haodf.com, which is the largest Chinese medical website that contains the hospital doctor information of each province. The doctors' data were collected as follows:

1. Data Crawler: A total of 19,000 doctor importations from 1,500 normal hospitals with eight specific attributes are manually selected as the data source. Specific data crawlers are developed for the normal hospitals and the doctor users. The normal hospital data crawler extracted a number of lists of followers of the normal department. The normal department data crawler also extracted several lists of followers of many doctors. Finally, a total of 19,000 doctor information is crawled.
2. Feature extraction: A number features are extracted for each doctor information, including the level of hospital, doctor title, academic title, curative effect, behavior, number of patient vote, number of thank-you letters, and working age.

3   System Overview

3.1   System Architecture

The system architecture of the smart health monitoring system mainly contains four parts as show in Fig. 1.

The Client, which can be mobile phones, laptops, PC, and so on, has the ability to access to the Cloud Gateway (CG) through fixed or mobile connection. The usage protocol is based on the web service between the website and the CG and the HTTP between the CG and the Client.

Remote service discovery and control use scenario can be achieved. The client and the website periodically upload/synchronize the list of information from the cloud platform.

The main technologies can explain as following two points which is introduced in detail in forth session:

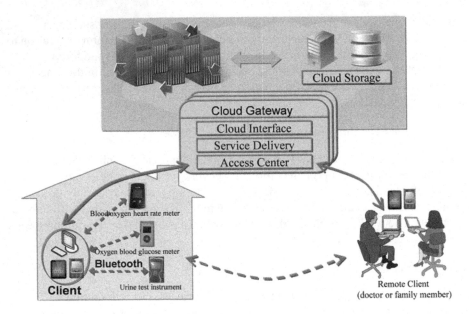

Fig. 1. System architecture

1. In the client app, combined with the similarity-based prediction and disease prediction based on causal bayes methods to provide the method of disease prediction.
2. In cloud storage side, we use the appropriate algorithm for doctors previously classified data and store it in the cloud, providing the doctor intelligent recommendation function.

3.2 Cloud-Based Storage

The traditional solution distributes data block uniformly in cloud platform servers. However, cloud servers may be distributed in different locations and network topologies, which may add transmission cost during access and data storage. Meanwhile, other family members often want to know the sign information of each other so that they can timely understand the physical condition of other family members. To optimize content distribution and reduce transmission cost, we suggest that multimedia data be distributed in a subset of cloud servers that are geographically near.

4 Methods

4.1 Similarity-Based Prediction

Similarity measurement between different samples is often performed when making classification by calculating the "distance" between samples (Distance). The geometric cosine of the angle between two vectors can be used to measure the difference in direction, borrowing the concept of machine learning to measure the difference between the sample vectors.

4.2 ELM-Based Service Recommendation

Figure 2 illustrates the basic concept of the proposed data processing model. In this solution, training data is converted into a series of feature vectors that consists of a set of formulated attribute values. These vectors construct the input value of a supervised machine learning algorithm. After training, a classification model is applied to distinguish whether the specific user belongs to excellent, good, and general or the poor.

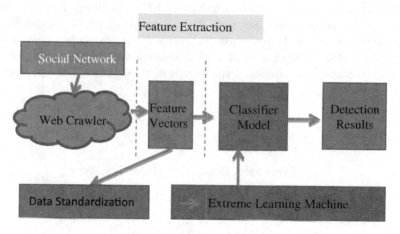

Fig. 2. Doctor recommended model

5   Evaluations

5.1 Prototype Implementation

We have implemented prototype in three components: Residential Media Server, Cloud Media Gateway, and Client. The programming language is based on JAVA.

Hadoop v 1.0.2 and HBase 0.92.0 are deployed in four PCs to build a cloud platform. Android v2.3.1 client testing simulator, Android v4.0 client testing simulator, and Huawei U8800 device (Android v2.2 system) are used as client testing tools.

Mobile client is capable to login CG, synchronous physical signs information and return the doctor recommended information. The network environment is based on campus network in Fuzhou University.

5.2 Performance Evaluation

5.2.1   Dataset Classification

In this paper, the doctor data classification is used as an example because this classification has different social features, which can distinguish the better ones. In this paper, a model based on the following 8 features is used: level of hospital, doctor title, academic title, curative effect, behavior, number of patient vote, number of thank-you letters, and working age of the doctor.

To evaluate the effectiveness of the experiment results, a confusion matrix illustrated is illustrated. TP (True Positive) represents the type of doctors correctly classified correctly classified, FN (False Negative) refers to the type of doctors misclassified as other type, FP (False Positive) expresses the type of doctors misclassified as other type, and TN (True Negative) is the type of doctors classified correctly. According to the confusion matrix, a set of metrics commonly evaluated in the machine learning field are introduced including precision (P), recall (R), and F-measure (F).

Table 1 shows a confusion matrix obtained by ELM classifiers. It shows that our proposed solution is quite efficient, with 99.32 % excellent, 98.97 % good, 99.06 % general and 99.65 % poor doctors classified correctly, leaving only a small fraction misclassified. Table 2 shows the value of evaluation metrics, in which precision, recall and F-measure are calculated for excellent, good, general and poor doctors respectively.

Table 1. Confusion matrix

	Predicted			
	Excellent	Good	General	Poor
Excellent	99.32 %	0.39 %	0.29 %	0 %
Good	0.82 %	98.97 %	0 %	0.21 %
General	0.71 %	0 %	99.06 %	0.23 %
Poor	0	0.26 %	0.09 %	99.65 %

Table 2. Classification evaluation

Evaluation	Precision	Recall	F-measure
Excellent	0.99070	0.99334	0.99202
Good	0.98760	0.98965	0.98862
General	0.99296	0.99063	0.99179
Poor	0.99823	0.99648	0.99735

5.2.2   Classification Result and Comparison

The training and testing times between SVM-based and ELM-based solutions are compared. The experiment results are illustrated in Table 3. The results indicate that the ELM-based solution is faster than the SVM-based solution, and, therefore, more efficient.

Table 3. Comparison between ELM and SVM

Classifier	Training time (s)	Testing time (s)
ELM	0.6094	0.0156
SVM	1.604	0.366

To further prove the effectiveness of the proposed doctor recommendation model, we consider two use scenarios, namely, data standardized and data non-standardized. The paper compares the training time together with testing accuracy under different activation functions (Sin and Sig) and different number of hidden nodes (L). The evaluation is illustrated in Figs. 3 and 4.

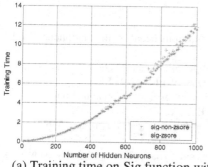

(a) Training time on Sig function with different number of hidden nodes

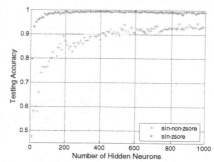

(b) Testing accuracy on Sig function with different number of hidden nodes

Fig. 3. Comparisons of training time and testing accuracy on Sig function

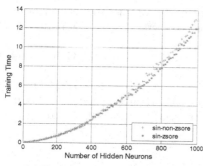

(a) Training time on Sin function with different number of hidden nodes

(b) Testing accuracy on Sin function with different number of hidden

Fig. 4. Comparisons of training time and testing accuracy under Sin function

6   Conclusions

In this paper, we proposed a system architecture for Smart Health Services and Applications. The system model resolves several issues, including new-style efficient data acquisition and transmission mode, a new method of predicting disease and intelligent doctor recommendation services. Through a set of experiments and evaluation work, these approaches guarantee feasibility and efficiency of system architecture.

Future works could include the extension of value-added service. Potential technical challenges may include availability and efficiency issue when a set of cloud servers negotiate and cooperate in the process of real-time transmission. The realization of this use scenario could greatly enhance user experience.

References

1. Lee, J.H.: Smart health; concepts and status of ubiquitous health with smartphone (2011). 978-1-4577-1268-5/11/ IEEE
2. Mathan Kumar, K., Venkatesan, R.S.: A design approach to smart health monitoring using android mobile devices (2014). ISBN No. 978-1-4799-3914-5/14/ IEEE
3. Chan, L.L., Celler, B.G., Lovell, N.H.: Development of a smart health monitoring and evaluation system (2006). 1-4244-0549-1/06/ IEEE
4. Huang, G.B., Zhu, Q.Y., Siew, C.K.: Extreme learning machine: theory and applications. Neurocomputing **70**(1), 489–501 (2006)
5. Rao, C.R., Mitra, S.K.: Generalized inverse of matrices and its applications. Wiley, New York (1971)
6. Ghanty, P., Paul, S., Pal, N.R.: NEUROSVM: an architecture to reduce the effect of the choice of kernel on the performance of SVM. J. Mach. Learn. Res. **10**, 591–622 (2009)
7. Huang, G.B., Ding, X., Zhou, H.: Optimization method based extreme learning machine for classification. Neurocomputing **74**(1), 155–163 (2010)
8. Zheng, X., Chen, N., Chen, Z., Rong, C., Chen, G., Guo, W.: Mobile cloud based framework for remote-resident multimedia discovery and access. J. Internet Technol. **15**(6), 1043–1050 (2014)
9. Hinton, G.E.: Learning multiple layers of representation. Trends Cogn. Sci. **11**(10), 428–434 (2007)
10. Bengio, Y.: Scaling up deep learning. In: Proceedings of the 20th ACM SIGKDD International Conference on Knowledge Discovery and Data Mining, p. 1966.1. ACM (2014)
11. Zhou, S., Chen, Q., Wang, X.: Active deep learning method for semi-supervised sentiment classification. Neurocomputing **120**, 536–546 (2013)

Author Index

Printed in the United States
By Bookmasters